Taxonomy Phytogeography and Evolution

Taxonomy Phytogeography and Evolution

Edited by

D. H. Valentine

Department of Botany
The University
Manchester, England

1972

ACADEMIC PRESS · LONDON · NEW YORK

ACADEMIC PRESS INC. (LONDON) LTD.
24/28 Oval Road,
London NW1

United States Edition published by
ACADEMIC PRESS INC.
111 Fifth Avenue
New York, New York 10003

Library of Congress Catalog Card Number: 72 84355
ISBN: 0 12 710250 7

PRINTED IN GREAT BRITAIN BY
W S COWELL LTD, 8 BUTTER MARKET, IPSWICH

List of Contributors

BAKER, H. G., *University of California, Berkeley, California, U.S.A.* (*p. 327*)

BÖCHER, T. W., *Institute of Plant Anatomy and Cytology, University of Copenhagen, Copenhagen, Denmark* (*p. 101*)

BRAMWELL, D., *Department of Botany, University of Reading, Reading, England* (*p. 141*)

CLAPHAM, A. R., *The Parrock, Arkholme, Carnforth, Lancashire, England* (*p. 397*)

CUTLER, D. F., *Jodrell Laboratory, Royal Botanic Gardens, Kew, England* (*p. 73*)

EXELL, A. W., *British Museum (Natural History), London, S.W.7., England* (*p. 307*)

FAVARGER, C., *Institut de Botanique, Université de Neuchâtel, Neuchâtel, Switzerland* (*p. 191*)

GILLETT, G. W., *University of California, Riverside, California, U.S.A.* (*p. 205*)

GREUTER, W., *The Goulandris Natural History Museum, Levidou 13, Kifissia, Greece* (*p. 161*)

HARA, H., *Department of Botany, University Museum, University of Tokyo, Japan* (*p. 61*)

JÄGER, E. J., *Sektion Biowissenschaften, Martin-Luther-Universität, Halle-Wittenberg, Germany (DDR)* (*p. 349*)

JARDINE, N., *King's College, Cambridge, England* (*p. 381*)

KORNAŚ, J., *Department of Botany, University of Zambia, Lusaka, Zambia* (*p. 37*)

LEWIS, H., *University of California, Los Angeles, California, U.S.A.* (*p. 179*)

MOORE, D. M., *Department of Botany, University of Reading, Reading, England* (*p. 115*)

MORLEY, B., *National Botanic Gardens, Glasnevin, Dublin, Ireland* (*p. 239*)

MORTON, J. K., *Department of Biology, University of Waterloo, Waterloo, Canada* (*p. 221*)

RAVEN, P. H., *Missouri Botanical Garden and Washington University, St. Louis, Missouri, U.S.A.* (*p. 259*)

ROSE, F., *Department of Geography, King's College, London, England (p. 363)*
SOLBRIG, O., *Department of Biology and Gray Herbarium, Harvard University, Cambridge, Massachusetts, U.S.A. (p. 85)*
STACE, C. A., *Botany Department, The University, Manchester, England (p. 307)*
STEBBINS, G. L., *Department of Genetics, University of California, Davis, California, U.S.A. (p. 7)*
STRID, A., *Department of Plant Taxonomy, University of Lund, Lund, Sweden (p. 289)*
VALENTINE, D. H., *Botany Department, The University, Manchester, England (p. 3)*
VAN STEENIS, C. G. G. J., *Rijksherbarium, Leiden, Netherlands (p. 275)*
WALTERS, S. M., *Botany School, University of Cambridge, Cambridge, England (p. 301)*

Preface

The idea of this conference was first suggested at a meeting of the Linnean Society. It was subsequently developed at a meeting of the Botanical Society of the British Isles, and later discussed at a meeting of the International Organisation of Plant Biosystematists held at Corvallis, Oregon, after the last International Botanical Congress.

To harmonize the aims of all three societies was a difficult task; but it was felt that the attempt was worth making, and a programme with a geographical theme was devised. The reaction of the speakers, all of whom were invited, was encouraging; and the Conference took place at the University of Manchester on September 9–11, 1971. This book presents the papers more or less as they were delivered and approximately in the order in which they were given. The main exception is Professor Kornaś's paper, which was written for the Conference, but could not be presented, as the author was called away to Zambia, and could not manage to travel back to Manchester in time. Nevertheless the paper is an important link in the proceedings of the Conference, and is included in its proper place. Dr. Jardine's paper was given, informally, as an evening talk, but is properly included as a full paper.

The sessions were chaired by Professor A. R. Clapham F.R.S. (Vice-President of the Linnean Society), Mr. David McClintock (President of the Botanical Society), Professor Harlan Lewis (President of the I.O.P.B.) and by Professor H. Merxmüller, Professor Reed C. Rollins and Dr. J. Heslop Harrison, F.R.S.

It would have been useful to include notes of the discussions which followed the papers. This was impracticable; but Professor Clapham's concluding paper, which includes responses invoked from the audience, deals with some of the points which arose.

The Conference was attended by 200 members and guests from 18 different countries. It was followed by an excursion to Derbyshire, in which moorland, heath, woodland and limestone grassland were visited. During the Conference, demonstrations by members of the Conference were on show in the University Department of Botany.

In addition to the societies named, many people and organizations contributed to the Conference. Financial help from the Royal Society, the Linnean Society and the B.S.B.I. made it possible for overseas contributors to come to Manchester, and this is gratefully acknowledged. I should like specially to thank Mr. J. C. Gardiner, until recently honorary treasurer of the B.S.B.I., who not only nursed the Conference in its formative stages, but also gave additional and very welcome financial help.

Thanks are due to the Vice-Chancellor and officers of the University, who kindly held a reception for the Conference. Dr. C. A. Stace and Miss J. Shore of the University Department of Botany headed the administrative and secretarial organization; the smooth running of the Conference (marred only by two minor crises with the coaches, but much favoured by the sunny weather) was due in large part to their hard work.

Readers of the book must judge if the Conference was timely and the contributions well chosen and arranged. There can be little doubt that geographical aspects of taxonomy and evolution have been somewhat neglected in recent symposia; and the papers given at Manchester certainly bring into focus some important problems of evolution. These problems are on the grand scale, in both the spatial and the temporal mode; and though the time is not yet ripe for a major synthesis, the way towards geographical treatment on a broad basis is beginning to appear.

I should like to acknowledge the assistance given by Academic Press and the prompt and efficient way in which they have seen the book through the press. I must also thank Dr. I. B. K. Richardson for preparing the index.

June 1972 D. H. Valentine

Contents

Section I

INTRODUCTION

1 | Introductory Remarks

D. H. VALENTINE

Department of Botany, The University, Manchester, England

This conference volume contains contributions from distinguished botanists from many parts of the world—a wide geographical spread which is matched by the range of the papers. Some aspects of the conference theme, it is true, had to be by-passed or are only lightly represented and there were few contributions from geographers or geologists; in fact the emphasis was taxonomic and evolutionary. But in a three-day conference, this restriction was no bad thing, and the discussions ranged widely enough.

My own interest in geographical distribution was first aroused when I attended the lectures of J. C. Willis in Cambridge many years ago. Certainly his views on evolution were not at all Darwinian; but he had gathered together a lot of data in a systematic way and had made a number of generalizations based on those data, which in the field of geographical distribution is not an easy thing to do. In classifying species into wides and endemics, and in demonstrating that in all larger taxa the number of endemic species greatly exceeded the number of wides, he posed questions which are still of fundamental importance. What makes a wide, wide? And why do endemics remain endemic? It is of course clear that wides are often the parents of endemics, which evolve in areas which are called centres of speciation; and it is equally clear that wides themselves must often begin as endemics, and have a centre of origin, from which they are able to spread without losing their individuality. A wide is thus a species which is so well buffered in respect of a series of environments and so well equipped in means of dispersal that, given a favourable opportunity, it can travel, apparently unchanged in essential characteristics, over considerable areas.

It is reasonable to suppose that this pattern of wides and endemics has existed at all times, in the past as well as at the present. A balance is

always maintained and there is an equilibrium; but it is a dynamic equilibrium. Environments are constantly changing, and vegetation and flora are in a state of constant flux, with constant migrations. As this happens, the range of a wide is fragmented and the two parts are separated, perhaps for ever. Some populations may be cornered, as it were, in a restricted area, and develop into a group or series of derived endemics. Here they may remain until they die out; or one of them may conceivably break out and set off along a new path and develop into a new wide. I think it is useful to consider evolution in this way against a very broad geographical background. In particular, we need to know where and how major evolutionary innovations arise. Maybe they are forged in the workshop of the local endemic, and then spread abroad in the form of a wide. Should this chance to meet another "original" wide, a crucial hybridization may occur, and new combinations may be made which are developed and worked out, as it were, in a new crop of endemics, until one of them enters again into the dispersal phase of a wide. These ideas are, I think, particularly relevant in thinking about major evolutionary problems, such as the origin of the angiosperms. One would not expect to find, in the fossil record, more than a few ancestral forms, in any one place; and the new evolutionary discovery may have been worked out, along somewhat different lines, in more than one geographical area.

In the papers presented at the conference, authors have considered, on the one hand the very large geographical areas, with their wides and series of vicariant taxa, and on the other hand, the smaller, more circumscribed and more or less isolated areas, in which clusters of endemic species have arisen. Both types of situation give rise to many problems, both taxonomic and evolutionary. A feature of geographical studies in the middle part of this century has been their investigation by biosystematic methods and the application to them of biosystematic concepts. It has thus become possible to study the flora of an area in a new way which is not entirely taxonomic, nor entirely chorological, but which combines the two with cytological and genetic information to form a comprehensive synthesis. A pioneer study of this kind was that of Á. and D. Löve (1956) on the flora of Iceland, itself based on the earlier paper of Á. Löve (1954) on corresponding taxa, and on the concepts of gradual and abrupt speciation (Valentine, 1949). These ideas have been developed in a number of ways, some of which are described in the Chapters below.

The general arrangement of the programme at the meeting was fairly straightforward. The first contributors looked mainly at evolution over major geographical regions, and wides were often the subject of discussion. Then the emphasis moved in the direction of endemics, and there were discussions of migration of both wild and cultivated plants. A third main topic was aspects of evolution and phytogeography in large or well-studied genera, including at least one apomictic genus. Papers on chorology, that is, on the classification of species into floral elements on a geographical basis, were unfortunately few in number. It was hoped at one time to have more papers of this kind but, for various reasons, the number had to be reduced. It is particularly regretted that it was not possible to have an up-to-date account of the chorology of the British Flora, which is now so well-known and well-mapped. As some compensation for this, the inclusion of Dr Jardine's contribution (given as an evening lecture) introduces some of the problems associated with the new techniques of computer mapping and analysis; accurate data on geographical distribution and efficient mapping are at the basis of discussion on phytogeography and evolution. An important project that might be mentioned here is the mapping of the whole European Flora, which is being organized by an International committee, based at the University of Helsinki.

Finally I would like to add a word about the sponsorship of the meeting. Three societies joined together as sponsors; much the oldest of these is the Linnean Society of London, famous for its Linnean collections, and its meetings at which biologists of all kinds can come together to discuss current theories and discoveries, particularly in the field of taxonomy. The Botanical Society of the British Isles can also boast a respectable history, going back in one form or another for more than a hundred years. As its title implies, its scope is less wide than that of the Linnean Society, but it, too, is famous both for the great skill of its members, especially its amateur members, who nowadays flourish perhaps more strongly than ever, and also for its increasing interest in modern developments in taxonomy and phytogeography. Last, and certainly the newest of the sponsoring societies was the International Organisation of Plant Biosystematists. It is only ten years old, and it has not yet produced a journal of its own, but through its occasional publications, its conferences and its world-wide membership, is able very effectively to promote the study of plant biosystematics. All three societies were represented at the meeting by their president or past-presidents; and in welcoming them and their members to Manchester, I expressed the hope that their co-operation would be fruitful and continuing.

REFERENCES

Löve, A. (1954). *Vegetatio* **8**, 212–220.
Löve, A. and Löve, D. (1956). *Acta Horti Gothoburg.* **20**, 65–290.
Valentine, D. H. (1949). *Acta Biotheor.* **9**, 75–88.

2 | Ecological Distribution of Centers of Major Adaptive Radiation in Angiosperms

G. LEDYARD STEBBINS

Department of Genetics, University of California, Davis, California, U.S.A.

INTRODUCTION

Geneticists and cytogeneticists concerned with the processes of evolution, with few exceptions, have confined their attention to the level of populations and species, and have paid little attention to evolutionary trends at the level of genera, families and higher categories. This focus is justifiable, since the problems of evolution above the species level cannot be attacked directly by experimental means. Nevertheless, plant evolution as a whole will never be understood until botanists become able to find out whether or not major trends of evolution can be entirely explained on the basis of processes that can be observed and studied at the level of populations and species, and if they can, what kinds of projections or extrapolations are necessary to establish connections between micro- and macroevolution.

The present discussion is based upon the hypothesis that evolution at all levels is guided by the same kinds of processes, and that the proper study of macroevolution is one that differs from experimental investigations of microevolution only in degree, with different emphasis placed upon certain aspects of the study. The macroevolutionist must place greater emphasis upon historical events, particularly the differences between past and present environments and consequent distributional patterns of biota; as well as extinction of formerly dominant floras. Furthermore, he can profit from the experimental method only by making certain assumptions. The first and most important of these is an hypothesis that can be called *genetic uniformitarianism*, since it corresponds to geological uniformitarianism, as this concept was first put forward by Hutton and Lyell, and formed the basis of the historical aspects of Darwin's theory.

THE HYPOTHESIS OF GENETIC UNIFORMITARIANISM

The hypothesis of genetic uniformitarianism states that the processes of evolution: mutation and genetic recombination as sources of variation in populations; natural selection as a guiding factor; and reproductive isolation as a basis of diversification and canalization—have operated in the past essentially as they do now, even though the genotypes and phenotypes upon which they operated, as well as the environmental conditions that created selection pressures, were different. In other words, processes of evolution are constant; phenotypes, genotypes and environmental conditions are variable and specific to certain places and certain times in the earth's history.

This hypothesis is implicit in the research of most modern evolutionists, both those whose research is at the level of populations and species, such as Dobzhansky (1970) and Mayr (1963), and those whose major contributions have been toward the understanding of macroevolutionary trends, such as Simpson (1953). On the other hand, many botanists have held to a philosophy that places great emphasis upon the morphology and geographic distribution of modern phenotypes that appear to them to be primitive and perhaps even ancestral, and pays little attention to evolutionary processes as understood by experimental evolutionists (Cronquist, 1968; Takhtajan, 1969, 1970; Smith, 1970).

The first corollary of the hypothesis of genetic uniformitarianism is that adaptive radiation, which can be demonstrated as the primary basis of diversification at the level of populations and species in recent times, was also the primary basis of the initial divergence of evolutionary lines that later became distinct genera and families. Its second corollary is that the origin of major categories, such as genera and families, does not require the evolution of differences with respect to characters or attributes that are qualitatively different from those that separate populations and species. Its third corollary is that the initial divergence that led to the origin of major categories took place under conditions similar to those that promote diversity of populations and speciation in the modern world. Its fourth corollary is that archaic forms, which among modern groups resemble most closely the putative ancestors of major taxonomic categories, are most likely to be preserved under conditions that are the least favorable for the origin of new, more efficient genotypes that would compete successfully with them and drive them out. This last corollary emphasizes the necessity for distinguishing clearly between biotic communities that function as evolutionary "laboratories" or "cradles," in which new adaptive complexes arise, and "museums," in which archaic forms are preserved.

A fifth corollary is that macroevolution is best studied not by focusing attention on relictual modern phenotypes that are believed, on morphological grounds to be primitive (but which certainly cannot be regarded as ancestral to any modern forms); but by analysing as carefully as possible differences between modern forms that are analogous to the differences that exist between major taxonomic groups.

ADAPTIVE RADIATION AT DIFFERENT LEVELS OF THE TAXONOMIC HIERARCHY

To what extent are the first four corollaries compatible with the existing variation pattern in angiosperms? The first two will be discussed in this section, since they are most directly related to evolutionary processes. The validity of both of them depends upon the demonstration that the characters which separate genera and families are not qualitatively different from those that differentiate related populations and species.

Some taxonomists and geneticists (Anderson, 1937) have maintained that the characters diagnostic for major categories are qualitatively different from those that separate populations and species. If one studies only a single group of plants, this conclusion can appear to be quite plausible. This is because if one confines one's attention to a single family, one finds that the characters used to separate that family from related families are usually different from the characters that are used to differentiate genera within the family, and that still different characters are the most useful for separating the species within any one genus.

If, however, one considers the entire class of angiosperms, one finds that there are no characters that are useful solely for distinguishing families or orders, and do not also vary occasionally at the level of genera or species. Table I illustrates this point. In order to keep this table relatively simple, only one out of numerous possible examples was chosen to represent each category.

Even the character that serves to give the two major subdivisions their names, cotyledon number, can sometimes vary at the level of genera or even species. For instance, most of the species of *Claytonia* (Portulacaceae) have seedlings possessing two normal cotyledons; yet *C. virginica* L.

TABLE I. Distribution of character differences at various hierarchical levels

Character Difference	Diagnostic at species level	Diagnostic at genus level	Diagnostic at family or order level
Woody *vs* herbaceous growth habit	*Mimulus longiflorus* (Nutt.) Grant *vs M. clevelandii* Brandg.	*Zanthorhiza vs Coptis* (Ranunculaceae)	Myrsinaceae *vs* Primulaceae
Compound *vs* simple leaves	*Ranunculus repens* L. *vs R. cymbalaria* Pursh	*Eschscholtzia vs Dendromecon* (Papaveraceae)	Oxalidaceae *vs* Linaceae
Capitate *vs* umbellate or other kind of inflorescence	*Arenaria congesta* Nutt. *vs A. macradenia* S. Wats.	*Trifolium vs Melilotus* (Leguminosae)	Dipsacaceae *vs* Valerianaceae; Asterales (Compositae) *vs* Campanulales
Bilateral (zygomorphic) *vs* radial (actinomorphic)	*Saxifraga sarmentosa* L. *vs* other Saxifraga spp.	*Tolmiea vs Heuchera* (Saxifragaceae)	Violaceae *vs* Cistaceae
Foliaceous *vs* awn-like or pappus-like calyx lobes	*Marrubium alysson* L. *vs M. vulgare* L.	*Dracocephalum vs Galeopsis* (Labiatae)	Dipsacaceae *vs* Caprifoliaceae; Asterales *vs* Campanulales
4-Merous *vs* 5-merous perianth	*Rhamnus crocea* Nutt. *vs R. californica* Esch.	*Ludvigia vs Jussiaea* (Onagraceae)	Cruciferae *vs* Moringaceae
Corolla lobes separate *vs* united	*Crassula zeyheriana* Schonl. *vs C. glomerata* Berg.	*Monotropa vs Pterospora* (Monotropaceae)	Pyrolaceae *vs* Ericaceae
Perianth biseriate *vs* uniseriate	*Sagina nodosa* (L.) Fenzl. *vs S. decumbens* (Ell.) T. & G.	*Agrimonia vs Sanguisorba* (Rosaceae)	Portulacaceae *vs* Chenopodiaceae
Carpels separate *vs* united	*Saxifraga lyallii* Engl. *vs S. arguta* D. Don	*Delphinium vs Nigella*	Dilleniaceae *vs* Actinidiaceae
Ovary superior (hypogynous) *vs* inferior (epigynous)	*Saxifraga umbrosa* L. *vs S. caespitosa* L.	*Tetraplasandra vs* other Araliaceae	Loganiaceae *vs* Rubiaceae
Placentae axial *vs* parietal	*Hypericum perforatum* L. *vs H. anagalloides* Cham. & Sch.	*Boykinia vs Heuchera* (Saxifragaceae)	Theaceae *vs* Cistaceae
Ovules numerous *vs* solitary	*Medicago sativa* L. *vs M. lupulina* L.	*Spiraea vs Holodiscus* (Rosaceae)	Campanulales *vs* Asterales

has only one cotyledon (Haccius, 1954). Several other examples exist of species having only one cotyledon, but belonging to normally dicotyledonous genera (Haskell, 1954).

In the other direction, species of the genus *Pittosporum* from Asia (*P. tobira* (Thunb.) Ait.f.) or Australia (*P. undulatum* Vent.) have embryos possessing two normal cotyledons, but in several species of New Zealand (*P. crassifolium* Putterl., Lubbock, 1892, pp. 200–204; *P. tenuifolium* Banks, personal observations; *P. rigidum* Hook.f., Cockayne, 1899; *P. anomalum* Laing & Gourl., *P. divaricatum* Cockayne, *P. crassicaule* Laing & Gourl., *P. lineare* Laing & Gourl., Laing and Gourlay, 1935) three to four cotyledons are always present (Fig. 1).

FIG. 1. Seedlings of *Pittosporum crassifolium*, showing three and four cotyledons. From specimens cultivated in Golden Gate Park, San Francisco, California.

CRITICAL ENVIRONMENTAL FACTORS, ECOLOGICAL DIVERSITY, AND ADAPTIVE RADIATION

The validity of the third corollary, that environmental conditions which favor diversification of populations and speciation in the modern world are similar to those that promoted the origin of modern genera and families at earlier periods in the earth's history, cannot be estimated until these environmental conditions have been clearly defined. Basic to such a definition is Sewall Wright's principle (Wright, 1931, 1950), that the most favorable population structure for rapid evolution is the division of a population into many small subpopulations, which are sufficiently isolated from each other so that they can become differentiated under the influence of differential selection pressures, but between which migration can occur often enough so that adaptive gene complexes arising in one subpopulation can spread to other subpopulations. A highly significant question is, therefore, what ecological conditions are most likely to maintain such partly subdivided populations, and to promote occasional migration between them, as well as the occasional fusion of different subpopulations?

An answer to this question can be deduced from a review of the climatic zones and ecosystems which support the highest proportion of species complexes consisting of a mosaic of allopatric or partly allopatric and narrowly endemic species and subspecies. An analysis of one region containing many such species complexes, California and Pacific North America in general, showed that mosaics of closely related endemics are most characteristic of semi-arid ecotones and borderline regions between mesic forests and truly arid deserts and steppes (Stebbins and Major, 1965). In other temperate regions that are noted for their high proportions of endemism and clusters of related species, such as southwestern Asia and the Cape region of South Africa, similar ecotones and borderline ecosystems prevail. Valentine (1970), on the basis of his investigations of the European flora, has also reached the conclusion that semi-arid border zones are regions of active speciation.

The data compiled and presented in Table II shed further light on this problem. These data were compiled in an attempt to answer the question: Are any habitats more favorable than others in promoting speciation? The flora of the Pacific Coast of the United States was selected partly because it is familiar to this author, and also because it offers a great variety of habitats, each of which (except for alpine habitats) covers an area not

TABLE II. Distribution by habitats of different sized genera in the Angiosperms of the Pacific Coast of the United States. Data from L. R. Abrams and R. S. Ferris, "Pacific Coast Flora". Genera are classified according to the habitat distribution of the majority of their species

Habitat	Number of genera and species in genera with:												Totals		
	1–2 species		3–5 spp.		6–10 spp.		11–20 spp.		21–30 spp.		over 30 spp.		Genera	Species	Spp./Genus
Lakes, streams, swamps, bogs, marshes	73	109	29	116	4	32	5	75	1	25	2	96	114	447	3·8
Mesic woodlands	81	122	37	148	17	136	10	150	1	25	1	47	146	628	4·3
Mesic open country	45	68	20	80	17	136	12	180	5	125	9	456	108	1045	9·7
Mesic alpine areas	15	23	6	24	0	0	0	0	1	25	0	0	72	22	3·3
Semi-xeric open woods, grasslands and shrub formations	114	171	40	160	40	320	37	555	9	225	23	1233	263	2664	10·1
Steppes and deserts	127	191	27	108	13	104	3	45	1	25	0	0	171	473	2·8

TABLE III. Comparisons between tropical and temperate Angiosperm floras as to numbers of families, genera and species of vascular plants

Flora	Number of families	Genera/ families	Number of genera	Species/ genera	Number of species
Philippine Islands (Merrill)	186	7·9	1468	5·4	7858
California (Munz)	147	6·9	1021	5·4	5470
Panama Canal Zone (Standley)	97	6·5	627	2·2	1397
Marin County, Calif. (Howell)	92	4·1	380	2·5	961
Brazilian cerrado (Rizzini) (Woody plants only)	70	3·5	242	1·8	537
California Chaparral (Woody plants only)	40	2·1	83	3·0	250

very different in extent from those of the others. The six generalized habitat categories represent the best classification permitted by the available data. In the case of many of the larger genera, such as *Astragalus*, *Eriogonum*, *Phacelia* and *Senecio*, most habitats contain at least a few of the numerous species belonging to them. These genera were classified according to the habitat distribution of the majority of their species. Three of the larger genera: *Carex*, *Veronica* and *Viola*, defied classification and were not tabulated.

As can be seen from Table III, the habitats fall neatly into two groups, with respect to the size of the genera that are principally adapted to them. In one group, the mean number of species per genus is low: from 2·8 to 4·3. This includes wet habitats, mesic woodlands, alpine habitats and desertic regions. The remaining two habitats are mesic open country: chiefly fields and meadows, with 9·7 species per genus, and semi-xeric open woods, grasslands and shrub formations, with 10·1 species per genus. The two habitats with high mean values are ecologically intermediate between the four having low values.

Pertinent comparisons between semi-arid and moist regions in the tropics are not known to me, and are probably not available. Nevertheless, comparisons between tropical and temperate regions as to numbers of families, genera and species do not support the hypothesis that the well known richness of tropical floras is due to extensive speciation in progress at the present time. Table III shows three such comparisons: Two large areas of comparable size: the Philippine Islands (from Merrill, 1923) and California (from Munz and Keck, 1959); Panama Canal Zone (from Standley, 1928) and Marin County, California (from Howell, 1949); and the Brazilian cerrado formation (from Rizzini, 1963) compared with the California chaparral (compiled personally from Munz and Keck, 1959). The most striking fact that emerges from this Table is that, while in each comparison the total number of species is greater in the tropical than in the temperate flora, the number of species per genus is in one comparison the same in the two regions, and in the two others is actually larger in the temperate, Californian example. The greater richness of the tropical floras is based chiefly upon the larger number of genera found in them, and to a lesser extent upon a larger number of families. When the fact is considered that many of the large genera of the tropical floras, as in the Orchidaceae, Gesneriaceae, and such genera as *Ficus*, have undergone speciation almost entirely in response to the diversity of animal pollinators with little or no dependence upon the diversity of the habitat, the conclusion can be reached that in the mesic or moist tropical forests speciation in response to habitat diversity and environmental changes is, if anything less active than it is in semi-arid temperate habitats. The large number of genera in the tropical floras could possibly reflect a greater amount of speciation in past geological epochs. More probably, however, it is due to the fact that in these floras there has been a lower rate of species extinction relative to addition of species by both immigration and speciation than there has been in temperate regions.

INDIVIDUAL EXAMPLES OF ADAPTIVE RADIATION

If semi-arid regions, both temperate and tropical, have been centers of diversification from which evolutionary lines have radiated into more mesic as well as more xeric habitats, examples should be available in which this radiation can be traced. Many groups, of course, do not show evidence of this radiation on the basis of their modern species, since the fragile nature of the ecosystems found in semi-arid habitats promotes a higher rate of extinction than in either mesic or truly arid habitats, so that in older groups that have exhibited this radiation in the past, the original forms are extinct and have no living counterparts. By examining floras of semi-arid regions with this concept in mind, I have identified a number of such examples, of which I shall present four, two at the level of a genus, and two at the family level.

The presentation is in the form of a device which I am calling an ecophyletic cross sectional chart. The phylogeny is represented in the manner recommended by Sporne (1956) and used by a number of authors, including myself (Stebbins, 1956). The conventional "tree" is dispensed with, since actual phylogenetic connections are unknown. Instead, each group (species or genus) is rated according to the number of advanced states that it possesses with respect to a series of characters for which generalized and advanced or specialized states can safely be postulated on the basis of a careful study of the group itself as well as its relatives. A symbol representing the group is then placed on the chart at a distance from its center that corresponds to the number of advanced states that the group possesses. When genera are shown, the size of the genus and the variability among its species with respect to states of advancement are represented by the size and shape of the symbol. Finally, the ecological adaptation or adaptations of the group are represented by conventional symbols similar to those used to identify rock formations on a geological map.

1. *The Genus* Antennaria

The first example is the genus *Antennaria*, of the Compositae, tribe Inuleae. Although the taxonomy of this genus is complex, due to the presence of apomixis, the apomicts can easily be recognized by the rarity or absence of male plants, and have been omitted in the construction of the chart. The relationships between the remaining sexual species are relatively simple (Fig. 2).

The distinctive characteristics of *Antennaria* compared to its immediate relatives, such as *Gnaphalium*, are the presence of stolons and the marked dimorphism of leaves that occurs in many of its species. These are, therefore, the principal specialized characters upon which the advanced condition of the species was based. There is a central group of two species in which horizontal stolons are not formed and the involucral phyllaries are pale in color. One of these is *A. geyeri* A. Gray, which most resembles other genera of Gnaphalinae in having basal leaves and stem leaves similar to each other in size and shape. Both it and *A. luzuloides* Torr. & Gray are adapted to dry, open woodlands or open country in the semi-arid climate of the western United States. The other three species that lack stolons, *A. carpathica* (Wahlb.) Bl. & Fing., *A. anaphaloides* Rydb. and *A. pulcherrima* (Hook.) Greene, all have dark brown or black involucral phyllaries and denser tomentum. They are adapted to mesic alpine or subalpine habitats.

Among the stoloniferous species, all of which have the stem leaves and basal leaves strongly differentiated from each other, only two of those shown, *A. microphylla* Rydb. and *A. argentea* Benth., are adapted exclusively to semi-arid habitats. Three others, *A. dimorpha* (Nutt.) Torr. & Gray, *A. flagellaris* A. Gray, and *A. rosulata* Rydb., are more xeric in their adaptation, having become adapted to *Artemisia* steppes, associated chiefly with a dwarf, cushion-like growth habit. The remaining thirteen are adapted to more mesic habitats. They include the complex of *A. alpina*, characterized by low stature and dark brown or black phyllaries and adapted to alpine meadows or arctic tundra; and several groups that have independently become adapted to open fields or dry woods in mesic climates. With respect to both complexity of growth pattern and divergence from other species of Gnaphalinae, the most specialized species of *Antennaria* is *A. solitaria* Rydb. It has long, prostrate stolons; large basal leaves that contrast strongly with the reduced cauline leaves and are differentially tomentose, being essentially glabrous above and heavily tomentose beneath; and its inflorescence is reduced to a single capitulum. It inhabits rich woods in the mesic climate of the eastern United States.

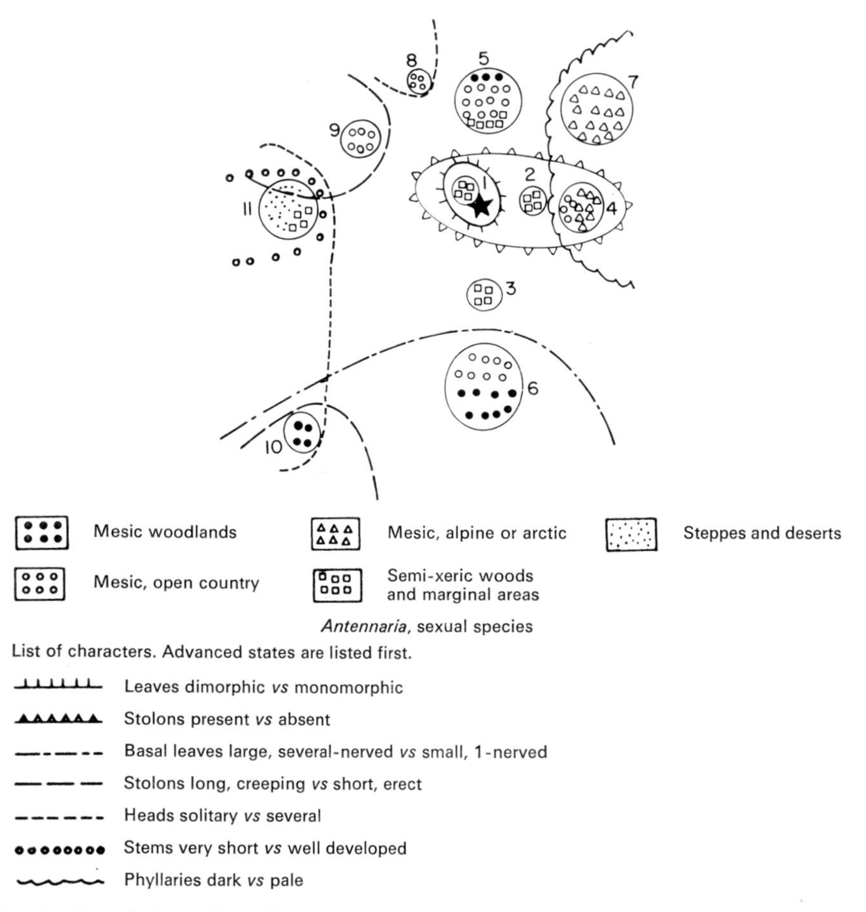

FIG. 2. Eco-phyletic chart showing the principal sexual species of the genus *Antennaria*. The species represented by the numbers are as follows:

1. *geyeri*
2. *luzuloides*
3. *argentea*
4. *carpathica*, *anaphaloides*, *pulcherrima*
5. *dioica* group (*dioica*, *virginica*, *microphylla*, *corymbosa*)
6. *plantaginifolia* group (*plantaginifolia*, *parlinii*, *racemosa*)
7. *alpina* group (*alaskana*, *media*, *monocephala*, *reflexa*)
8. *suffrutescens*
9. *neglecta*
10. *solitaria*
11. *dimorpha*, *flagellaris*, *rosulata*

2. *The Genus* Ceanothus

The second genus selected, *Ceanothus* subg. *Euceanothus*, consists almost entirely of shrubs, as do the great majority of other genera belonging to its family, the Rhamnaceae. The great bulk of the modern species are adapted to the semi-arid, Mediterranean type climate of California. It has been well monographed by McMinn (1942). For the sake of simplicity, only the larger of its two subgenera, *Euceanothus*, is included in Fig. 3. Its species are all closely related to each other, and hybrids between them are frequent. In the opinion of Nobs (1963), they might all be regarded as subspecies of a single, highly polymorphic genetic species. On the other hand, my observations in several localities indicate that most of the hybrids between sympatric species of *Euceanothus* exist for only short periods of time in disturbed habitats, and that under stable ecological conditions sympatric species may remain edaphically isolated from each other and thus retain their identity for indefinite periods of time.

The evaluation of the character states has been made by the present author on the following basis. The prostrate growth habit is an obvious specialization that has arisen only occasionally. Rigid branches that end in spines have been evolved repeatedly in the Rhamnaceae, and are one adaptation against browsing predators in a climate where recovery from browsing is slow, due to the rigors of periodic drought. Branchlets that are angular in cross section are peculiar to a few species of subg. *Euceanothus*, and are not found elsewhere in the Rhamnaceae. The 3-veined leaf, with essentially palmate venation, is also regarded as a more specialized condition than typical pinnate venation, since it is found elsewhere in the Rhamnaceae only in relatively specialized genera, such as *Paliurus* and *Zizyphus*. Deciduous leaves are justly regarded by botanists as a specialized condition, associated with the evolution of specific physiological mechanisms involving particular hormones. The compoundly cymose inflorescences found in the larger species of subg. *Euceanothus* are unique to the subgenus, being absent from other genera of Rhamnaceae, and are not paralleled in genera of the related and perhaps more primitive family Celastraceae. They are, therefore, to be regarded as secondary aggregations of simple corymbs. Finally, blue flowers are unknown elsewhere in the Rhamnaceae, and must, therefore, be regarded as specialized in the genus *Ceanothus*.

On the basis of these considerations, the most generalized living species of *Ceanothus* is *C. ochraceus* Suesseng., endemic to the semi-arid mountains of northern Mexico. The great majority of the remaining species, having

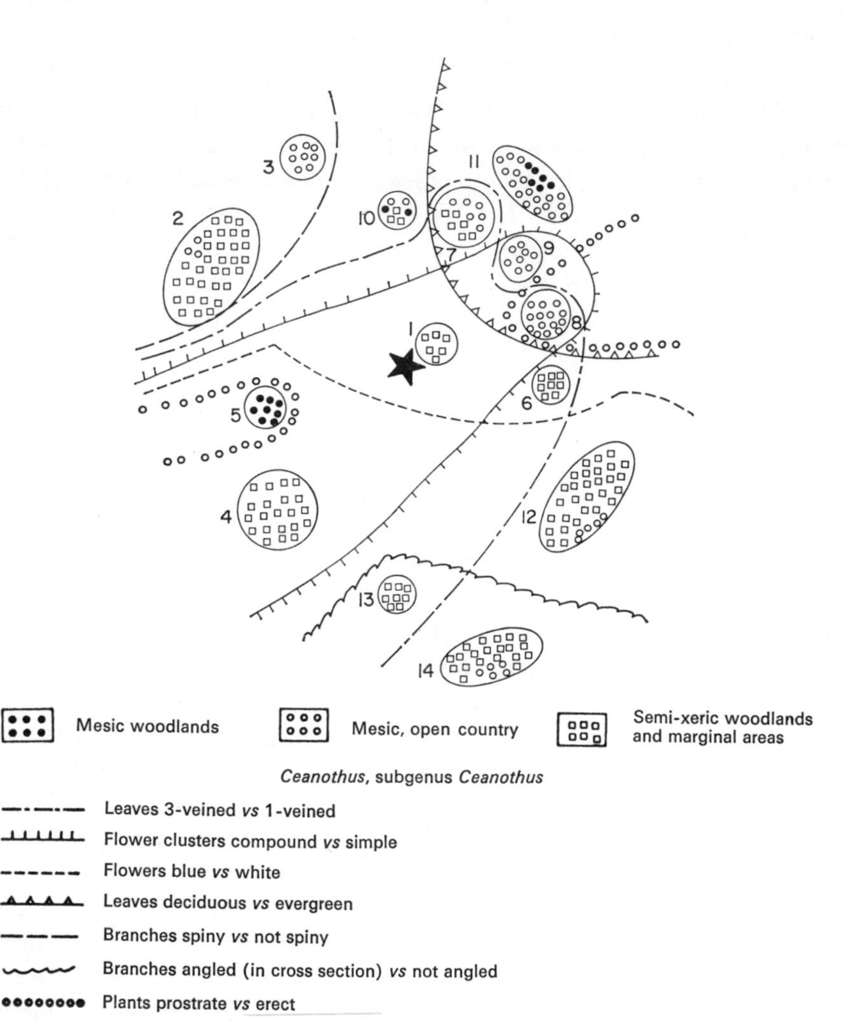

FIG. 3. Eco-phyletic cross section of *Ceanothus* sect. *Euceanothus* Species:

1. *ochraceus*
2. *buxifolius*, *cordulatus*, *depressus*, *fendleri*, *leucodermis*
3. *incanus*
4. *dentatus*, *foliosus*, *impressus*, *lemmonii*, *papillosus*
5. *diversifolius*
6. *spinosus*
7. *integerrimus*, *palmeri*
8. *microphyllus*, *serpyllifolius*
9. *ovatus*
10. *velutinus*
11. *americanus*, *sanguineus*, *Martinii*
12. *arboreus*, *coeruleus*, *oliganthus*, *sorediatus*, *tomentosus*
13. *parryi*
14. *cyaneus*, *griseus*, *thyrsiflorus*

intermediate degrees of specialization, are endemic to the semi-arid Mediterranean-type climatic region of California. Several species have, however, become adapted to more mesic climates. Each of these has acquired certain specializations. *Ceanothus incanus* Torr. & Gray, adapted to the redwood border forest, retains the spiny branches of its more xeric relatives, *C. leucodermis* Greene, *C. cordulatus* Kell. and *C. fendleri* A. Gray, and differs from them chiefly in its more vigorous growth, larger leaves and more numerous flowers. An obvious specialization of *C. diversifolius* Kell., adapted to dense forests in the central Sierra Nevada, is its prostrate growth habit. In the same habitat occurs *C. integerrimus* Hook. & Arn., which has compound inflorescences and 3-veined, deciduous or partly deciduous leaves. The most widespread of the truly deciduous species, *C. ovatus* Desf., *C. americanus* L. and *C. sanguineus* Pursh, are all adapted to mesic climates in the eastern United States and the Pacific Northwest.

The second subgenus of *Ceanothus*, *Cerastes*, contains a larger number of xeric adaptations than subg. *Euceanothus*, particularly the thick leaves provided with elaborate stomatal crypts (Nobs, 1963). It has not evolved any species adapted to truly mesic habitats. If the entire genus is considered, therefore, *Ceanothus* agrees with *Antennaria* and *Crepis* in having evolved both more xeric and more mesic species than the original progenitors.

3. *The Family Hydrangeaceae*

Eco-phyletic cross-sectional charts have been attempted for two woody families that are relatively unspecialized, the Hydrangeaceae and the Dilleniaceae.

The Hydrangeaceae were selected because they are entirely woody, they are well enough defined so that most taxonomists interested in phylogeny agree on their limits, and they occupy a wide range of ecological habitats. Furthermore, I am reasonably familiar with all of the genera that are regarded as primitive.

The sixteen genera of this family are shown in Fig. 4. The smallest number of specialized characters (1 out of 10) is found in *Carpenteria*, a monotypic genus confined to a restricted portion of the semi-arid foothills of the Sierra Nevada in California. The genus *Fendlera*, which is second in this respect, has three species that occupy similar habitats in the southwestern United States and northern Mexico. The monotypic *Jamesia*, with three specialized characters out of ten, occurs in dry forests of the southwestern United States, but extends also to subalpine and alpine cliffs. The large genus *Philadelphus* contains many species with deciduous leaves

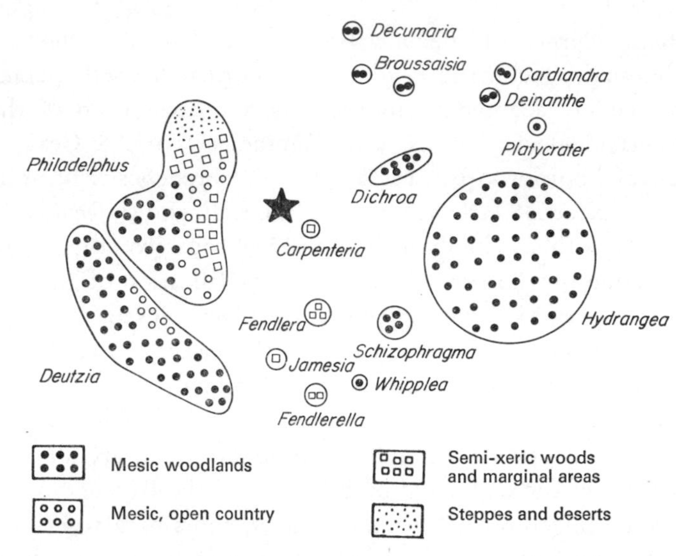

FIG. 4. Hydrangeaceae: List of characters. Advanced states are listed first:

1. Stems herbaceous, trailing or twining *vs* shrubby or arboreal.
2. Leaves deciduous *vs* persistent.
3. Inflorescence bracteate and cymose or paniculate *vs* leafy and corymbose or flowers solitary.
4. Showy or otherwise differentiated marginal sterile flowers present *vs* absent.
5. Perianth 4-merous or 7 to 10-merous *vs* 5-merous.
6. Stamen number = no. perianth parts or less, *vs* stamens more numerous.
7. Ovary fully inferior *vs* ovary superior or half inferior.
8. Styles united *vs* styles free.
9. Ovary locules and/or styles ("carpels") fewer in number than petals *vs* equal in number.
10. Ovary with one locule and parietal placentation *vs* several locules and axial placentation.
11. Ovules 1–2 per locule *vs* more numerous.

Genera and their advanced states. Numbers in parentheses refer to characters that are found only in some of the species of the genus.

1. *Carpenteria*, 1 sp.: 8
2. *Philadelphus*, 50–60 spp.: 2, 5, (7), (8)
3. *Jamesia*, 1 sp.: 2, 6, 9, 10
4. *Fendlera*, 3 spp.: 2, 5, 6
5. *Deutzia*, 40 spp.: 2, 3, 7, 9
6. *Fendlerella*, 2 spp.: 2, 3, 6, 7, 11
7. *Whipplea*, 1 sp.: 1, 3, 6, 9, 11
8. *Deinanthe*, 2 spp.: 1, 2, 3, 4, 8
9. *Cardiandra*, 2 spp.: 1, 2, 3, 4, 7, 9, 10
10. *Platycrater*, 1 sp.: 1, 2, 3, 4, 5, 7, 9
11. *Hydrangea*, 80 spp.: (1), 2, 3, 4, (5), 6, 7, (9)
12. *Schizophragma*, 4 spp.: 1, 2, 3, 4
13. *Dichroa*, 9 spp.: 3, 6, (9), 11
14. *Pileostegia*, 2 spp.: 3, (5), 6, 7, 8
15. *Broussaisia*, 2 spp.: 3, 6, 7, 8
16. *Decumaria*, 2 spp.: 1, 2, 3, 7, 8

that inhabit mesic forests in temperate regions. A whole group of small-leaved species inhabits the arid southwestern United States, and another group, regarded by Hu (1954–1956) as the least specialized in the genus, is confined to tropical and subtropical Mexico and Central America, in both mesic and semi-arid situations. The remaining twelve genera are all more specialized and occur exclusively in mesic habitats.

4. *The Family Dilleniaceae*

The Dilleniaceae were selected because they are the most primitive family of angiosperms that exhibits a wide range of ecological preferences. Other families of apocarpus woody angiosperms are either very small, or, as in the Magnoliaceae, Winteraceae, Annonaceae and Monimiaceae, occur only in mesic habitats. Information on the Dilleniaceae has been obtained in part from the treatment by Gilg and Werdermann (1925), and in part from the studies of Dickison (1967, 1968, 1969) and Kubitzki (1968).

The most remarkable feature of the Dilleniaceae is the extraordinary diversity of its largest genus, *Hibbertia*, with respect to both growth habit and floral structure (Fig. 5). There is probably no other genus of angiosperms that exhibits such a high degree of variation in those floral characteristics that are often regarded as "fundamental" and are usually associated with the separation of genera or even higher categories, and at the same time is equally diverse with respect to the size, growth habit and ecological adaptations of its species. The majority of its species are, however, low growing shrubs with slender stems and small to medium sized, more or less xeromorphic leaves. Some of them, belonging to the section *Cyclandra* and inhabiting the western and southern portions of Australia, have a primitive combination of characters with respect to wood and nodal anatomy, leaf variation, inflorescence (flowers solitary or in few flowered leafy corymbs), and floral parts (five sepals, five petals, many stamens regularly distributed, five separate carpels with two or more ovules per carpel, dehiscing ventrally). Other species are more highly specialized with respect to leaves, inflorescences, floral parts or all of these, although all of those investigated apparently have relatively unspecialized wood and nodal anatomy (Dickison, 1967, 1969). The specializations of their leaves are adaptations to the dry climate of Australia; those of their flowers will not be understood until careful investigations have been made of their pollination biology. The small genus *Pachynema* of North Australia is an extreme xeromorphic derivative of *Hibbertia*.

The isolated, relictual genus *Acrotrema*, found in Ceylon and Malaya, has

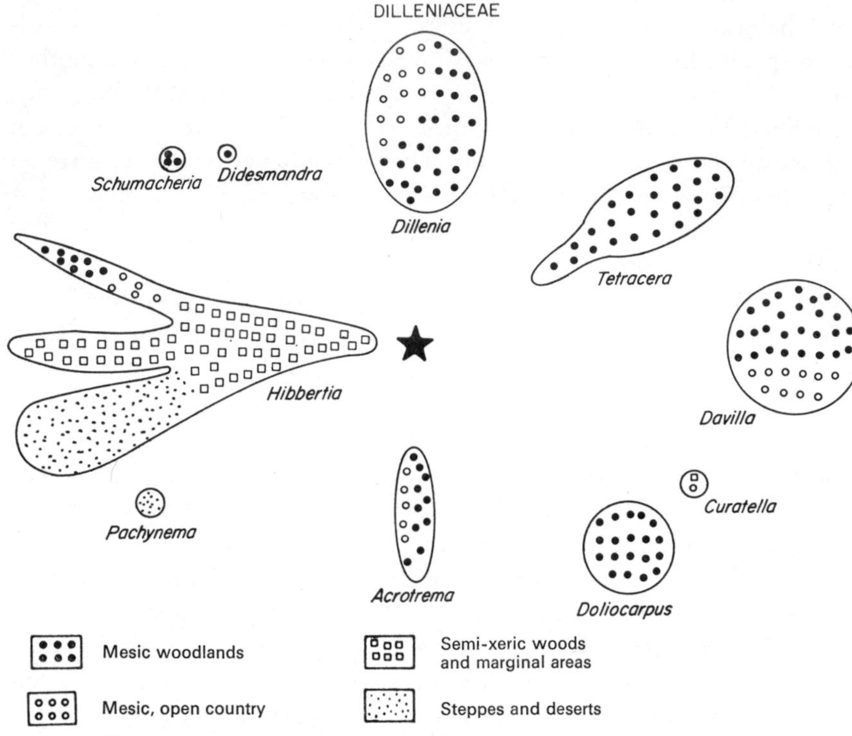

FIG. 5. Dilleniaceae. List of characters. Advanced states are listed first.

1. Herbaceous *vs* woody.
2. Plants climbing (lianas) *vs* not climbing.
3. Simple vessel perforations present *vs* absent.
4. Xylem rays of type IIA present *vs* absent.
5. Xylem rays uniseriate, homogeneous *vs* heterogeneous.
6. Multilacunar nodes present *vs* absent.
7. 1-lacunar nodes present *vs* absent.
8. Leaves scale-like *vs* normal in appearance.
9. Inflorescence spicate, racemose or paniculate *vs* cymose and leafy bracted or flowers solitary.
10. Inflorescence lateral or axillary *vs* terminal.
11. Flowers asymmetrical *vs* actinomorphic.
12. Stamens fewer or as many as perianth segments *vs* stamens more numerous.
13. Carpels fewer than petals *vs* the same number or more numerous.
14. Ovules 1–2 per carpel *vs* ovules more numerous.
15. Calyx enlarged and fleshy in fruit *vs* not so.
16. Carpels united at base *vs* completely free.
17. Sepals unequal (dissimilar) *vs* equal and similar to each other.
18. Non-methylated flavonols present *vs* absent.
19. Methylated flavonols present *vs* absent.

a smaller number of specialized characteristics than any other genus of the family except for the more unspecialized species of *Hibbertia*, in spite of its herbaceous growth habit. It is a distinct mesophyte.

The remaining seven genera consist chiefly of trees, climbing shrubs or lianas, and nearly all of them are confined to moist tropical forests. They have intermediate to high numbers of specialized characters, the highest number being in the genus *Davilla*, which occurs chiefly in Brazil.

The Dilleniaceae provide an admirable example of ancient adaptive radiation from intermediate habitats both toward more xeric and more mesic adaptations. Other tropical and subtropical families, if they are carefully studied both morphologically and ecologically, may provide similar examples. I have seen suggestive indications of this nature in the Ochnaceae, Guttiferae (Clusiaceae and Hypericaceae), Capparidaceae, Tiliaceae, Sterculiaceae, Mimosaceae, Caesalpiniaceae, Rutaceae, Oleaceae and Apocynaceae.

INCREASED SPECIALIZATION THROUGH TEMPORARY ADAPTATION TO SEVERE ENVIRONMENTS

One reason why plant groups that are adapted to transitional, semi-arid regions are most likely to become diversified is that, as a result of climatic fluctuations that occur through different geological epochs, they have been subjected alternatively to different selective pressures of an opposite nature. One result of such alternating selective pressures would be the evolution of more efficient adaptations to a favorable environment via an intermediate stage of adaptation to a severe environment, that would exert strong selective pressures on the evolving line. An example of this kind of evolution is in the Compositae, tribe Cichoricae. The two New World genera *Microseris* and *Agoseris* resemble each other in size, leaf

---

Advanced states found in the genera. (Numbers in parentheses refer to characters found only in some of the species of the genus.)

1. *Tetracera*, 30 spp.: (2), 3, 4, 6, 9, (13), (14), (18), (19)
2. *Davilla*, 35 spp.: (2), 3, 4, 6, 9, (10), 13, 14, 15, 17, 18
3. *Curatella*, 2 spp.: 3, 4, 6, 9, 13, 14, 16, 18
4. *Doliocarpus*, 20 spp.: 3, 4, 6, 9, 10, 13, 16, 18
5. *Hibbertia*, 110 spp.: (2), (7), (8), (9), (10), (11), (12), (13), (14), (16), (18)
6. *Pachynema*, 4 spp.: 5, 7, 8, 10, 12, 13, 14, 18
7. *Acrotrema*, 12 spp.: 1, 5, 13, (16), (18), (19)
8. *Schumacheria*, 3 spp.: 2, 6, 9, 10, 11, 13, 14, 18
9. *Didesmandra*, 1 sp.: 6, 9, 11, 12, 13, 14, 18?
10. *Dillenia*, 55 spp.: 3, 5, (10), (14), 15, 16, (18), (19).

outlines, large capitula, yellow flower color, relatively short, broad style branches, a peculiar orange color of their pollen and in the number ($x = 9$), size and morphology of their chromosomes (Stebbins, 1953; Stebbins *et al.*, 1953). They are clearly more closely related to each other than either of them is to any other of the larger genera in the tribe. For both of them, the center of distribution is Pacific North America, and in the dry forests of northern California, species of the two genera are regularly sympatric.

The extreme forms; on the one hand *Microseris laciniata* (Hook.) Sch.-Bip. and *M. paludosa* (Greene) J. T. Howell and on the other hand *Agoseris grandiflora* (Nutt.) Greene and *A. retrorsa* (Benth.) Greene (the annual species *A. heterophylla* (Nutt.) Greene is somewhat apart) differ principally with respect to their stems, achenes, and pappus. In all of these respects, *Agoseris* is distinctly more specialized than *Microseris* (Fig. 6). Nevertheless, as shown in Fig. 6, these extreme forms are connected to each other by a more or less continuous series of intermediates.

The habitats occupied by the members of this series are of particular interest. *Microseris laciniata* and *M. paludosa*, have the primitive characters: leafy, branched stems; beakless achenes; and a paleaceous pappus of which the segments resemble most their homologues, calyx lobes. They occupy mesic woodlands or open meadows along the Pacific coast, in a climate that is cool in summer and mild in winter, with a long growing season. From species essentially similar to these, three lines of adaptive radiation can be detected. One of them led to a group of annuals inhabiting the warm, dry regions of California, that developed leafless, unbranched flowering stems, but retained the paleaceous pappus. A second line became adapted to moister habitats, and also developed unbranched, leafless stems. Its termini are two rather isolated, swamp or bog inhabiting species; *M.* (*apargidium*) *borealis* (Bung.) Sch.-Bip., which has a capillary pappus and occurs along the coast from northern California to Alaska; and the epappose *Phalacroseris bolanderi* A. Gray of the Sierra Nevada.

The third and most successful radiating line is the most interesting one for this discussion. It starts with *Microseris nutans* (Hook.) Sch.-Bip., which occurs at high altitudes in the mountains, and differs from *M. laciniata* in its depressed stature and short lower internodes. Next come two similar species, *M. cuspidata* Sch.-Bip. and *M. troximoides* A. Gray, which are widespread in the cold, arid steppes of the Great Basin, Rocky Mountain region and western Great Plains. Probably in response to a short growing season, they have evolved an unbranched, leafless scape bearing a single capitulum, a condition which persists in all of the remaining members of this evolutionary line. In response to the great adaptive advantage of wind dispersal in the open, windy places where they grow, they have evolved a pappus consisting of many slender, papery paleae, that offer a large amount of wind resistance. They are closely related to *M. alpestris* Q. Jones ex Cronquist, which differs only in its somewhat smaller size, and in having more numerous pappus parts, that appear like bristles but have broadened, palea-like bases. The habitat of *M. alpestris* is more mesic, but being alpine

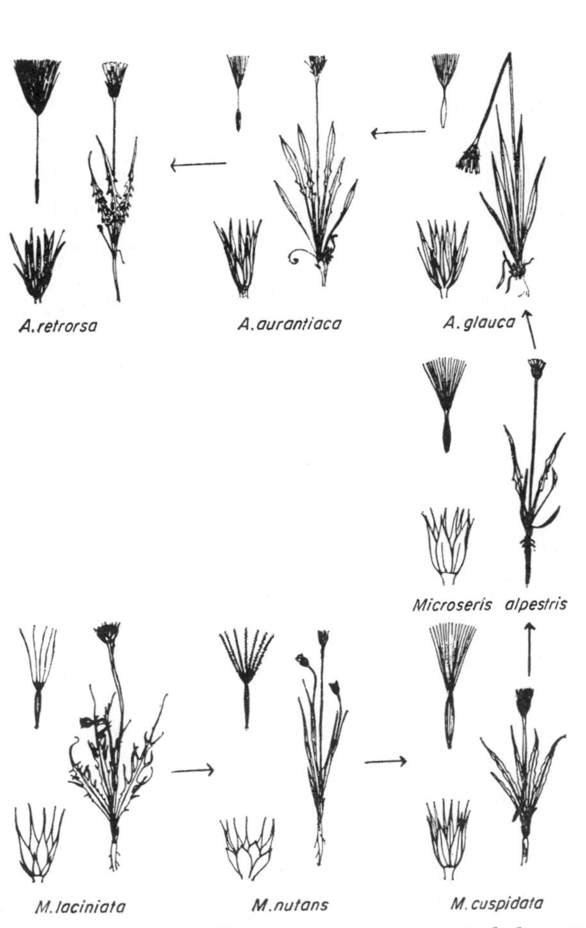

FIG. 6. Diagram showing species of the genera *Microseris* (including *Nothocalais*) and *Agoseris*, which exhibit a series of advancing specialization. The species on the left hand side of the diagram are adapted to mesic or semi-arid woodlands; those on the right hand side are adapted to more severe conditions: steppes and alpine slopes.

or subalpine, has an even shorter growing season. All of the differences between the two species could be explained as adaptations to this different habitat. Capillary pappus bristles, which are fewer celled than are paleae, can develop more rapidly. *M. alpestris* has been placed by various well recognized authorities in any one of the three genera: *Microseris* (Quentin Jones); *Nothocalais* (Chambers); or *Agoseris* (Greene and Gray, as *Troximon*).

The most primitive species of *Agoseris*, *A. glauca* (Pursh) Raf., occupies a geographic and ecological distribution about equal to that of *M. troximoides* and *M. cuspidata* combined. It has three subspecies, and includes both diploid and tetraploid cytotypes. It differs from *Microseris alpestris* chiefly in its completely capillary pappus and achenes bearing a short, stout beak. Next come a group of three species, all of them more mesic in habitat than *A. glauca*, and differing from that species in the longer, more slender beaks of their achenes. Two of them, *Agoseris aurantiaca* (Hook.) Greene and *A. elata* (Nutt.) Greene, inhabit mountain meadows; while the third, *A. apargioides* (Less.) Greene, occurs along the mild, foggy coastal strip of California, being therefore sympatric with *Microseris laciniata*. The ability of its ancestors to descend from the mountains and re-invade the coastal region may well have been based upon the production of a very large number of small, light, wind borne achenes. The two remaining species groups of *Agoseris*, which probably evolved from ancestors resembling *A. apargioides*, are the perennial *A. grandiflora* and *A. retrorsa*, and the annual *A. heterophylla*. The former two species are distinctive in their very large heads, containing numerous florets, and long beaked achenes. They have reached the ultimate degree of seed production, ease of dispersal, and of establishment, since *A. grandiflora* tends to become weedy. *Agoseris heterophylla* is the smallest and most ephemeral of all of the species in this complex of genera. These most advanced groups have acquired a very wide distribution, both geographically and ecologically.

In summary, the most specialized species of *Agoseris* occur sympatrically with the least specialized species of *Microseris*, but the former are much more common and widespread than the latter. The distribution of the intermediate species groups suggests that the specialized characteristics of *Agoseris* evolved in response to the strong selective pressures for rapid development and efficient seed dispersal that were exerted in the severe steppe and arid montane habitats where these intermediate species occurred, followed by selection for fecundity under more favorable conditions.

In the Old World, the evolution of the genus *Taraxacum* from its probable ancestors in the genus *Launaea* probably took a similar course. The most primitive species of *Taraxacum*, which resemble *Agoseris glauca* in both growth habit and the morphology of their achenes, occur in semi-arid mountain slopes in the Middle East (Handel-Mazzetti, 1907). They are linked to the most advanced species, such as the common dandelion (*T. officinale* Weber) of western Eurasia and the circumpolar *T. cerato-phorum* (Ledeb.) DC. complex, by species such as *T. kok-saghyz* Rodin, which inhabit moist mountain meadows in central Asia, and so are both morphologically and ecologically comparable to *Agoseris elata* and *A. aurantiaca*.

Table IV lists a number of similar examples in various plant families. Careful studies of other groups with this possibility in mind would doubtless reveal many more.

A CRITIQUE OF THE "EVIDENCE" PRESENTED BY J. W. BEWS

Admittedly, the line of thought followed in this paper is sharply at variance with the opinions that have been expressed by most botanists. The opinion of Bews (1927), that angiosperms originated and underwent their major diversification in moist tropical forests, is still dominant. At first sight, the evidence which he advanced to support his point of view appears impressive. Upon closer scrutiny, however, many flaws in it can be found. A case in point is the list of 70 groups in his "Appendix" on pp. 39–41. Twelve of these (nos 2, 3, 10, 17, 18, 21, 35, 48, 50, 58, 69, 70) are relevant only to the postulate that temperate groups are derived from tropical or subtropical groups, since the groups listed as "advanced" are temperate and mesic. I agree completely with Bews on this point, and question only his belief that among tropical and subtropical groups, plants adapted to semi-arid conditions are all derived from the inhabitants of mesic rain and cloud forests.

Of the remaining 58 groups, 22 are poorly classified (nos 1, 11, 13–16, 22–26, 32, 36–45). Although the "relatively primitive" families or tribes listed opposite these numbers are characterized as "hygrophilous or mesophytic," actually they contain many semi-xeric forms or even xerophytes, as in evergreen Fagaceae, Phytolaccaceae, Nyctaginaceae, Capparidaceae, Mimoseae, Caesalpinioideae, Sophoreae, Euphorbiaceae subf. Crotonoideae, and most tropical species of *Euphorbia*. In another five (nos 46, 49,

TABLE IV. Sympatric species of mesic habitats having different degrees of specialization, and intermediates found in more severe habitats

Family	Less specialized	More specialized	Intermediate	Habitat
Ranunculaceae	*Isopyrum* *Anemonella*	*Thalictrum* spp.	*T. petaloideum* L. *T. sultanabadense* Stapf	Semi-arid steppes
Primulaceae	*Lysimachia vulgaris* L.	*Primula veris* L., *P. elatior* (L.) Hill	*Primula* spp., *Androsace* spp.	Alpine and subalpine
Ericaceae	*Arbutus menziesii* Pursh	*Arctostaphylos andersonii* A. Gray *A. columbiana* Piper	*Arctostaphylos insularis* Greene *et al.*	Semi-xeric Mediterranean
Saxifragaceae	*Peltiphyllum peltatum* (Torr.) Engl.	*Boykinia elata* (Nutt.) Greene	*Bolandra californica* Gray *Suksdorfia ranunculifolia* Engl.	Wet rocks, cold climate
Hydrangeaceae	*Philadelphus lewisii* Pursh	*Whipplea modesta* Torr.	*Fendlerella utahensis* (S. Wats.) Heller	Semi-xeric
Rosaceae	*Potentilla* spp.	*Poterium sanguisorba* L.	*Poterium* spp.	Semi-xeric, Mediterranean
Leguminosae	*Genista anglica* L. *et al.*	*Ulex europaeus* L., *U. gallii* Planch., *U. minor* Roth	*Ulex* spp.	Semi-xeric, Mediterranean
Fagaceae	*Alnus* spp.	*Betula*, most spp.	*Betula nana* L., *B. pumila* L., *B. glandulosa* Michx	Arctic, subarctic
Onagraceae	*Clarkia* spp.	*Heterogaura heterandra* (Torr.) Coville	*Gaura* spp.	Semi-xeric
Compositae	*Helianthus*, other Heliantheae	*Ambrosia* spp. *Xanthium* spp.	*Iva* spp.	Semi-xeric, halophytic
Compositae	*Launaea*, *Sonchus* spp.	*Taraxacum* spp.	*Taraxacum primigenium* Hand.-Mazz.	Subalpine

53, 64, 68), the "advanced" groups are not regarded by modern botanists interested in phylogeny as derived from forms similar to those listed as "primitive" in the same example. For instance, Anacardiaceae, Celastraceae and Staphyleaceae are not regarded as advanced derivatives of Icacinaceae and Sapindaceae, nor are Onagraceae believed to be specialized derivatives of Myrtaceae. In another six of his examples (nos 20, 27, 29, 32, 34, 59), the groups designated as "advanced" are, with respect to floral structure and other characteristics, less specialized than those designated as "primitive". For instance *Clematis*, having its carpels reduced to one-seeded indehiscent achenes bearing elaborate plumose styles, is certainly not the most primitive genus in the Ranunculaceae. In the Rosales, the Saxifragaceae are not more advanced than the Pittosporaceae. The Hamamelidaceae, if they can be placed in the Rosales at all, would certainly qualify as one of its most advanced rather than primitive families. In the "Rutales" (or Sapindales), the Meliaceae are definitely not the most primitive family, since they have the stamens joined into a column, and the seeds with little or no endosperm. Among the Rutaceae themselves, the most generalized floral structure, with carpels almost separate and seeds having copious endosperm, is found in many shrubby forms that inhabit semi-arid habitats, while subfamilies such as the Flindersoideae, which consist entirely of mesic trees, have more specialized floral structure. Still another series of five examples (nos 5–9) involve specialized parasites or hemiparasites, and so are peripheral to the problem of major evolutionary trends. The remaining twenty examples are also peripheral, since even the groups designated as "primitive" are already rather specialized, and in several instances (nos 12, 28, 30, 31, 33, 54) represent relatively minor examples of adaptive radiation, which would be compatible with any theory of ecological evolution of angiosperms.

Since the work of Bews, botanists who support the hypothesis of angiosperm origins in the moist tropics use as their examples principally the Magnoliales or "woody Ranales," which in many respects are obviously primitive. Most of the members of this order are, however, relictual representatives of mono- or digeneric families that are only distantly related to each other (Bailey and Nast, 1945). The larger families are the Annonaceae, which with respect to anatomical characteristics are relatively specialized; the Monimiaceae and Lauraceae, which are relatively specialized with respect to both vegetative and reproductive characteristics; the Magnoliaceae; and the Winteraceae. The last two families consist chiefly or entirely of genera having a polyploid origin (Ehrendorfer *et al.*, 1968), and their diploid ancestors are extinct. Since polyploids very often invade regions having different ecological conditions from those to which their diploid ancestors were adapted, we are by no means justified in concluding that the diploid ancestors of these families were adapted to moist tropical forests.

The last argument brings us back to the point brought up at the beginning of the paper. If modern phenotypes are used as the principal guide to hypotheses concerning the conditions under which angiosperms first diversified during the Mesozoic era, the hypothesis of mesic origins receives some support. If, on the other hand, the principal of genetic uniformitarianism is regarded as more significant than the distribution of modern relictual phenotypes, the hypothesis of semi-xeric origins appears more plausible.

SUMMARY

The principle of genetic uniformitarianism is set forth, as follows. The processes of evolution have operated in the past essentially as they do now, but on different phenotypes. Hence macroevolution is best studied not by focusing attention on relictual phenotypes that are believed to be primitive, but on differences between modern, obviously related forms that are analogous to the differences between major taxonomic groups. Analogies of this kind can be made because every major morphological difference that separates families or orders can be found, in some groups, to separate related species or genera. The hypothesis of genetic uniformitarianism also has the corollary that the ecological conditions that promoted the initial differentiation of higher taxa in past geological periods must have been similar to those that promote a high rate of speciation in modern floras. Following this lead, numbers of species per genus were found to be highest in those genera of Pacific North America that inhabit semi-arid ecotones between forest and steppe or desert. Moreover, several examples of both genera and families (*Antennaria*, *Ceanothus*, Hydrangeaceae, Dilleniaceae) indicate that lines of specialization have radiated in many directions from initial generalized types that are adapted to semi-arid regions. The more specialized types include extreme xerophytes, extreme mesophytes, and adaptations to alpine conditions. In several examples, particularly *Microseris* and *Agoseris* (Compositae), two genera, one of them relatively generalized and the other more specialized, occur sympatrically in relatively favorable habitats. Intermediates between them occur in more severe habitats. This indicates that the morphological specialization has been brought about by an initial stage of selection imposed by a more severe habitat, followed by a second stage of selection for rapid growth, fecundity, vigor and greater ease of seed dispersal, fostered by a more favorable habitat.

REFERENCES

ANDERSON, E. (1937). *Am. Nat.* **71**, 223–235.

BAILEY, I. W. and NAST, C. G. (1945). *J. Arnold Arbor.* **26**, 37–47.

BEWS, J. W. (1927). "Studies in the Ecological Evolution of Angiosperms". New Phytologist Reprint no. 16.

COCKAYNE, L. (1899). *Trans. N.Z. Inst.* **31**, 354–398.

CRONQUIST, A. (1968). "The Evolution and Classification of Flowering Plants". Houghton Mifflin, Boston.

DICKISON, W. C. (1967). *J. Arnold Arbor.* **48**, 1–23.

DICKISON, W. C. (1968). *J. Arnold Arbor.* **49**, 317–329.

DICKISON, W. C. (1969). *J. Arnold Arbor.* **50**, 384–400.

DOBZHANSKY, T. (1970). "Genetics of the Evolutionary Process". Columbia University Press, New York.

EHRENDORFER, F., KRENDL, F., HABELER, E. and SAUER, W. (1968). *Taxon* **17**, 337–353.

GILG, E. and WERDERMANN, E. (1925). *In* "Die Naturliche Pflanzenfamilien" (A. Engler, ed.), 2nd ed., Vol. 21, pp. 7–36.

HACCIUS, B. (1954). *Öst. Bot. Z.* **101**, 285–303.

HANDEL-MAZZETTI, H. VON. (1907). "Monographie der Gattung *Taraxacum*". Leipzig und Wien.

HASKELL, G. (1954). *Phytomorphology* **4**, 140–152.

HOWELL, J. T. (1949). "Marin Flora". University of California Press, Berkeley.

HU, S-Y. (1954–1956). *J. Arnold Arbor.* **35**, 275–333; **36**, 52–109, 325–368; **37**, 15–90.

KUBITZKI, K. (1968). *Ber. dtsch. bot. Ges.* **81**, 238–251.

LAING, R. M. and GOURLAY, H. W. (1935). *Trans. R. Soc. N. Z.* **65**, 44–62.

LUBBOCK, J. (1892). "A Contribution to Our Knowledge of Seedlings". Vol. 1, London.

MCMINN, H. E. (1942). *In* "Ceanothus" pp. 131–279. Publications Santa Barbara Botanic Garden.

MAYR, E. (1963). "Animal Species and Evolution". Harvard University Press, Cambridge, Massachusetts.

MERRILL, E. D. (1923). "An Enumeration of Philippine Flowering Plants". 4 Volumes. Manila, Bureau of Printing.

MUNZ, P. A. and KECK, D. (1959). "A California Flora". University of California Press, Berkeley, California, U.S.A.

Nobs, M. A. (1963). *Carnegie Inst. of Wash. Publ.* no. 623, Washington, D. C.
Rizzini, C. de T. (1963). *In* "Simposio sôbre o Cerrado" (Editora de Universidade de Sao Paulo), pp. 125–177. Brazil.
Simpson, G. G. (1953). "The Major Features of Evolution". Columbia University Press, New York.
Smith, A. C. (1970). *Harold L. Lyon Arboretum Lecture Number* **1**, 1–26.
Sporne, K. R. (1956). *Biol. Rev.* **31**, 1–29.
Standley, P. C. (1928). *Contr. U. S. natn Herb.* **27**, 1–417.
Stebbins, G. L. (1953). *Madroño* **12**, 33–64.
Stebbins, G. L. (1956). *Am. J. Bot.* **43**, 890–905.
Stebbins, G. L., Jenkins, J. A. and Walters, M. S. (1953). *Univ. Calif. Publs Bot.* **26**, 401–430.
Stebbins, G. L. and Major, J. (1965). *Ecol. Monogr.* **35**, 1–35.
Takhtajan, A. (1969). "Flowering Plants: Origin and Dispersal". Oliver and Boyd, Edinburgh.
Takhtajan, A. L. (1970). "Proiskhozhdenie i Rasselenie Tsvetkovykh Rastenii". Akademia Nauk, Leningrad, U.S.S.R.
Valentine, D. H. (1970). *Repert. Spec. Nov. Regni. Veg.* **81**, 33–39.
Wright, S. (1931). *Genetics* **16**, 97–159.
Wright, S. (1950). *In* "Moderne Biologie, Festschrift für Hans Nachtsheim" (F. W. Peters, ed.), pp. 275–287. Berlin.

Section II

MAJOR GEOGRAPHICAL DISJUNCTIONS IN RELATION TO EVOLUTION AND MIGRATION

3 | Corresponding Taxa and their Ecological Background in the Forests of Temperate Eurasia and North America

JAN KORNAŚ

Institute of Botany, Jagellonian University, Kraków, Poland

INTRODUCTION

European botanists visiting temperate North America, especially its eastern part, are always impressed by the extent to which the forest floras of both continents look alike. The European–American affinities are very clear in the boreal zone* of coniferous forests, as well as in the nemoral zone of broad-leaved deciduous forests. In each zone, however, they manifest themselves in a different manner. Similar conditions occur in analogous latitudes of eastern Asia. The present paper attempts to analyse this phenomenon from phytogeographical, historical and evolutionary points of view. We shall first of all consider the ecological background, on the basis of the European situation which is by far the best known.

ECOLOGICAL BACKGROUND

In Central Europe, and especially in its northern part, two principal types of forests occur side by side: acidophil coniferous forests and subneutrophil broad-leaved deciduous forests. The coniferous type, poor in species and limited to more or less oligotrophic sites, has much affinity with the boreal taiga. The deciduous type, much more rich floristically, occupies eutrophic or mesotrophic sites and is optimally developed in the nemoral zone. Phytosociologists early recognized these two groups of forests as syntaxonomic units of very high rank (Vaccinio–Piceetea and Querco–Fagetea silvaticae; cf. Braun-Blanquet, 1964; Ellenberg, 1963; Medwecka-Kornaś *et al.*, 1966).

The differences between the classes are fundamental. They concern the

* Names of vegetation zones according to Sjörs (1963) and Walter (1968).

structure of vegetation, the dominant life forms of plants and their periodicity, as well as the influence of the community on microclimate, especially light conditions, and on soil forming processes. The structure of coniferous forests is very simple; their canopy is normally one-layered, the larger shrubs are mostly absent, the field layer is dominated by low shrubs, and the lowest layer of mosses and lichens is very well developed. In the deciduous forests the canopy is often composed of 2–3 layers of trees, the undergrowth is normally rather rich, in the field layer herbaceous plants prevail, and mosses and lichens are absent or very scarce. The trees dominating the coniferous woods are evergreen; in the broad-leaved forests they shed their leaves in autumn. Therefore the light conditions in the coniferous stands are rather uniform the whole year long, but in the deciduous forests there are two basically different light periods, one when the trees are leafless, another when they are covered with leaves. The annual rhythms of lesser plants very well correspond to these conditions. In the coniferous forests the seasonal aspects of the ground flora are rather obscure; an important role is played by evergreen plants. In the deciduous forests a very impressive spring aspect of flowering is observed, and a pronounced summer depression of activity of the ground flora. Most nemoral herbs have very short seasonal cycles, especially the numerous spring geophytes. The coniferous forests produce acid litter, which gives origin to unsaturated humus of "mor" type. This results in a more or less advanced podzolization of the soils. In deciduous forests a neutral saturated humus of "mull" type accumulates, which favours the formation of brown forest soils.

Forest communities of each class are further differentiated into subordinate units. Within the coniferous group, differences should be especially stressed between the pine forests occupying poor and dry soils or forming the initial stages of forest succession (e.g. after fire), and the spruce (and spruce-fir) forests which are usually more stable and require better sites. Among the deciduous forests there are also permanent zonal communities, as well as seral communities and local types of special habitats (e.g. flood plain forests).

In those regions where both the coniferous and the broad-leaved forests occur together, mixed communities of intermediate composition are met with. There are many reasons to believe that they are young derivative types which originated rather recently by intermingling of elements of the coniferous and the deciduous types. Putting aside these mixed communities, the floristic limits between the two principal classes of central

European forests prove to be surprisingly sharp. Very few plants have such a vast ecological amplitude that they are able to occur in typical coniferous and typical broad-leaved stands. Neither are there pairs of closely related corresponding species, one in the coniferous woods, and another in the broad-leaved deciduous types. The conclusion from these facts is that the floras of two principal classes of central European forests derive from different sources and—in spite of occurring side by side for a long time—still preserve their original individuality.

Similar differentiation of forests is also found in climatically analogous regions of eastern North America (Dansereau, 1959; Medwecka-Kornaś, 1961; Grandtner, 1966; Curtis, 1959; Küchler, 1964) and in eastern Asia (Lavrenko and Sochava, 1956; Hou *et al.*, 1956; Hara, 1959; Suzuki, 1966). The parallels existing concern not only the main classes of forest vegetation—coniferous and broad-leaved deciduous—but also their subordinate units (e.g. pine forests, spruce and spruce-fir forests, flood-plain forests, etc.). Not only are the habitat conditions, the vegetation structure and the plant life forms in corresponding communities similar; there are also distinct floristic affinities between them. These facts also are expressed in phytosociological classifications; the coniferous forests of all sectors of the Holarctic realm have been included in one syntaxonomic unit of the highest rank (class Vaccinio–Piceetea *sensu lato*—Braun-Blanquet *et al.*, 1939; group of classes Vaccinio–Piceetea—Braun-Blanquet, 1959), which is opposed to another unit of the same rank, comprising Holarctic deciduous forests of nemoral type (group of classes Aceri–Fagetea—Suzuki, 1966). Compare also Knapp, 1957, 1959, 1965; Medwecka-Kornaś, 1961.

HISTORICAL BACKGROUND

The principal central European classes of forest communities show remarkable differences in their phytogeographical connections. The coniferous forests are dominated by northern species ("boreale Arten" *sensu* Meusel *et al.*, 1965) which have very extensive, most often circumpolar distributions (Hultén, 1962, 1968). This makes the central European conifer woods very similar to the north-European, Siberian and North American taiga. The broad-leaved deciduous forests are composed chiefly of more southern "boreomeridional" species ("temperate Arten" *sensu* Meusel *et al.*, 1965) with rather narrow, mostly European ranges. Their affinities with eastern Asia and North America are very clear but mostly concern taxa of supraspecific rank (sections, genera), and only very few species.

Such a situation may be easily explained by what is known from fossil evidence about the late Tertiary and Quaternary history of the Holarctic forest vegetation (Cain, 1944; Walter and Straka, 1970). There is no doubt that in the Lower Tertiary, and even in Miocene times, a continuous circumpolar zone of nemoral forests existed throughout the whole Northern Hemisphere. It has been pushed gradually southwards, as the climate of northern latitudes has become more and more cool. The broad-leaved deciduous forests of this zone, most often called Arcto-tertiary, have been much more rich and more uniform than their present derivatives. This is an evidence of a long-lasting floristic exchange between all three Holarctic continents. One of its proofs is the presence of numerous nemoral species, which are today eastern-Asiatic or North American only, in the Pliocene fossil floras of Europe (Szafer, 1946, 1947, 1954; Walter and Straka, 1970). The general decrease of temperature and the resulting cataclysm of Pleistocene glaciations have divided the formerly continuous zone of nemoral forests into three refugia (in eastern Asia, eastern North America and southeastern Europe—Transcaucasia), the geographical isolation of which has resulted in development of numerous series of corresponding taxa. Unfortunately, we are not able to find the exact date at which the exchange of floras between nemoral refugia was definitely broken. We have to content ourselves with the statement that this must have happened not later than in the Pliocene.

Less distant, but also much less well-known, is the origin of the boreal zone of coniferous taiga. In this respect we still depend upon hypotheses, the best founded among them being that of Tolmachev (1954). According to this author, the taiga type of vegetation was formed very early, at least in the Miocene period, in the mountains of the temperate zone, surrounded by lowland vegetation of the nemoral type. The cradle of the montane taiga should be looked for most probably in the Pacific parts of Asia and North America. The zonal taiga of the northern lowlands is considerably younger; its expansion has been stimulated only by the deterioration of climate and the retreat of nemoral vegetation from northern latitudes in late Pliocene and Pleistocene. It is not possible to say whether during the maximal extensions of the Pleistocene ice-sheets the boreal forest zone preserved its continuity or not. At any rate, the geographical isolation of various sectors of this zone is not only less complete than in the nemoral zone, but also much more recently developed, and not earlier than the Pleistocene.

IDENTITY AND CORRESPONDENCE WITHIN THE CONIFEROUS FOREST FLORA

To define the phytogeographical connections within the Holarctic coniferous forests the method of distributional spectra of communities (Braun-Blanquet, 1964), has been used. Such spectra provide information on the role of various chorological groups of species in the community analysed, and make it possible to compare different communities with one another. The chorological groups of species are dealt with here on a broad basis. They have been distinguished according to the division of the Holarctic realm into large longitudinal sectors and arranged in order of their decreasing areas (Table I). Only thus has it been possible to emphasize the most general transcontinental and transoceanic affinities of the communities discussed. To simplify still more the comparison of spectra, all chorological groups have been fused into three higher categories of pluricontinental, bicontinental and unicontinental species. Thirty-two acidophil coniferous forests have been studied, representing the following types: (1) zonal lowland communities of the boreal taiga; (2) extrazonal communities, closely related to the preceding group, but occurring in the lowlands of the boreo-nemoral zone; (3) montane coniferous forests with boreal affinities, found at lower geographical latitudes. In each case, a collective spectrum of the whole community has been calculated, as well as separate spectra of each of its vegetation layers, viz., the tree (and shrub) layer, the field layer, and the moss layer. A detailed discussion of the materials and methods* used, and of the results obtained will be published elsewhere. In the present paper only some principal conclusions are summarized.

Distributional spectra of coniferous forests from various sectors of the Holarctic realm are rather similar. Widely distributed, pluricontinental (mainly circumboreal) and bicontinental species predominate; local (unicontinental) species are rather scarce. The spectra of individual vegetation layers of the same community are different from each other; the lower the layer, the broader the geographical range of its components.

The canopy of all forests studied is composed of local (unicontinental) European, Asiatic or North American species (and in the lowlands of

* Adapted from Kornaś (1970) and only slightly modified.

TABLE I. Chorological groups of species adopted in the present paper

Groups of chorological types	Chorological types	Europe	Asia: W. Siberia	Asia: E. Siberia and Pacific Asia	North America: Pacific North America	North America: Centr. and E. North America
Pluri-continental spp.	Cosmopolitan (cosm)	– – –	– – –	– – –	– – –	– – –
	Circumboreal (c-bor)	– – –	– – –	– – –	– – –	– – –
Bi-continental spp.	European–Asiatic Transcontinental (eur-as)	– – –	– – –	– – –		
	European–Western–Asiatic (eur-w-as)	– – –	– – –			
	European–American (eur-am)	– – –			– – –	– – –
	Asiatic–American (as-am)		– – –	– – –	– – –	– – –
Uni-continental spp.	European (eur)	– – –				
	Asiatic Transcontinental (as)		– – –	– – –		
	Western–Asiatic (w-as)		– – –			
	Eastern–Asiatic (e-as)			– – –		
	North-American Transcontinental (am)				– – –	– – –
	Western North-American (am pacif)				– – –	
	Eastern North-American (am atl)					– – –

eastern Europe also of bicontinental, European–Asiatic species). It has been already pointed out more than once, that in the Holarctic boreal forests there are no circumboreal trees (Table II). The dominant genera of conifers—*Pinus*, *Picea*, *Abies*, *Larix*, etc.—are differentiated into very typical series of corresponding taxa (Tolmachev, 1954); the same is also true of accessory trees, e.g. *Betula*, *Alnus* and *Sorbus* (Clausen, 1968). The rank of all these taxa is rather low, and many of them do not show any sterility barriers within the whole group, thus sharing a common gene-pool. Clausen, who called these groups "species clusters", suggested that their components should be reduced to subspecific status. Whether we accept this suggestion or not, the fact of geographical differentiation of the taiga trees is unquestionable. This is rather surprising, because it suggests that the evolutionary rate has been more rapid in trees than in low shrubs and herbs growing in their shade. Tolmachev (1954) tries to explain this paradox by pointing out that the canopy trees in the boreal forests must have been much more influenced by the vicissitudes of climate than the lesser vegetation.

The mosses and lichens in coniferous woods are almost exclusively circumboreal or cosmopolitan species, widely distributed in the whole Northern Hemisphere (Table III, group D). These taxa must have been very stable for a long time, because they do not show any evolutionary divergence throughout the enormous area they occupy.

Most interesting and manifold are the phytogeographical connections of the field layer of coniferous woods, which is somewhat richer in species (Fig. 1). In all the communities studied, circumboreal species play a very important role. Many are strictly limited to the coniferous forests ("characteristic species" of the phytosociologists). Numerous circumboreal taiga plants are, so far as we know, remarkably uniform, or they are only very slightly differentiated into infraspecific taxa with obscure geographical ranges (Table III, group A). Other species—by far less numerous—have differentiated into corresponding varieties or subspecies (group B). Series of corresponding species are surprisingly infrequent in the ground flora of coniferous woods, and their components are very close to each other (group C). The local (unicontinental) species in this habitat are mostly strangers which have invaded from other community types (sands, rocks, tall forb communities, etc.), where most probably they originated. These data seem to prove that the main stock of the flora of the taiga formation is a uniform group which has had a common origin and history, and which has not yet differentiated into corresponding taxa.

Table II. Presence classes of some trees in various coniferous forests of the Holarctic realm*

Trees	Chorological types	Mountains							E
		Pyrenees		Alps			Carpathia		
		1	2	3	4	5	6	7	8
Pinus uncinata Mill. ex Mirb. in Buff.	eur	5	.	.	.	.	.	.	.
Pinus silvestris L.	eur-as	.	.	.	2	.	.	.	.
Pinus contorta Dougl.	am pacif	.	.	.	.	.	.	.	.
Pinus banksiana Lamb.	am	.	.	.	.	.	.	.	.
Pinus cembra L. ssp. *cembra*	eur	.	.	1	2	.	.	.	.
Pinus cembra L. ssp. *sibirica* (Rupr.) Kryl.	as	.	.	.	.	.	.	.	.
Pinus pumila (Pall.) Rgl.	e-as	.	.	.	.	.	.	.	.
Pinus strobus L.	am atl	.	.	.	.	.	.	.	.
Pinus monticola Dougl.	am pacif	.	.	.	.	.	.	.	.
Abies alba Mill.	eur	.	5	.	.	2	.	5	1
Abies sibirica Ldb.	eur-w-as	.	.	.	.	.	.	.	.
Abies veitschii Lindbl.	e-as	.	.	.	.	.	.	.	.
Abies mariesii Mast.	e-as	.	.	.	.	.	.	.	.
Abies lasiocarpa Nutt.	am pacif	.	.	.	.	.	.	.	.
Abies balsamea Mill.	am	.	.	.	.	.	.	.	.
Abies grandis Lindl.	am pacif	.	.	.	.	.	.	.	.
Picea excelsa Link.	eur	.	.	5	5	5	5	5	5
Picea obovata Ldb.	eur-as	.	.	.	.	.	.	.	.
Picea yezoënsis (Sieb. et Zucc.) Carr.	e-as	.	.	.	.	.	.	.	.
Picea glauca Voss.	am	.	.	.	.	.	.	.	.
Picea mariana Britt.	am	.	.	.	.	.	.	.	.
Picea engelmannii Engelm.	am pacif	.	.	.	.	.	.	.	.
Larix decidua Mill. ssp. *decidua*	eur	.	.	3	4	.	.	.	.
Larix leptolepis Gord.	e-as	.	.	.	.	.	.	.	.
Larix laricina (Du Roi) K. Koch	am	.	.	.	.	.	.	.	.
Larix occidentalis Nutt.	am pacif	.	.	.	.	.	.	.	.
Tsuga diversifolia (Maxim.) Mast.	e-as	.	.	.	.	.	.	.	.
Tsuga heterophylla (Rafin.) Sarg.	am pacif	.	.	.	.	.	.	.	.
Thuja standishii Carr.	e-as	.	.	.	.	.	.	.	.
Thuja occidentalis L.	am atl	.	.	.	.	.	.	.	.
Thuja plicata D.Don.	am pacif	.	.	.	.	.	.	.	.
Sorbus aucuparia L.	eur-w-as	4	3	3	4	3	1	+	4
Sorbus sibirica Hedl.	eur-as	.	.	.	.	.	.	.	.
Sorbus conmixta Hedl.	e-as	.	.	.	.	.	.	.	.
Sorbus decora (Sarg.) Britt.	am atl	.	.	.	.	.	.	.	.
Sorbus americana Marsh.	am atl	.	.	.	.	.	.	.	.
Sorbus scopulina Greene	am pacif	.	.	.	.	.	.	.	.

* See footnotes on pp. 48–9.

Lowlands ntral rope 12	13	N. Eur. 14	Ural and Siberia 15	16	17	18	Asia Mountains of Japan 19	20	21	22	23	24	North America Lowlands W. part 25	26	E. part 27	28	29	Mountains (Pacific NW.) 30	31	32
.	.	.	.	.	.	.	.	.	.	.	.	.	.	.	.	.	.	.	.	.
5	5	.	.	.	.	.	.	.	.	.	.	.	.	.	.	.	.	.	.	.
.	.	.	.	.	.	.	.	.	.	.	.	.	2	.	.	.	.	.	.	1
.	.	.	.	.	.	.	.	.	.	.	.	.	.	1	1	.	3	.	.	.
.	.	.	.	.	.	.	.	.	.	.	.	.	.	.	.	.	.	.	.	.
.	.	.	5	4	5	4	.	.	.	.	.	.	.	.	.	.	.	.	.	.
.	.	.	.	.	.	.	5	.	.	.	.	.	.	.	.	.	.	.	.	.
.	.	.	.	.	.	.	.	.	.	.	.	.	.	.	2	1	.	.	.	.
.	.	.	.	.	.	.	.	.	.	.	.	.	.	.	.	.	.	4	2	1
.	.	.	.	.	.	.	.	.	.	.	.	.	.	.	.	.	.	.	.	.
.	.	2	5	5	5	5	.	.	.	.	.	.	.	.	.	.	.	.	.	.
.	.	.	.	.	.	.	.	5	5	.	.	.	.	.	.	.	.	.	.	.
.	.	.	.	.	.	.	+	.	.	4	5	4	.	.	.	.	.	.	.	.
.	.	.	.	.	.	.	.	.	.	.	.	.	3	2	.	.	.	.	.	.
.	.	.	.	.	.	.	.	.	.	.	.	.	.	.	5	5	5	.	.	.
.	.	.	.	.	.	.	.	.	.	.	.	.	.	.	.	.	.	.	4	5
5	5	.	.	.	.	.	.	.	.	.	.	.	.	.	.	.	.	.	.	.
.	.	5	5	5	5	1	.	.	.	.	.	.	.	.	.	.	.	.	.	.
.	.	.	.	.	.	.	.	.	.	1	3	.	.	.	.	.	.	.	.	.
.	.	.	.	.	.	.	.	.	.	.	.	.	5	5	5	5	2	.	.	.
.	.	.	.	.	.	.	.	.	.	.	.	.	3	5	4	5	5	.	.	.
.	.	.	.	.	.	.	.	.	.	.	.	.	.	.	.	.	.	.	1	2
.	.	.	.	.	.	.	.	.	.	.	.	.	.	.	.	.	.	.	.	.
.	.	.	.	.	.	.	.	5	5	.	.	.	.	.	.	.	.	.	.	.
.	.	.	.	.	.	.	.	.	.	.	.	.	.	1	.	.	1	.	.	.
.	.	.	.	.	.	.	.	.	.	.	.	.	.	.	.	.	.	5	1	1
.	.	.	.	.	.	.	.	4	3	1	5	.	.	.	.	.	.	.	.	.
.	.	.	.	.	.	.	.	.	.	.	.	.	.	.	.	.	.	5	.	1
.	.	.	.	.	.	.	.	.	.	.	1	.	.	.	.	.	.	.	.	.
.	.	.	.	.	.	.	.	.	.	.	.	.	.	.	2	.	.	.	.	.
.	.	.	.	.	.	.	.	.	.	.	.	.	.	.	.	.	.	5	5	.
.	2	5	.	.	.	.	.	.	.	.	.	.	.	.	.	.	.	.	.	.
	.	.	5	5	5	5	.	.	.	.	.	.	.	.	.	.	.	.	.	.
	.	.	.	.	.	.	.	5	5	3	4	4	.	.	.	.	.	.	.	.
	.	.	.	.	.	.	.	.	.	.	.	.	.	.	5	5	4	.	.	.
	.	.	.	.	.	.	.	.	.	.	.	.	.	.	2	1	3	.	.	.
	.	.	.	.	.	.	.	.	.	.	.	.	.	.	.	.	.	+	1	3

TABLE III. Presence classes of some field layer and ground layer plants in various coniferous forests of the Holarctic realm

Species	Chorological types	Mountains								Eu
		Pyrenees		Alps			Carpathian			
		1	2	3	4	5	6	7	8	9
A. Circumboreal species of the field layer:										
Gymnocarpium dryopteris (L.) Newm.	c-bor	1	1	5	2	2	2	2	1	2
Pyrola secunda L.	c-bor	+	2	4	4	4	.	1	.	2
Dryopteris dilatata (Hoffm.) Gray	c-bor	1	3	3	.	3	5	2	5	3
Lycopodium annotinum L. ssp. *annotinum*	c-bor	.	.	4	2	1	5	2	1	2
Vaccinium myrtillus L.	c-bor	5	5	5	5	3	5	5	5	5
Moneses uniflora (L.) Gray	c-bor	.	1	5	2	2	.	3	.	1
Thelypteris phegopteris (L.) Slosson	c-bor	.	1	.	.	1	3	1	+	.
Goodyera repens (L.) R.Br.	c-bor	.	2	1	1	1	.	.	.	.
Athyrium filix-femina (L.) Roth.	c-bor	1	2	1	.	4	2	4	+	5
Listera cordata (L.) R.Br.	c-bor	.	.	3	1	.	2	2	.	.
Equisetum silvaticum L.	c-bor	.	.	.	.	.	1	.	.	.
Dryopteris spinulosa (Mill.) Ktze.	c-bor	.	.	.	.	.	1	4	1	3
Lycopodium selago L. ssp. *selago*	c-bor	.	.	1	+	2	4	2	1	2
Circaea alpina L.	c-bor	.	.	.	.	2	.	1	.	.
Streptopus amplexifolius (L.) DC.	c-bor	+	.	.	.	.	1	.	2	.
Corallorhiza trifida Châtelain	c-bor	.	.	3	1	+	.	1	.	.
Blechnum spicant (L.) Roth.	c-bor	1	2	.	.	1	3	4	1	1
Pteridium aquilinum (L.) Kuhn.	cosm	.	.	.	.	2	.	.	.	.
Galium triflorum Michx.	c-bor	.	.	.	.	.	.	.	.	.
Lycopodium complanatum L. *s.str.*	c-bor	.	.	.	.	.	.	.	.	.
Pyrola chlorantha Sw.	c-bor	.	.	.	.	1	.	.	.	.
Pyrola minor L.	c-bor	1	1	1	.	.	.	1	.	.
Lycopodium clavatum L.	cosm	.	.	.	.	.	.	1	.	.
Pyrola rotundifolia L.	c-bor	.	.	.	.	.	.	.	.	.
Monotropa hypopitys L.	c-bor	.	1	.	+	1	.	1	.	.
Empetrum nigrum L. s.l.	c-bor	.	.	.	.	.	.	.	.	.
Arctostaphylos uva-ursi (L.) Spreng.	c-bor	.	.	.	.	.	.	.	.	.
B. Circumboreal species of the field layer with corresponding infraspecific taxa:										
Oxalis acetosella L.										
ssp. *acetosella*	eur-as	3	5	5	4	5	5	5	1	5
ssp. *griffithii* (Hook.f. et Edgew.) Hara.	e-as	.	.	.	.	.	.	.	.	.
ssp. *montana* (Raf.) Hult.	am atl	.	.	.	.	.	.	.	.	.
Vaccinium vitis-idaea L.										
ssp. *vitis-idaea*	eur-as	.	.	5	5	1	.	1	1	.
ssp. *minus* (Lodd.) Hult.	as-am	.	.	.	.	.	.	.	.	.
Rubus idaeus L.										
ssp. *idaeus*	eur-w-as	1	1	.	.	.	2	3	3	3
ssp. *melanolasius* (Dieck) Focke	as-am	.	.	.	.	.	.	.	.	.
Linnaea borealis L.										
ssp. *borealis*	eur-as	.	.	5	3	.	.	.	.	.
ssp. *americana* (Forbes) Hult.	am	.	.	.	.	.	.	.	.	.
ssp. *longiflora* (Torr.) Hult.	am pacif	.	.	.	.	.	.	.	.	.

Asia													North America							
Lowlands			Ural				Mountains						Lowlands					Mountains		
ntral rope		N. Eur.	and Siberia				of Japan						W. part		E. part			(Pacific NW.)		
12	13	14	15	16	17	18	19	20	21	22	23	24	25	26	27	28	29	30	31	32
.	1	4	4	5	5	3	.	1	.	.	.	.	.	1	5	5	2	2	3	.
4	3	3	1	1	3	.	.	3	5	.	.	.	5	4	4	4	4	5	3	2
1	.	2	4	5	5	.	.	1	1	5	.	.	.	.	5	5	1	.	2	.
.	5	4	5	2	3	4	.	.	.	.	.	.	1	1	5	4	3	.	.	.
5	5	5	5	2	2	3	.	.	.	.	.	.	.	.	.	.	.	4	.	.
.	1	+	1	.	3	.	.	.	.	.	.	.	4	1	4	4	.	.	.	+
.	.	4	3	4	4	1	.	1	1	+	1	.	.	.	1	3	.	.	.	.
3	.	3	3	3	2	3	.	1	2	.	.	.	2	2	4	3	2	.	.	.
.	2	.	.	.	.	1	.	.	.	.	.	.	.	.	.	1	.	.	4	2
.	2	1	.	.	.	.	.	3	1	.	1	.	.	1	.	3	4	.	1	.
.	.	5	2	3	3	1	.	.	.	.	.	.	.	1	3	1	4	.	.	.
1	5	3	.	.	.	3	.	.	.	.	.	.	.	.	.	.	.	.	.	.
.	+	.	.	.	.	.	.	.	.	.	.	.	.	.	.	.	.	.	.	.
.	.	+	.	.	.	1	.	.	.	.	.	.	.	.	3	.	.	1	2	1
.	.	.	.	.	.	.	.	.	.	.	.	.	.	.	.	4	1	.	2	1
.	1	.	.	.	.	.	.	.	.	.	.	.	3	1	.	.	.	.	.	.
.	.	.	.	.	.	.	.	.	.	.	.	.	.	.	.	.	.	.	.	.
2	.	.	.	.	.	.	.	.	.	.	.	.	.	.	3	.	1	2	2	1
.	.	+	.	.	.	1	.	.	.	.	.	.	.	.	5	1	.	1	5	5
+	.	+	.	.	.	1	.	.	.	.	.	.	.	.	.	1	1	1	.	.
.	.	.	.	.	.	.	.	.	.	.	.	.	1	4	2	.	.	4	.	1
.	.	.	.	1	.	1	.	.	.	.	.	.	.	.	.	.	.	.	.	.
+	.	.	.	.	.	.	.	.	.	.	.	.	.	.	2	.	1	.	.	.
1	1	1	1	.	3	.	.	.	.	.	.	.	.	.	.	.	.	.	.	.
1	.	.	.	.	.	.	.	1	.	.	.	.	.	.	.	.	.	.	.	.
.	.	.	.	.	.	.	.	.	.	.	.	.	3	4	.	.	.	.	.	.
+	.	.	.	.	.	.	.	.	.	.	.	.	2	.	.	.	.	.	.	.
4	3	5	5	5	5	4	2	.	.	5	4	4	.	.	.	.	.	.	.	.
.	.	.	.	.	.	.	.	.	.	+	.	.	.	.	.	.	.	.	.	.
.	.	.	.	.	.	.	.	.	.	.	.	.	.	.	3	5	2	.	.	.
5	5	5	4	1	1	4	.	.	.	.	.	.	.	.	.	.	.	.	.	.
.	.	.	.	.	.	.	3	5	.	.	.	.	4	5	.	.	.	.	.	.
.	3	3	+	5	3	.	.	.	.	.	.	.	.	.	.	.	.	.	.	.
.	.	.	.	.	.	.	.	.	.	.	.	.	2	.	4	4	3	.	.	.
.	.	5	4	5	5	4	.	.	.	.	.	.	.	.	.	.	.	.	.	.
.	.	.	.	.	.	.	.	.	.	.	.	.	5	4	5	5	5	5	.	.
.	.	.	.	.	.	.	.	.	.	.	.	.	.	.	.	.	.	.	2	4

Table III. Presence classes of some field layer and ground layer plants in various coniferous forests of the Holarctic realm—*continued*

Species	Chorological types	Eu								
		Mountains								
		Pyrenees		Alps			Carpathian			
		1	2	3	4	5	6	7	8	9
Chimaphila umbellata (L.) Barton										
ssp. *umbellata*	eur-as	.	.	.	.	.	.	.	.	.
ssp. *cisatlantica* (Blake) Hult.	am atl	.	.	.	.	.	.	.	.	.
ssp. *occidentalis* (Rydb.) Hult.	am pacif	.	.	.	.	.	.	.	.	.
C. Circumboreal clusters of corresponding species of the field layer:										
Majanthemum bifolium (L.) F.W.Schm.	eur-as	.	.	2	2	3	2	4	+	2
Majanthemum dilatatum (How.) Nels. et Macbr.	as-am	.	.	.	.	.	.	.	.	.
Majanthemum canadense Desf.	am	.	.	.	.	.	.	.	.	.
Trientalis europaea L.	c-bor	.	.	.	.	.	.	.	1	.
Trientalis borealis Raf.										
ssp. *borealis*	am atl	.	.	.	.	.	.	.	.	.
ssp. *latifolia* (Hook.) Hult.	am pacif	.	.	.	.	.	.	.	.	.
Rubus saxatilis L.	eur-as	.	.	2	5	1	.	.	.	.
Rubus pubescens Raf.	am	.	.	.	.	.	.	.	.	.
D. Mosses:										
Hylocomium proliferum (L.) Lindb.	c-bor	4	5	5	5	5	1	4	+	2
Entodon schreberi (Willd.) Mönkem.	c-bor	1	.	5	5	1	5	5	3	3
Rhytidiadelphus triqueter (L.) Warnst.	c-bor	3	4	5	5	5	1	4	+	.
Dicranum scoparium (L.) Hedw.	c-bor	3	1	5	5	3	5	5	4	4
Polytrichum attenuatum Menz.	c-bor	1	2	1	.	3	4	5	5	3
Ptilium crista-castrensis (L.) de Not.	c-bor	.	1	3	.	.	.	.	1	.
Polytrichum commune L.	cosm	.	.	.	.	.	2	1	1	.
Sphagnum girgensohnii Russ.	c-bor	.	.	.	.	.	4	4	1	.
Dicranum fuscescens Turn.	c-bor	.	.	.	.	.	1	.	1	.
Dicranum undulatum Ehrh.	c-bor	.	.	.	.	.	.	.	.	.
Hylocomium umbratum Bryol. eur.	c-bor	.	.	.	.	.	1	1	.	.
Rhytidiadelphus loreus (Hedw.) Warnst.	eur-am	.	1	.	.	.	5	4	2	2
Plagiothecium undulatum (L.) B.S.G.	eur-am	.	.	.	.	.	4	3	5	1

Footnotes to Tables II and III.

Chorological types as in Table I.

Presence classes: 5 = 81–100%, 4 = 61–80%, 3 = 41–60%, 2 = 21–40%, 1 = 1–20%.

Plant community numbers:

1. Rhododendro–Pinetum uncinatae typicum (Rivas-Martinez, 1968b);
2. Galio rotundifolii–Abietetum albae (Rivas-Martinez, 1968a);
3. Piceetum (excelsae) subalpinum myrtilletosum, var. *Linnaea* (Braun-Blanquet *et al.*, 1954);
4. Piceetum (excelsae) subalpinum vaccinietosum, var. *Peltigera–Hylocomium* (Braun-Blanquet *et al.*, 1954);
5. Piceetum (excelsae) montanum galietosum, var. *Eurhynchium striatum–Picea* (Braun-Blanquet *et al.*, 1954);
6. Piceetum (excelsae) tatricum subnormale (Medwecka-Kornaś, 1955);
7. Piceetum (excelsae) tatricum abietetosum (Medwecka-Kornaś, 1955);
8. Piceetum (excelsae) carpaticum myrtilletosum + filicetosum (Stuchlik, 1968);
9. Abieti–Piceetum (excelsae) montanum (Stuchlik, 1968);
10. Piceetum (excelsae) oricarpaticum normale (Pawłowski and Walas, 1949);
11. Vaccinio myrtilli–Pinetum (silvestris) typicum, var. typ. (Sokołowski, 1965);
12. Vaccinio myrtilli–Pinetum (silvestris) typicum, var. *Carex digitata* (Sokołowski, 1966);
13. Sphagno girgensohnii–Piceetum (excelsae), "sous-ass. moussue" (Czerwiński, 1966);

[L]owlands [Cen]tral [Euro]pe	N. Eur.	Asia: Ural and Siberia				Asia: Mountains of Japan						North America: Lowlands, W. part		North America: Lowlands, E. part			North America: Mountains (Pacific NW.)		
13	14	15	16	17	18	19	20	21	22	23	24	25	26	27	28	29	30	31	32
.	.	.	.	.	.	.	.	.	.	.	.	.	.	.	.	.	.	.	.
.	.	.	.	.	.	.	.	.	.	.	.	.	.	.	.	.	.	.	.
.	.	.	.	.	.	.	.	.	.	.	.	.	.	.	.	.	5	2	3
3	5	5	5	5	5	.	.	.	.	.	.	.	.	.	.	.	.	.	.
.	.	.	.	.	.	3	5	5	+	1	5	.	.	.	.	.	.	.	.
.	.	.	.	.	.	.	.	.	.	.	.	.	1	5	5	4	.	.	.
1	5	4	5	4	4	.	2	4	.	.	.	.	.	.	.	.	.	.	.
.	.	.	.	.	.	.	.	.	.	.	.	.	.	5	5	4			
.	.	.	.	.	.	.	.	.	.	.	.	.	.	.	.	.	.	.	+
1	2	1	.	1	.	.	.	.	.	.	.	.	.	.	.	.	.	.	.
.	.	.	.	.	.	.	.	.	.	.	.	.	2	5	4	.	.	.	.
5	5	5	5	5	5	5	...	...	1	4	4	...	...	...	...	...	5	...	...
5	5	5	5	5	4	5	...	...	3	4	5	...	...	...	...	...	5	...	...
2	4	3	3	5	1	2	...	...	.	.	.	...	...	...	...	...	4	...	...
1	+	.	.	.	.	.	...	...	.	.	.	...	...	...	...	...	3	...	...
3	.	.	.	.	.	.	...	...	.	.	.	...	...	...	...	...	.	...	...
.	2	5	5	5	4	5	...	...	.	.	.	...	...	...	...	...	5	...	...
2	5	4	2	1	3	.	...	...	.	.	.	...	...	...	...	...	.	...	...
4	5	1	.	.	1	3	...	...	.	.	5	...	...	...	...	...	.	...	...
.	.	5	5	4	.	3	...	...	2	2	.	...	...	...	...	...	5	...	...
2	1	3	1	.	1	.	...	...	.	.	.	...	...	...	...	...	1	...	...
.	.	3	1	3	.	.	...	...	.	.	.	...	...	...	...	...	.	...	...
.	.	.	.	.	.	.	...	...	.	.	.	...	...	...	...	...	.	...	...
.	.	.	.	.	.	.	...	...	.	.	.	...	...	...	...	...	.	...	...

14. Piceetum (obovatae) sorbosum, filicetosum, equisetosum et humidum, var. div. (Korchagin, 1929);
15. Abietetum (sibiricae) myrtillosum (Gorchakovsky, 1954);
16. Abietetum (sibiricae) austriaci–dryopteridetosum (Gorchakovsky, 1954);
17. Abietetum (sibiricae) aconitosum (Gorchakovsky, 1954);
18. "Tyomnokhvoynaya taiga" (Krasnoborov, 1965);
19. Vaccinio–Pinetum pumilae (Miyawaki *et al.*, 1969a);
20. Abietetum mariesii, *Vaccinium vitis-idaea* subass., var. typ. (Miyawaki *et al.*, 1969b);
21. Abietetum mariesii, *Cacalia adenostyloides* subass., var. *Larix leptolepis* (Miyawaki *et al.*, 1969b);
22. *Abies mariesii* ass., *Cacalia adenostyloides* subass. (Miyawaki *et al.*, 1967);
23. *Abies mariesii* ass., *Tsuga diversifolia* subass. (Miyawaki *et al.*, 1967);
24. Abietetum mariesii, *Shortia soldanelloides* subass. (Miyawaki *et al.*, 1968);
25. *Picea glauca*, *Populus/Salix/Shepherdia*, *Solidago–Empetrum* community (La Roi, 1967);
26. *Picea glauca*, *Abies–Populus//Rosa/Mertensia*, *Corylus/Diervilla/Aster–Anemone* community (La Roi, 1967);
27. *Picea glauca*, *Abies/Acer–Pyrus//Clintonia–Oxalis–Gaultheria* community (La Roi, 1967);
28. *Picea mariana*, *Rosa–Ribes/Mitella–Mertensia*, *Salix//Empetrum* community (La Roi, 1967);
29. *Picea mariana*, *Kalmia*, *Picea glauca–Populus//Rubus chamaemorus* community (La Roi, 1967);
30. *Thuja plicata–Tsuga heterophylla* forest (Kornaś, 1970);
31. *Thuja plicata–Pachistima myrsinites* ass. (Daubenmire and Daubenmire, 1968);
32. *Abies grandis–Pachistima myrsinites* ass. (Daubenmire and Daubenmire, 1968).

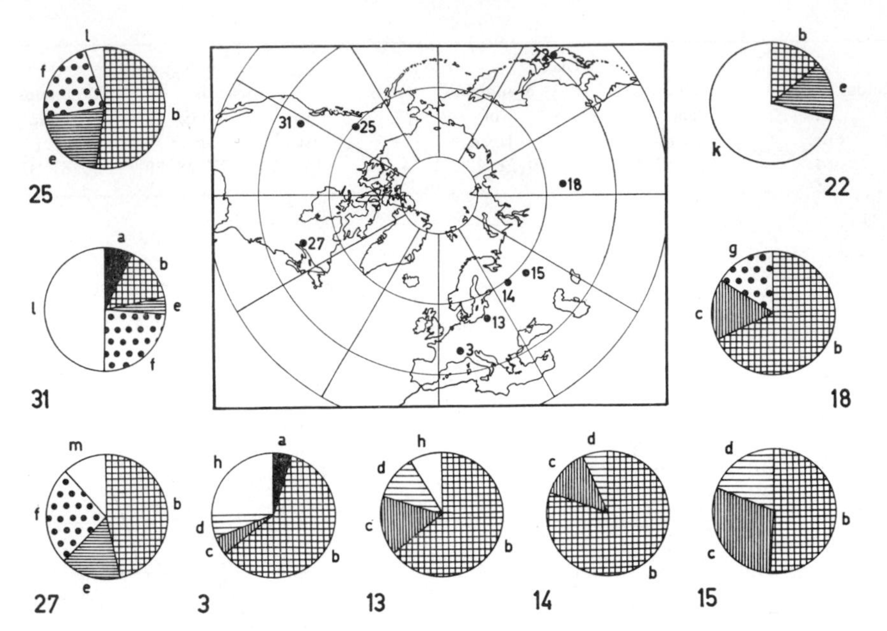

FIG. 1. Distributional spectra of field layer plants in 9 Holarctic coniferous forest communities. Points on the map show the regions where the communities occur. Community numbers as in Table II. Chorological groups: a, b–pluricontinental species (a–Cosmopolitan, b–Circumboreal), c–e–bicontinental species (c–European–Asiatic Transcontinental, d–European–Western–Asiatic, e–Asiatic–American), f–m–unicontinental species (f–North–American Transcontinental, g–Asiatic Transcontinental, h–European, k–Eastern–Asiatic, l–Western–North-American, m–Eastern–North-American).

However, the distributional spectra of various Holarctic coniferous forests are not quite uniform. They display some interesting regional features (Fig. 1), which may be summarized in the following points:

(1) Local species are especially scarce in the zonal boreal taiga of Euro-Sibiric and North American lowlands.

(2) Montane coniferous forests are phytogeographically more distinct than the lowland types.

(3) Montane forests of the lower latitudes in Pacific North America and eastern Asia have a peculiar character, because widely distributed species, especially of circumpolar range, are relatively scarce in them and local species prevail (Kornaś, 1970). It should be especially emphasized that these local species belong to groups which are strictly attached to the coniferous type of vegetation, and represent most often the same genera which include the common circumboreal taiga plants: *Pyrola*, *Chimaphila*,

Monotropa, *Vaccinium*, *Listera*, *Corallorhiza*, *Goodyera*, etc. Such facts indicate that there is a rich genetic centre of the coniferous forest flora in the mountains on both sides of the northern Pacific. Nearly all the widely distributed coniferous forest species occur there, too.

Thus the results of a phytogeographical analysis support the hypothesis of Tolmachev (1954) on the origin and development of the taiga formation. It seems very probable (and more and more paleobotanical data are in favour of such an interpretation) that the boreal coniferous forests are a very old plant formation, dating back perhaps as far as the lower Tertiary. This explains their primitive characteristics. These are (a) the dominance of ancient genera of conifers (*Picea*, *Abies*, *Pinus*, *Larix*, etc.), (b) the presence in the field layer of peculiar, isolated, monotypic or oligotypic genera and subgenera (*Linnaea*, *Trientalis*, *Majanthemum*, *Cornus* subgen. *Arctocrania*, *Vaccinium* subgen. *Vitis-idaea*, etc.), or even taxa of still higher rank, including one family (Pyrolaceae), (c) their archaic morphophysiological characters (e.g. the dominance of evergreens), (d) their advanced ecological specialization (e.g. in mycotrophic Orchidaceae and Monotropoideae), etc. In the supposed cradle of taiga formation in the northern Pacific mountains, some links between temperate coniferous forest plants and their probable subtropical ancestral types are still preserved. On the other hand, even in that place, there are no perceptible affinities between coniferous and broad-leaved deciduous formations. This confirms the supposition, that the two principal types of temperate forest vegetation in the Northern Hemisphere have had an independent origin and have been developing side by side for thousands and millions of years without mingling their components, most probably due to ecological (especially edaphic) isolation.

CORRESPONDING TAXA IN THE BROAD-LEAVED DECIDUOUS FORESTS

Phytogeographical connections of the Holarctic broad-leaved deciduous (nemoral) forests are quite different from those of the coniferous woods. In all main vegetation layers (tree, shrub and herb layers), local (unicontinental) species prevail (though in lowland forests of the eastern part of Central Europe, they occur together with bicontinental European–western-Asiatic species). Because of this, the distributional spectra of nemoral communities based upon lists of species do not reveal the mutual connections, but strongly emphasize the differences between various sectors of the Holarctic realm (Fig. 2). However, remarkable similarities may be demonstrated when the coefficients of generic affinity of the communities have been calculated.

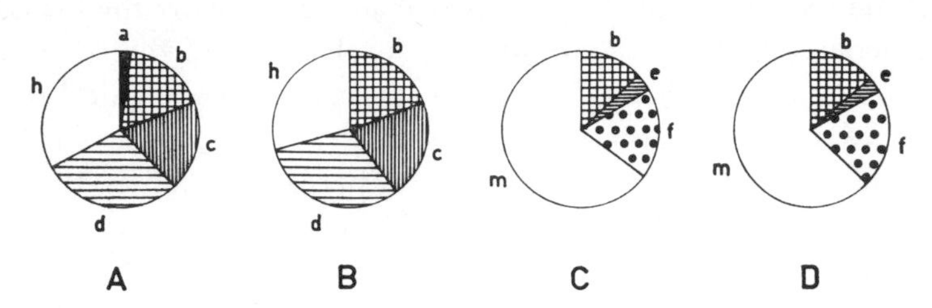

FIG. 2. Distributional spectra of field layer plants in 4 broad-leaved deciduous forest communities in Poland (A, B) and southeastern Canada (C, D). A–Tilio–Carpinetum (= Querco–Carpinetum), var. *Ranunculus cassubicus* (Medwecka-Kornaś, 1952); B–Fraxino–Ulmetum, var. *Ulmus campestris* (Wojterski, 1960); C–Caryo–Aceretum (Grandtner, 1966); D–Aceretum sacchari laurentianum (Grandtner, 1966). Chorological groups as in Fig. 1.

Both methods have been used by the present author to compare some few nemoral communities of Central Europe and eastern North America. The coefficients of generic similarity have been calculated according to the method applied by Szymkiewicz (1947) to compare the whole floras of various regions with one another. However, cosmopolitan genera have not been excluded. On both sides of the Atlantic similar regions have been selected: Southeastern Canada and Poland. They have been deeply affected by Pleistocene glaciations and definitively colonized by forests not before the Holocene. The results of this analysis, in spite of being scarce and preliminary, do make it possible to draw some general conclusions.

There are very few species in common in the nemoral forests of southeastern Canada and Poland. Mostly they are Pteridophytes (*Athyrium filix-femina* (L.) Roth., *Dryopteris spinulosa* (Müll.) Ktze., *Equisetum silvaticum* L., *Gymnocarpium dryopteris* (L.) Newm., *Thelypteris phegopteris* (L.) Slosson, etc.), and only exceptionally Phanerogams (*Milium effusum* L.). However, the generic affinities between the deciduous forests of the two regions compared are very clear (Table IV). This is due above all to the presence of numerous pairs of corresponding taxa of trees, shrubs and herbs. Some of them are as follows:

In Southeastern Canada	In Poland
Trees	
Carpinus caroliniana Walt.	*C. betulus* L.
Fagus grandifolia Ehrh.	*F. silvatica* L.
Fraxinus nigra Marsh.	*F. excelsior* L.
Padus serotina (Ehrh.) Agardh	*P. avium* Mill.
Quercus alba L.	*Q. robur* L.
Tilia americana L.	*T. cordata* Mill.
Ulmus americana L.	*U. laevis* Pall.
Ulmus rubra Muhl.	*U. carpinifolia* Gled., etc.
Shrubs	
Sambucus racemosa L.	
ssp. *pubens* (Michx) House	ssp. *racemosa*
Viburnum trilobum Marsh.	*V. opulus* L. etc.
Herbs	
Actaea rubra (Ait.) Willd.	*A. spicata* L.
Asarum canadense L.	*A. europaeum* L.
Circaea quadrisulcata (Maxim.) Franch. & Sav.	*C. lutetiana* L.
Hepatica acutiloba DC.	*H. triloba* Gilib.
H. americana (DC.) Ker.	
Impatiens capensis Meerb.	*I. noli-tangere* L.
Sanicula trifoliata Bicknell	*S. europaea* L., etc.

If a larger territory, including never-glaciated areas in Europe and North America, is taken into consideration, the number of such examples increases considerably, e.g. in representatives of the genera *Aesculus*, *Ostrya*, *Platanus; Cornus*, *Corylus*, *Evonymus*, *Staphylea; Anemone*, *Dentaria*, *Isopyrum*, *Luzula* sect. *Pterodes*, *Polygonatum*, *Waldsteinia*, and many others.

Nearly all series of corresponding taxa listed above are represented also in eastern Asia. It would be extremely interesting to know the degree of possible genetic isolation within them, and which are the centres where the most primitive forms occur, e.g. in regard to their karyotypes. Some attempts made (Hara, 1952, 1956; Löve, 1954; Kawano and Koyama, 1965) seem to indicate, that the situation in this respect is rather varied and complicated.

TABLE IV. Coefficients of generic affinity between four nemoral forest communities from Poland and southeastern Canada

		T.-C.	F.-Ulm.	C.-Ac.	Ac.s
Poland	Tilio-Carpinetum var. *Ranunculus cassubicus* (Medwecka-Kornaś, 1952)	100,0	77,5	35,9	35,1
Poland	Fraxino-Ulmetum var. *Ulmus campestris* (Wojterski, 1960)		100,0	39,4	35,2
S.E. Canada	Caryo-Aceretum (Grandtner, 1966)			100,0	77,5
S.E. Canada	Aceretum sacchari laurentianum (Grandtner, 1966)				100,0

The phytogeographical affinities of the nemoral flora, as stated above, agree again very well with the geological history of this type of vegetation. The processes of differentiation into corresponding taxa are much more advanced here than in the coniferous forests, but the connections which prove the origin of all nemoral floras from a common Arctotertiary ancestral stock are still clearly visible.

Beside similarities there are also some interesting phytogeographical differences between European and North American nemoral communities. In Europe there are many bicontinental species—a group which is scarcely represented in North America. Nearly all of them occur in Asia; they are either in a few places within isolated, relic deciduous forests in the mountains, especially the Ural and Altai Mountains (e.g. *Actaea spicata* L., *Asarum europaeum* L., *Lathyrus vernus* (L.) Bernh., *Sanicula europaea* L., *Stellaria holostea* L., etc.), or scattered through the whole of Siberia, thus connecting the European and eastern Asiatic centres of the nemoral flora (e.g. *Asperula odorata* L., *Melica nutans* L., *Paris quadrifolia* L., *Viola mirabilis* L., etc.) (Lipmaa, 1938; Gorchakovsky, 1968, 1969).

Very typical for the European deciduous forests are also numerous representatives of the ancient Mediterranean genetic element (e.g. the genera *Arum*, *Galanthus*, *Helleborus*, *Leucoium*, *Phyteuma*, *Symphytum*, etc.). They are remnants of a former autochthonous flora which developed in Europe before the Miocene invasion of the Arctotertiary nemoral flora.

In North America, especially in the never glaciated territories, a most remarkable feature is the presence of residual Arctotertiary genera which occur also in the eastern Asiatic refugium, but are completely absent in Europe and western Asia (e.g. *Liriodendron*, *Magnolia; Chionanthus*, *Hamamelis; Caulophyllum*, *Diphylleia*, *Menispermum*, *Phryma*, *Trautvetteria*, etc.). Another interesting group, also missing in Europe, are the representants of genera (or even families) of tropical origin (Anonaceae: *Asimina*; Bignoniaceae: *Bignonia*, *Campsis*, *Catalpa*; Ebenaceae: *Diospyros*; Lauraceae: *Sassafras*; Symplocaceae: *Symplocos*; etc.). Their presence may be explained by better conditions for survival of Tertiary tropical relics (Fernald, 1931), as well as by more recent contacts between the subtropical and temperate floras in North America. Similar conditions are found in eastern Asia.

ECOLOGICAL CORRESPONDENCE OF TAXA

The most fascinating fact discovered when comparing temperate forest floras of North America and Europe is that the corresponding taxa within them are always "ecologically corresponding" (Medwecka-Kornaś, 1961; Lebrun, 1961; Grandtner, 1962). This means, that "both counterparts have similar ecological ranges of tolerance, and consequently also phytosociological values, i.e. they occur on analogical habitats and in analogical plant communities on both continents" (Kornaś, 1965). As far as we know, there are no cases which are at variance with this rule (i.e. no temperate European forest species, the American counterparts of which grow in a different type of plant community from the European plant). The conclusion is that the ecological constitution of such taxa must be very rigid. Medwecka-Kornaś (1961) was the first to point out the evolutionary aspect of this problem: "The pairs of North American and European corresponding taxa may probably be descendants of common ancestral forms, growing in the Tertiary Holarctic forest formations. The similar phytosociological characters of these taxa which can now be observed on both continents suggest that they have preserved the same ecological affinities from very ancient times, when they were not yet isolated and not differentiated morphologically". Thus ecological and phytosociological data on present-day corresponding taxa may be used to reconstruct the "prototypic plant associations" which once existed in the Tertiary forests of the Northern Hemisphere. These data also throw very interesting light on the evolutionary processes in the forest plants themselves; the ecological characters of plants which we would expect to be flexible, have remained nearly unchanged since the Pliocene or even Miocene, i.e. at least some 10–15 million years (Löve, 1967), in spite of distinct morphological divergence. Can speciation processes really affect in this case only those characters which are not essential to the life-strategy of the plants?

QUESTIONS FOR FUTURE INVESTIGATIONS

Corresponding taxa of the temperate Holarctic forest flora are certainly excellent material for ecological–evolutionary studies. For the moment this material has been only very little exploited; its further examination may give answers to many questions of general importance, such as:

(1) Of what nature was the original differentiation of the Holarctic Tertiary flora into "prototypic plant associations"?

(2) What are the rates of evolution, the degree of divergence and the possible sterility barriers in corresponding taxa of various systematic groups?

(3) What is the karyological and genetic background of this situation?

(4) How can the ecological conservatism of temperate forest plants be reconciled with their morphological evolution?

To solve these questions, phytosociological methods may be especially useful. Bringing to light the ecological correspondence of taxa, they make it possible to find the most appropriate subjects for biosystematical investigation. To this end, comparative phytosociological studies on the forests of the whole Holarctic realm are indispensable; and such studies can only be carried out through long-term co-operation between plant ecologists of Europe, Asia and North America.

SUMMARY

The forest floras of temperate Eurasia and North America show striking similarities. In the boreal coniferous forests this is due to the presence of numerous circumpolar species occurring on all three continents of the Holarctic realm. Series of corresponding taxa are rather infrequent there, and their components are only slightly different from one another. On the other hand, in the nemoral broad-leaved deciduous forests, circumpolar species are very rare, but numerous series of corresponding species exist, closely related to each other, though clearly separated taxonomically. This situation is easy to explain by the different origin and different geological history of the two principal types of the Holarctic forest vegetation. As wide-ranging zonal types, the boreal coniferous forests are rather young. Geographical isolation is only slightly marked and relatively recent (of late Pleistocene or Holocene date). The nemoral deciduous forests are remnants of a formerly continuous temperate Arctotertiary zone, which has been pushed southwards and divided up into isolated refugia by the deterioration of climate of the northern latitudes in Pliocene and early Pleistocene. This long-lasting and complete isolation has resulted in advanced evolutionary divergence of nemoral floras.

Nearly all series of closely related taxa in the forests of temperate Eurasia and North America consist of ecologically corresponding components which have similar ranges of tolerance, occupy similar habitats and grow in strictly analogous plant communities. The ecological constitution of such groups must be very ancient and very rigid. Even those nemoral taxa which are separated into morphologically distinct species have undergone no noticeable ecological changes since their isolation some 10–15 million years ago.

REFERENCES

BRAUN-BLANQUET, J. (1959). *Vistas Bot.* **1**, 145–171.

BRAUN-BLANQUET, J. (1964). "Pflanzensoziologie. Grundzüge der Vegetationskunde". 3 Aufl. Springer Verlag, Wien–New York.

BRAUN-BLANQUET, J., SISSINGH, G. and VLIEGER, J. (1939). *Prodromus der Pflanzengesellschaften* **6**, 1 123.

BRAUN-BLANQUET, J., PALLMANN, H. and BACH, R. (1954). *Ergebn. wiss. Unter. schweiz Natn. Parks* 4(N.F.) **28**, 1–200.

CAIN, S. A. (1944). "Foundations of Plant Geography". Harper Bros., New York.

CLAUSEN, J. (1968). *Yb. Carnegie Instn Wash.* **66**, 234–243.

CURTIS, J. T. (1959). "The Vegetation of Wisconsin. An Ordination of Plant Communities". Univ. of Wisconsin Press, Madison.

CZERWIŃSKI, A. (1966). *Bull. Soc. Amis Sci. Lett. Poznań*, Sér. D. (Sci. Biol.) **7**, 15–36.

DANSEREAU, P. (1959). *Contr. Inst. bot. Univ. Montréal* **75**, 1–147.

DAUBENMIRE, R. and DAUBENMIRE, J. B. (1968). *Bull. Wash. agric. Exp. Stn* **60**, 1–104.

ELLENBERG, H. (1963). "Vegetation Mitteleuropas mit den Alpen in kausaler, dynamischer und historischer Sicht". E. Ulmer, Stuttgart.

FERNALD, M. L. (1931). *Rhodora* **33**, 25–63.

GORCHAKOVSKY, P. L. (1954). *Zap. ural. Otd. geogr. Obshch.* **2**, 12–77.
GORCHAKOVSKY, P. L. (1968). *Tr. Inst. Ekol. Rast. Zhivotn. Ural. Filial Akad. Nauk. SSSR* **59**, 1–207.
GORCHAKOVSKY, P. L. (1969). *Tr. Inst. Ekol. Rast. Zhivotn. Ural. Filial Akad. Nauk. SSSR* **66**, 1–286.
GRANDTNER, M. (1962). *Bull. Soc. r. for. Belg.* **1962**, 413–436.
GRANDTNER, M. (1966). "La végétation forestière du Québec méridional". Presses Univ. Laval, Québec.
HARA, H. (1952). *J. Fac. Sci. Tokyo Univ., Sect.* 3 (*Bot.*) **6**(**2**), 29–96.
HARA, H. (1956). *J. Fac. Sci. Tokyo Univ., Sect.* 3 (*Bot.*) **6**(**7**), 343–391.
HARA, H. (1959). *In* "Distribution Maps of Flowering Plants in Japan" (H. Hara and H. Kanai), Vol. 2, pp. 1–96. Inoue Book Co., Tokyo.
HOU, H.-Y., CHEN, CH.-T. and WANG, H.-P. (1956). "The Vegetation of China with Special Reference to the Main Soil Types". Rep. Sixth. Int. Congr. Soil Sci., Soil Soc. China, Peking.
HULTÉN, E. (1962). *K. svenska Vetensk.-Akad. Handl.*, 4 *Ser.*, **8**(**5**), 1–275.
HULTÉN, E. (1968). "Flora of Alaska and Neighbouring Territories. A Manual of Vascular Plants." Stanford Univ. Press, Stanford, California.
KAWANO, S. and KOYAMA, T. (1964). *Can. J. Bot.* **42**, 859–884.
KNAPP, R. (1957). *Geobot. Mitt.* **4**, 1–63.
KNAPP, R. (1959). *Geobot. Mitt.* **7**, 1–27.
KNAPP, R. (1965). "Die Vegetation von Nord- und Mittelamerika und der Hawaii-Insel." G. Fischer, Stuttgart.
KORCHAGIN, A. A. (1929). K voprosu o tipakh lesa po issledowaniyu v Totemskom uyezde Vologodskoy gubernii. V kn.: Ocherki po fitosociologii i fitogeografii, 287–327. Nowaya Derevnya, Moskva.
KORNAŚ, J. (1965). *Fragm. flor. geobot.* **11**, 307–338.
KORNAŚ, J. (1970). *Fragm. flor. geobot.* **16**, 123–136.
KRASNOBOROV, I. M. (1965). *In* Rastitelny pokrov Krasnoyarskovo kraya **2**, 24–62. Novosib. Otd. Akad. Nauk. SSSR., Novosibirsk.
KÜCHLER, A. W. (1964). *Spec. Publ. Am. geogr. Soc.* **36**, 1–38, 1–116.
LA ROI, G. H. (1967). *Ecol. Monogr.* **37**, 229–253.
LAVRENKO, E. M. and SOCHAVA, V. B. (red.) (1956). Rastitelny pokrov SSSR. Poyasnitelny tekst k "Geobotanicheskoy karte SSSR" m. 1:4 000 000. I, II. Izd. Akad. Nauk., Moskva-Leningrad.
LEBRUN, J. (1961). *Vegetatio* **10**, 25–41.
LIPMAA, T. (1938). *Acta Inst. Hort. bot. Tartu.* **6**(**2**), 1–152.
LÖVE, A. (1954). *Vegetatio* **5–6**, 212–224.
LÖVE, A. (1967). *Taxon* **16**, 324–333.
MEDWECKA-KORNAŚ, A. (1952). *Ochr. Przyr.* **20**, 133–236.
MEDWECKA-KORNAŚ, A. (1955). *Ochr. Przyr.* **23**, 1–111.
MEDWECKA-KORNAŚ, A. (1961). *Bull. Acad. pol. Sci. Sér. Sci. Biol.* **9**(**6**), 255–260.

MEDWECKA-KORNAŚ, A., KORNAŚ, J. and PAWŁOWSKI, B. (1966). *In* "The Vegetation of Poland" (W. Szafer, ed.), pp. 294–509. Pergamon Press, Oxford, Warszawa.

MEUSEL, H., JÄGER, E. and WEINERT, E. (1965). "Vergleichende Chorologie der zentraleuropäischen Flora". I. Text und Karten. Fischer Verlag, Jena.

MIYAWAKI, A., ITOW, S. and OKUDA, S. (1967). *In* "Sci. Rep. on the Mt. Aizukomagatake, Tashiroyama and its vicinity, Fukushima Prefecture" pp. 15–43. Nat. Conservation Soc. Japan, Tokyo.

MIYAWAKI, A., OHBA, T., OKUDA, S., NAKAYAMA, K. and FUJIWARA, K. (1968). *In* "Sci. Rep. on the Echigo-Sanzan, Okutadami and its Vicinity, Niigata and Fukushima Prefecture" pp. 57–152. Nat. Conservation Soc. Japan, Tokyo.

MIYAWAKI, A., OHBA, T. and OKUDA, S. (1969a). *Rep. natn. Conserv. Soc., Japan* **36**, 50–103.

MIYAWAKI, A., SUGAWARA, H., HAMADA, T. and IZUKA, M. (1969b). *In* "Scientific Studies on the North Slope of Mt. Fuji, Yamanashi Prefecture" pp. 1–43. Yamanashi Prefecture, Kafu.

PAWŁOWSKI, B. and WALAS, J. (1949). *Bull. Acad. pol. Sci., Cl. Sci. Math.-Nat., Sér. B*1. *(Bot.)* **1948**, 117–180.

RIVAS-MARTÍNEZ, S. (1968a). *Publnes Inst. Biol. apl.* **45**, 81–105.

RIVAS-MARTÍNEZ, S. (1968b). *Publnes Inst. Biol. apl.* **44**, 5–44.

SJÖRS, H. (1963). *In* "North Atlantic Biota and their history" (A. and D. Löve, eds), pp. 109–125. Pergamon Press, Oxford.

SOKOŁOWSKI, A. W. (1965). *Fragm. flor. geobot.* **11**, 97–119.

SOKOŁOWSKI, A. W. (1966). *Pr. badaw. Inst. badaw. Leśn.* **305**, 71–105.

STUCHLIK, L. (1968). *Fragm. flor. geobot.* **14**, 441–484.

SUZUKI, T. (1966). *Sinrin-Ritti (Forest Site)* **8(1)**, 1–12.

SZAFER, W. (1946). The Pliocene Flora of Krościenko in Poland. I. General part. *Rozpr. Wydz. mat. przyr. pol. Akad. Umiejet. (Dział B, Nauki Biol.)* **72(1)**, 1–162.

SZAFER, W. (1947). The Pliocene Flora of Krościenko in Poland. II. Descriptive part. *Rozpr. Wydz. mat. przyr. pol. Akad. Umiejet. (Dział B, Nauki Biol.)* **72(2)**, 1–213.

SZAFER, W. (1954). *Inst. Geol. Prace* 11 : 1–238.

SZYMKIEWICZ, D. (1947). *Bull. Acad. pol. Sci., Cl. Sci. Math.-Nat.* Sér. B l Bot. **1946**, 1–29.

TOLMACHEV, A. I. (1954). K istorii wozniknovaniya i razvitiya temnokhvoynoy taigi. Izd. Akad. Nauk., Moskva-Leningrad.

WALTER, H. (1968). "Die Vegetation der Erde in ökophysiologischer Betrachtung. Bd. II. Die gemässigten und arktischen Zonen". G. Fischer Verlag, Jena.

WALTER, H. and STRAKA, H. (1970). "Arealkunde. Floristisch-historische Geobotanik". E. Ulmer, Stuttgart.

WOJTERSKI, T. (1960). *Pr. Kom. biol. Poznań* **23(3)**, 1–231.

4 | Corresponding Taxa in North America, Japan and the Himalayas

HIROSHI HARA

Department of Botany, University Museum, University of Tokyo, Tokyo, Japan

INTRODUCTION

As Asa Gray first pointed out in 1840 and 1846, the flora of Japan has a striking resemblance to that of eastern North America. Attention was drawn to another resemblance, that between the temperate flora of Sikkim in the Himalayas and that of Japan, by J. D. Hooker in 1855; and he cited the genera *Helwingia*, *Aucuba*, *Stachyurus*, and *Enkianthus* as conspicuous instances of this similarity. These classical examples of disjunct distributions have attracted the attention of many botanists, and corresponding taxa in those regions have been studied by various authors.

Fortunately I have had the opportunity to visit both North America and the Eastern Himalayas several times, and I have examined corresponding taxa of various plant groups in the field, as well as the herbarium. I have partly enumerated the corresponding taxa between Japan and North America in 1952 and 1956, and between Japan and the Eastern Himalayas in 1966 and 1971. The present climates of these three regions are similar to some extent; but in the Himalayas the dry season (October–May) and the rainy reason (June–September) are much more distinct than in the other regions, and the annual range of temperature is much smaller. So it is often very difficult to cultivate in Japan the Himalayan plants which have been adapted to such a climate, and they are liable to die in severe frost, or because of the hot summer or by disease.

As I pointed out in 1956 and 1962a, in the taxa common to these regions, different patterns of differentiation are shown by different plant groups, even when they have lived through similar geohistorical changes. Here I wish to give several examples of those taxa which are interesting from an evolutionary point of view.

SPECIES COMMON TO THE HIMALAYAS, JAPAN AND NORTH AMERICA

Representative examples of species which are distributed from the Himalayas through China to Japan, and then across the Pacific Ocean to North America, are *Phryma leptostachya* L. (Phrymaceae), *Monotropa uniflora* L. (Pyrolaceae) and *Symplocarpus foetidus* (L.) Nuttall (Araceae).

In the case of *Phryma leptostachya*, the Himalayan, Japanese and eastern North American plants agree well with each other, and certainly belong to a single species, although slight differences can be found in the size of the flowers, the length of the spinose calyx-lobes and in the pubescence; they have the same chromosome number ($2n = 28$). The Himalayan and American plants grow well at Karuizawa in middle Japan where wild plants are also found. A form with crowded leaves in the middle part of the stems was named var. *confertifolia* Fernald in North America, but a similar form is also found in Japan.

As regards *Monotropa uniflora*, the Japanese plants tend to have thicker stems and broader scale-leaves, are snowy white, and flower from September to October, but they are not clearly distinguishable from the American plants (Hara, 1956). It is noteworthy that the genus *Monotropastrum*, which strikingly resembles *Monotropa uniflora* in general appearance, occurs in Japan, China and the Himalayas, but not in North America. *Monotropastrum* differs clearly from *M. uniflora* in having a smooth unilocular ovary with 6–13 parietal placentas, anthers opening by an elliptical lid, an indehiscent berry and oval seeds without appendages. It is very variable in the size of the stems, flowers and fruits, in the shape of the scaly leaves, tepals and ovary and in the number of floral parts; but it seems to belong to a single species, *M. humile* (D. Don) Hara (Hara, 1965). As compared with the Himalayan plants, the Japanese plants have thicker petals, villose filaments and distinctly papillose anthers.

In these examples, the populations, which are widely separated in East Asia and North America, have undergone little change, although they have been isolated for at least 5 million years. But in some other genera such as *Adenocaulon* (Compositae) and *Antenoron* (*Tovara*, *Sunania*) (Polygonaceae) (Hara, 1962b and 1965), the North American plants are generally treated as a different species from the Asiatic.

CORRESPONDING SPECIES IN JAPAN AND THE SINO-HIMALAYAN REGION

The genus *Trillium* (Liliaceae), comprising about 40 species, is distributed in temperate Asia and North America, and both Japanese and American species have been cytologically studied in detail; but in the Himalayas only two species have been reported. One species, *T. tschonoskii* Maximowicz, occurs widely from Japan to the Eastern Himalayas. I have cultivated in Japan the plants from Bhutan, and noticed that they are so similar to the Japanese plants in external morphology and chromosomal characters that they must be considered to be conspecific. But the Bhutanese plants tend to have more rhombic leaves, shorter peduncles, elliptic sepals which are less acuminate at the apex, stamens shorter than the pistil, shorter anthers, and sometimes smaller purple petals. This Himalayan race, var. *himalaicum* Hara, is allotetraploid and the karyotype is the same as that of typical *T. tschonoskii* of Japan (Hara, 1969c).

The other species, *T. govanianum* Wallich ex Royle, endemic in the Himalayas, has shortly petiolate leaves, small flowers, and lanceolate petals nearly as long as the sepals, and is also allotetraploid ($2n = 20$) with one pair of C chromosomes satellited. In these characters, the species somewhat resembles the allied genus *Paris*.

The genus *Paris* is distributed in temperate Eurasia and consists of 8 species. Among them, *P. japonica* (Franch. & Sav.) Franch. of northern Japan is singular in having 7–10 white petaloid sepals and 40 somatic chromosomes, and *P. polyphylla* Smith of the Sino–Himalayan region in having an angular ovary and a capsular, dehiscent fruit which exposes the seeds enveloped in bright red arils.

P. polyphylla is an extremely polymorphic species, and is variable in the size, shape and number of leaves, tepals, stamens, and styles; it also includes two cytotypes, diploid and tetraploid. Each of the morphological characters changes almost continuously and somewhat independently, without definite correlation between the characteristics, and they vary from one population to another. Thus extreme populations may look very different from one another; and more than 25 races with different character-combinations have been described as independent species.

The plants of this species-group are protogynous, but the stigma is functional for a long time, and self-pollination seems to take place frequently. I have confirmed that a tetraploid plant collected at Darjeeling and cultivated in Japan produces good, fertile seeds by self-pollination.

It would appear that individuals in one population are mostly propagated either by rhizomes or by inbreeding, and the populations are thus homogeneous. But cross-pollination occasionally takes place between individuals of different populations; and extreme populations are connected by a series of transitional populations with intermediate characters or different combinations of characters. Similar populations found in the Eastern Himalayas, West China and Taiwan may be the result of parallel mutation in different regions. In this group, the various races have not yet, in my opinion, been established as species and those with different combinations of variable characters can be treated as infra-specific taxa (Hara, 1969a).

However, many other plant groups distributed from Japan to the Himalayas have been differentiated into three or more species. A typical example is seen in the genus *Aucuba* (Cornaceae); it comprises three principal species, one in Japan (*A. japonica* Thunb.), one in China (*A. chinensis* Benth.) and one in the Eastern Himalayas and Northern Burma (*A. himalaica* Hook. f. & Thoms.). *A. himalaica* is distinguishable from *A. japonica* in having densely hairy young leaves and inflorescences, elongate oblong leaves with minute serrations and distinctly reticulately-impressed veins above, and caudate–acuminate petals. Typical *A. japonica* together with its var. *borealis* and several cultivars has been studied cytologically in detail, and it is known as a tetraploid with $2n = 32$ (Meurman, 1929; Yamamoto, 1937; Viinikka, 1970). Later, *A. himalaica* from the Eastern Himalayas and *A. chinensis* from Taiwan were reported as diploid ($2n = 16$). Recent investigation (Kurosawa, 1971b) has shown that a diploid *Aucuba* also occurs in southernmost Japan. The diploid plants from the Ryukyus, Amami-Ôshima, and southern Kyushu are morphologically similar to *A. japonica* and have been referred to this species; they are slightly different from one another, especially in the location of the secondary constriction of the chromosomes. It is presumed that the genus *Aucuba* was once distributed continuously from the Himalayas to Japan, and differentiated morphologically to some extent at the diploid level. Tetraploid *A. japonica* may have originated in western Japan and extended its distribution northwards, differentiating into var. *borealis* with decumbent stems in the northernmost area, where it has become adapted to the Japan-Sea climate, with very heavy snows in winter. Typical tetraploid *A. japonica*, including various cultivars, is the most vigorous species in the genus, and is very widely cultivated.

The genus *Helwingia* (Cornaceae) (Fig. 1), which is peculiar in having epiphyllous flowers, shows similarities to the genus *Aucuba* which has just been described.

H. japonica (Thunb.) Dietrich of Japan has roundish leaves acuminate at the apex, male flowers 4–6 mm in diameter, and black fruits, while *H. himalaica* Hook. f. & Thoms. has more elongate leaves with a long, caudately acuminate apex, male flowers 3 mm in diameter, and dull red fruits. *H. chinensis* Batalin seems to be more closely related to *H. himalaica* than to *H. japonica*. Cytologically *H. japonica* has been reported as a high polyploid with about 114 somatic chromosomes. In 1965 *H. himalaica* was found to be diploid with 38 somatic chromosomes (Hara and Kurosawa, 1965). Recently *H. formosana* Kanehira & Sasaki from Taiwan and *H. liukiuensis* Hatusima from the Ryukyus were also proved to have $2n = 38$, but they have black fruits and are morphologically more similar to *H. japonica*. Moreover, it is noteworthy that *H. japonica* var. *parvifolia* Makino of south-western Japan, which has smaller thicker and lustrous leaves seems to have 38 somatic chromosomes (Kurosawa, unpublished). It would appear that in this case, too, the group has become differentiated at the diploid level, and a high polyploid race, which represents the type of *H. japonica*, has later spread widely in Japan.

Another example is the genus *Panax* (Araliaceae) disjunctly distributed in temperate East Asia and eastern North America. The American plants are regarded as two distinct species, *P. quinquefolius* L. and *P. trifolius* L., but the Asiatic plants, which include *P. ginseng*, a very famous Chinese drug, have been given widely different taxonomic treatments (Hara, 1970).

Panax pseudo-ginseng Wallich of Nepal is characterized, in its typical form, by fleshy carrot-like roots which are sometimes fascicled; very short rhizomes; obovate-oblong leaflets which are sometimes long caudate-acuminate at the apex, doubly incise-serrate with cuspidate close teeth, and conspicuously bristly on the nerves of the upper surface; distinct stipules at the base of the petioles; slender smooth pedicels; inconspicuous bracts; and large flattish seeds; it proved to have $2n = 24$. The species is common in the Himalayas; it differs from typical Nepalese plants in having long creeping rhizomes with elongate internodes and thickened nodes, no stipules, minutely scabrous pedicels and smaller ovoid seeds; it is now referred to subsp. *himalaicus* Hara. The shape of the rhizome is variable even in the same population, and is influenced to some extent by ecophysiological conditions; plants growing at higher

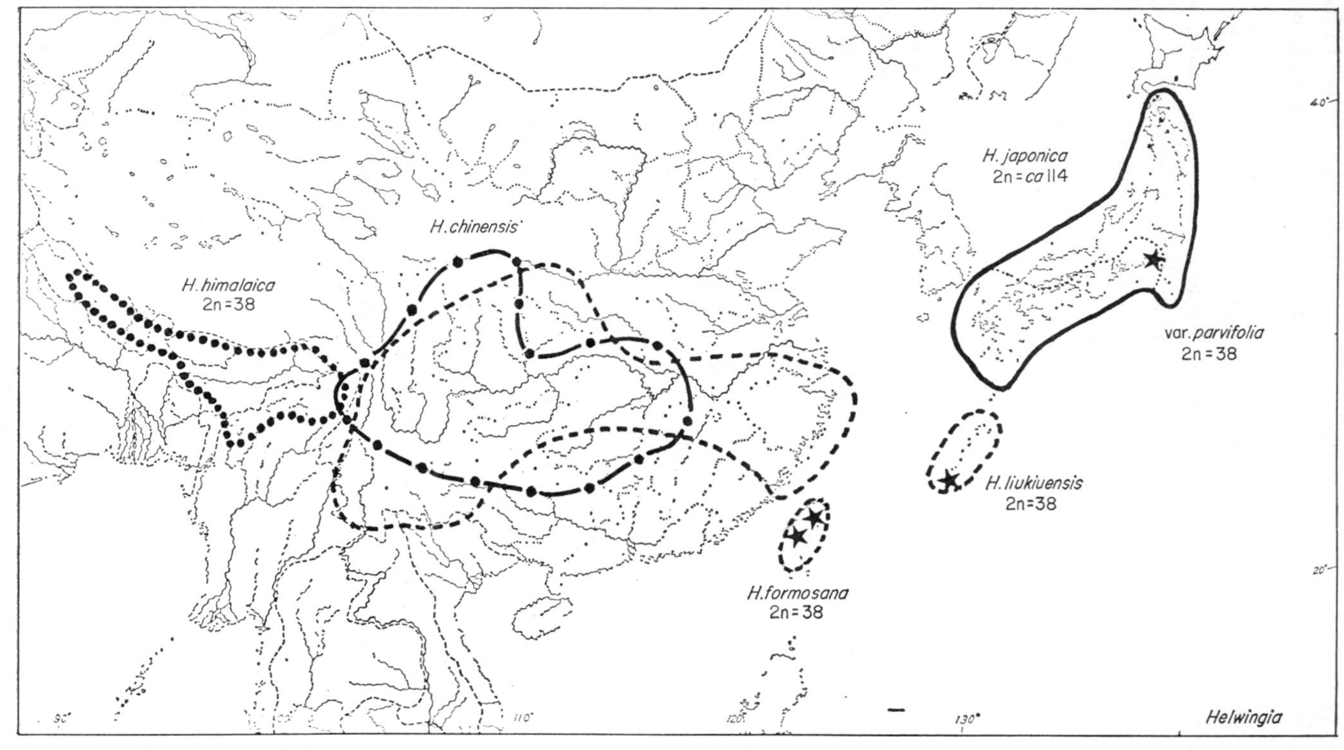

FIG. 1. Distribution map of the genus *Helwingia*.

altitudes tend to have slender nodulose rhizomes. The shape of the leaflets is also variable, but is stable in the same population. The somatic chromosome number seems to be mostly 24.

The Japanese plants (subsp. *japonicus* (Meyer) Hara) differ from those of the Himalayas in having thick creeping rhizomes; elliptical to obovate leaflets which are more shortly acuminate at the apex, irregularly serrate with shorter-pointed teeth and glabrescent or sparsely bristly; no stipules; minutely scabrous pedicels; numerous minute bracts; smaller ovoid seeds; and 48 somatic chromosomes.

Typical *P. ginseng* C. A. Meyer, which has been known only in North Korea, S. and E. Manchuria, and Ussuri in the wild, but is cultivated in China, Korea and Japan for medicinal use, is similar to typical *P. pseudo-ginseng* in the shape of roots and seeds, but differs in having glabrescent leaflets shortly acuminate at the apex and shortly serrate, no stipules, scabrous pedicels, pollen grains coarsely reticulate on the surface, and 44 or 48 somatic chromosomes.

These plants are polygamo-dioecious or polygamo-monoecious and are considered to be pollinated by insects. But the individuals in one population are usually uniform, and seem to be propagated by branching of rhizomes or inbreeding; and the plants are variable from population to population, as out-breeding also occurs occasionally as in the case of *Paris polyphylla*. The Asiatic races which have been mentioned above cannot be separated from one another by any single character, but the range of variations is not the same in different geographical regions.

Recently, detailed studies on saponins and sapogenins contained in the roots and rhizomes of the genus *Panax* have been made by Professor S. Shibata and his associates at the University of Tokyo. Roots and rhizomes of *P. ginseng* and *P. quinquefolius* contain sapogenins consisting of a large quantity of dammarane-type triterpenes (20s-protopanaxadiol and 20s-protopanaxatriol) which are the main effective chemical constituents of *Panax*, and a small quantity of oleanane-type triterpene (oleanolic acid). On the other hand, all Japanese and Himalayan races contain sapogenins consisting of a very large quantity of oleanane-type triterpene and only a very small quantity of dammarane type triterpenes. Slight differences were observed in the quantity of dammarane-type sapogenins from race to race, and typical *P. pseudo-ginseng* has only a trace of it, although the roots resemble those of *P. ginseng* in shape.

In most of the cases mentioned above, polyploidy seems to play a role in the differentiation of the group. However, in the genus *Stachyurus*, belonging to the monotypic family Stachyuraceae, *S. himalaicus* Hook. f. & Thoms. of the Eastern Himalayas, *S. chinensis* Franch. of China, and *S. praecox* Sieb. & Zucc. of Japan are differentiated morphologically, but all races found in the Eastern Himalayas, Taiwan and Japan have 24 somatic chromosomes (Kurosawa, 1971a).

There are also some examples in which polyploidy is correlated with other essential characters. In 1958, I reported the correlation between polyploidy and the pattern of the pollen grains in the *Anemone hepatica* L. group (Hara and Kurosawa, 1958). The group occurs in three isolated areas namely Europe, eastern North America and East Asia, and all the European, two of the North American, and the common Japanese races are diploid ($2n = 14$) with 3-colpate pollen grains (Huynh, 1970), although variable in the shape of the leaves and sepals and the colour of the flowers. Only in 6 localities in western Japan, mostly on calcareous mountains, have we found var. *pubescens* Hiroe with $2n = 28$ and 6- to 10-rugate pollen grains. It is reasonable to consider that the tetraploid race with poly-rugate pollen has been derived from a diploid race with tricolpate pollen. This is the first example in which the pattern of the pollen grains has been shown to vary in relation to polyploidy within a single species.

Differentiation in the *Rubia cordifolia* group which is distributed from Japan to the Himalayas is also interesting. The roots of this group have been used as a red dye from ancient times in Japan and the Himalayas, and all the plants have generally been called *R. cordifolia*. Now it became apparent that the plants are roughly divided into two groups by the mode of germination. The Himalayan plants have narrowly ovate epigeous cotyledons, although they are variable in the shape and hairiness of the leaves, and the colour of the flowers and fruits. *R. manjith* Roxburgh, common at altitudes less than 2000 m, has reddish young leaves and nodes, long-acuminate leaves, small, dull orange flowers, and 66 somatic chromosomes. *R. wallichiana* Decaisne, generally found at higher altitudes, has green stems and leaves, broader leaves, and larger greenish or dark purple flowers, and it includes at least two cytotypes, with 44 and 132 somatic chromosomes.

On the other hand, the two species found in Japan, *R. cordifolia* L. and *R. akane* Nakai, have consistently hypogeous cotyledons which remain underground enclosed in the seed-coat. *R. cordifolia* has 4–10 whorled leaves, and fruits which are yellowish brown or orange at first and purplish black when fully ripe; it is diploid ($2n = 22$). *R. akane*, which is very common in Japan, has always 4-nate, broad leaves, cordate at the base, and black fruits, and is tetraploid ($2n = 44$).

In the *R. cordifolia* group it is reasonable to suppose that a species with hypogeous germination was derived from an ancestral diploid species with epigeous germination; polyploidy then occurred in parallel in the Himalayas and in the Far East (Hara, 1969b).

DIFFERENTIATION AND POLYPLOIDY IN JAPAN

In Japan, where there have been complicated geohistorical and climatic changes several times since the Tertiary, polyploidy and dysploidy seem to have occurred locally in some plant groups.

The genus *Chionographis* (Liliaceae), a genus endemic to Japan and South China, with sessile ebracteate flowers and very unequal tepals, is most closely allied to the genus *Chamaelirium* ($2n = 24$) of eastern North America, and must be of very ancient origin (Hara, 1968). Two Japanese species, *C. japonica* (Willd.) Maximowicz and *C. koidzumiana* Ohwi, have 24 somatic chromosomes, although variable in the shape and size of the leaves and tepals.

However, *C. japonica* has two subspecies, subsp. *hisauchiana* (Okuyama) Hara and subsp. *minoensis* Hara, which occur in very limited localities near the eastern edge of its distribution. They are smaller than the typical subspecies, and have 42 somatic chromosomes. If the basic number of the genus is presumed to be 6, then $2n = 24$ can be interpreted as tetraploid, and 42 as heptaploid. But, in both subspecies, the reduction division is regular with 21 chromosomes in pollen mother-cells; the pollen grains are normal, and the capsules produce fertile seeds freely. These facts suggest that these subspecies with 42 somatic chromosomes may be amphiploids of ancient origin, derived from an ancestral form with the basic number 6 or 7, and that at the time they became established, they already had the basic number 21.

Iris setosa Pallas, distributed from Siberia to Alaska and southwards to Japan, has been reported to have $2n = 38$, and its var. *canadensis* Foster of north-eastern North America has the same chromosome number. Near the southern border of the distribution of the species are found two very local races, var. *nasuensis* Hara and var. *hondoensis* Honda (Hara and Kurosawa, 1963). As compared with the parent species, both races have larger violet inner tepals, sterile fruits, pollen grains of high sterility which are more minutely reticulate on the surface, and about 54 somatic chromosomes, but they can be distinguished by several morphological characters. The two races are not directly related phylogenetically, and seem to have originated from a triploid or a hybrid.

THE GENUS *Chrysosplenium*

The genus *Chrysosplenium* (Saxifragaceae) has been monographed by Hara (1957). About 14 species occur in Japan, 23 in China, 12 in the Himalayas, 5 in North America, 2 in South America, and 2 in Europe.

In the Series Alternifolia, *C. tetrandrum* (N.Lund) Th.Fries, *C. wrightii* Franch. & Sav. and also *C. japonicum* (Maxim.) Makino of Japan have 24 somatic chromosomes, *C. alternifolium* L. of Europe 48, *C. rosendahlii* Packer of North America 96, and *C. iowense* Rydb. also of North America about 120; they form a polyploid series with the basic number 6, and all the chromosomes are small. It is striking that *C. lanuginosum* Hook.f. & Thoms. of the Eastern Himalayas with alternate leaves is diploid with 14 very large somatic chromosomes (Kurosawa, 1966) (Fig. 2).

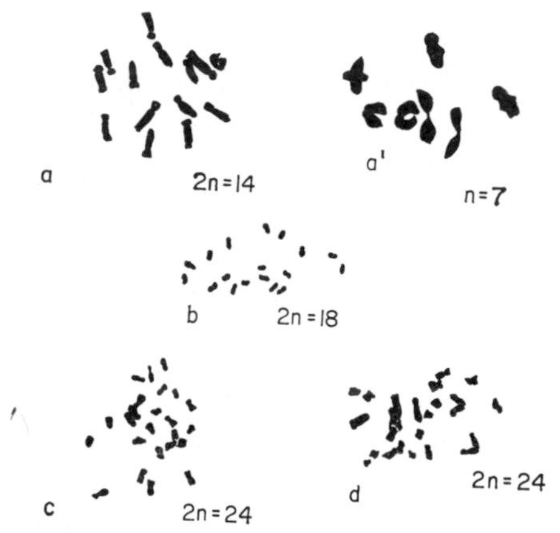

FIG. 2. Chromosomes of *Chrysosplenium*. a, a'. *C. lanuginosum*. b. *C. americanum*. c. *C. japonicum*. d. *C. ramosum*.

In the groups with opposite leaves also, the chromosome number 24 prevails, though the numbers 22 and 42 have been reported for a few species. *C. ramosum* Maximowicz of Japan ($2n = 24$) has a different karyotype from the others in having both large and small chromosomes (Kurosawa, unpublished).

It is noteworthy that the basic number of the American species with opposite leaves seems to be 9. *C. macranthum* Hooker, which occurs in a very isolated area in southern South America, and which retains some archaic characters, has 36 somatic chromosomes. *C. glechomaefolium* Nuttall ex Torrey & Gray of western North America was reported to have 18 somatic chromosomes. *C. americanum* Schweinitz ex Hooker of eastern North America, collected in the Shenandoah National Park of Virginia, was found to have 18 somatic chromosomes, although previously reported to have 24 chromosomes.

Thus the basic numbers of the genus *Chrysosplenium* seem to be 6, 7, 9 and 11, and we expect that more detailed cytotaxonomical studies on fuller material will reveal interesting interrelationships between the various species of the genus. It is clear that there still remains much interesting work to do, even in genera which have long been studied by many authors.

CONCLUSION

The data which we have presented illustrate that the mode and tempo of evolution vary greatly from group to group. The analysis of the relationships between corresponding taxa found in disjunct areas should always be based on ample phytogeographic and biosystematic data, and on experimental work wherever possible.

REFERENCES

HARA, H. (1952). *J. Fac. Sci. Univ. Tokyo (Bot.)* **6(2)**, 29–96.
HARA, H. (1956). *J. Fac. Sci. Univ. Tokyo (Bot.)* **6(7)**, 343–391.
HARA, H. (1957). *J. Fac. Sci. Univ. Tokyo (Bot.)* **7(1)**, 1–90.
HARA, H. (1962a). *Am. J. Bot.* **49**, 647–652.
HARA, H. (1962b). *J. Jap. Bot.* **37**, 326–332.
HARA, H. (1965). *J. Jap. Bot.* **40**, 97–103.
HARA, H. (ed.) (1966). "Flora of Eastern Himalaya" pp. 1–744. University of Tokyo.
HARA, H. (1968). *J. Jap. Bot.* **43**, 257–267.
HARA, H. (1969a). *J. Fac. Sci. Univ. Tokyo (Bot.)* **10(10)**, 141–180.
HARA, H. (1969b). *Am. J. Bot.* **56**, 732–737.
HARA, H. (1969c). *J. Jap. Bot.* **44**, 373–378.
HARA, H. (1970). *J. Jap. Bot.* **45**, 197–212.
HARA, H. (ed.) (1971). "Flora of Eastern Himalaya, Second Report" pp. 1–393. University of Tokyo.
HARA, H. and KUROSAWA, S. (1958). *J. Jap. Bot.* **33**, 265–275.

HARA, H. and KUROSAWA, S. (1963). *J. Jap. Bot.* **38**, 113–116.
HARA, H. and KUROSAWA, S. (1965). *J. Jap. Bot.* **40**, 36–40.
HOOKER, J. D. and THOMSON, T. (1855). "Flora Indica" Vol. 1, pp. 178–183. London.
HUYNH, K.-L. (1970). *Pollen Spores* **12**, 329–364.
KUROSAWA, S. (1966). *In* "Flora of Eastern Himalaya" (H. Hara, ed.), pp. 658–670. Tokyo.
KUROSAWA, S. (1971a). *In* "Flora of Eastern Himalaya, Second Report" (H. Hara, ed.), pp. 355–363. Tokyo.
KUROSAWA, S. (1971b). *J. Jap. Bot.* **46**, 231–238.
MEURMAN, O. (1929). *Hereditas* **12**, 179–209.
VIINIKKA, Y. (1970). *Ann. Bot. Fenn.* **7**, 203–211.
YAMAMOTO, Y. (1937). *Cytologia*, Fujii Jubil. Vol., 181–187.

5 | Vicarious Species of Restionaceae in Africa, Australia and South America

D. F. CUTLER

Jodrell Laboratory, Royal Botanic Gardens, Kew, England

INTRODUCTION

Botanists in the northern hemisphere are, in the main, unfamiliar with the Restionaceae. This is not unexpected since its members are rush-like, lack attractive flowers, and all but one grow south of the equator.

Hutchinson (1959) placed Restionaceae with Juncaceae, Thurniaceae and Centrolepidaceae in the order Juncales. Anatomical studies support this grouping (Cutler, 1969).

The plants have a creeping or tufted rhizome. Very few have persistent leaves, although some have much reduced leaf blades which fall early. The leaf sheaths are conspicuous, encircle the culm by about $1\frac{1}{3}$ turns and are split to the base; they are usually scarious and are enlarged in some genera. New culms or flowering stems arise annually from the rhizome. Culms have become adapted to take over functions of photosynthesis and transpiration. Culms in most species reach about 20–50 cm in height and 0·5–3 mm in thickness; in some they may be up to 1–2 m high and 4–8 mm thick. The plants are, thus, xeromorphic.

Flowers are small, dioecious, occasionally monoecious or rarely hermaphrodite. They are borne in individual spikelets or spikelets in lax inflorescences. Male inflorescences of different species can be so similar that in some instances it is difficult to decide from which genus a male plant comes, and within a genus, which species is represented. Male and female inflorescences of the same species are also often different in appearance. Most key features are consequently based on details of the female inflorescence. This has hindered correct classification and when using characters of gross morphology alone the identification of a large proportion of the male plants is a problem.

Current works on the family define about 29 genera, but there are probably more than this, as anatomical evidence has shown (Cutler, 1969). It will be appreciated that the total number of species is not known with any degree of accuracy, but there may be about 300. Over $\frac{1}{3}$ of the genera are very small, indeed, 10 are monospecific (*Anthochortus*, *Chaetanthus*, *Coleocarya*, *Dielsia*, *Harperia*, *Hopkinsia*, *Meeboldina*, *Onychosepalum*, *Phyllocomos* and *Sporadanthus*). *Restio* and *Leptocarpus* are the largest, with about 125 and 40 species respectively. Taxonomic confusion was extreme in the past and at present many generic boundaries, particularly among S. African species, are undoubtedly incorrect. Pillans (1928, 1950) who carried out the best and most comprehensive studies of the S. African genera and species lists, for example, 10 synonyms for one of the species and four of these are from different genera! It is not uncommon for a species to have 4–6 synonyms. This is not entirely due to difficulties of identification or definition. Part of the trouble has been caused by enthusiastic taxonomists who have split or lumped—most having no additional evidence when they committed their views to the nomenclatural record.

Restionaceae were studied anatomically for inclusion in the series "Anatomy of the Monocotyledons" (C. R. Metcalfe, ed.) and the results of the study were reported in Volume IV, "Juncales" (Cutler, 1969). The Juncales appear to have anatomical affinities with Cyperaceae and Gramineae.

It was necessary to have anatomical information about a large proportion of the species in Restionaceae before reasonable suggestions could be made concerning probable interrelationships. In the following sections I shall present the evidence for, and develop the arguments leading to, the conclusions that no single genus is represented both in S. Africa and Australia and that the only readily detectable vicarious species occur in *Leptocarpus sensu stricto*. It is of course possible that other genera have vicarious species within their continental ranges, but this has not yet been demonstrated.

DISTRIBUTION OF RESTIONACEAE

The two main centres of the present day distribution are S. Africa and Australia (including Tasmania). Species also occur in New Zealand, the Chatham Islands, Malaysia and South Vietnam. One species is recorded for Basutoland, one for Madagascar and one for tropical Africa in Malawi. The disjunct distribution is further accentuated by the presence of one species in Chile and Patagonia, South America.

Most species grow in poor soils which are moist during the growing period and then dry out, but some are found in wetter situations and some in saline conditions. In S. Africa most species grow in Cape Province in seasonally wet places where up to 60 inches of rain falls per annum. Certain species grow where only 10–20 inches per annum falls, and in Australia there are occasional species which will grow in even drier conditions. Open habitats are favoured; light shade is tolerated by some species.

Australia has two major, discontinuous areas of distribution, the south-west where many interesting endemics occur, and the east. One species only is known to me which grows in both south-west and east, and that is *Leptocarpus tenax* (Labill.) R.Br.

Perhaps this distribution results from the separation of east and west Australia by sea in miocene times, rather than a lack of suitable present-day habitats to link the two groups.

Three species grow in New Zealand, among them *Leptocarpus similis* Edgar. *Leptocarpus* is also the only genus found in Malaysia, South Vietnam and South America. The genus *Restio* occurs mainly in S. Africa but is also represented in Madagascar and Malawi.

CURRENT TAXONOMIC TREATMENT OF RESTIONACEAE

The current taxonomic treatment of the family obscures both the probable interrelationships of species and an understanding of the distribution of species and genera. It was not until the anatomy of members of the family had been reported in detail that numbers of discontinuities were conclusively demonstrated (Cutler, 1969). Several of the larger genera need to be revised, a revision which should wait for a truly synthetic approach when data from palynology and chemistry among others can be added to those from morphology and anatomy. A reiteration of the guidelines for such a revision derived from the anatomical survey is relevant here. The most important observation is that no single genus is represented in both S. Africa and Australia. Consequently *Restio*, *Leptocarpus* and *Hypolaena* must be revised. *Restio* occurs in S. Africa only; there are at present no generic names suggested for the Australian species thought of until recently as *Restio*. There are some names in the literature that might have to be revived. Previous authors using some anatomical data in their studies (Gilg, 1890; Gilg-Benedict, 1930) suggested that *Leptocarpus* was an Australian genus and S. African plants called *Leptocarpus* correctly constitute *Calopsis* Beauv. ex Desv. Gilg-Benedict gave the name *Mastersiella* to S. African plants previously in *Hypolaena*, leaving *Hypolaena* for Australian species. New evidence shows that Gilg's and Gilg-Benedict's ideas were basically sound, but *Mastersiella* and *Calopsis* are anatomically very similar and their retention as separate genera must rest on strong morphological data.

So, details apart, we have a situation where all the S. African species and genera are distinct from all others (apart from one species of *Restio* in Malawi and one in Madagascar). Most Australian species and genera are distinct from all others in the world (except for 3 genera in N. Zealand), and the genus *Leptocarpus sensu stricto* is found also in Malaysia, S. Vietnam and S. America.

BASIC CULM ANATOMY

The simplest form of culm organization is found in some Australian species from the wetter habitats (Fig. 1A). In cross-section of an internode, working from the outside inwards, the culm has an epidermis with paracytic stomata, one layer of palisade-like cells, a complete cylinder or sheath of parenchymatous cells, a moderately strong sclerenchyma cylinder of fibres, in the outer part of this cylinder are embedded one ring of small vascular bundles and to the inner side of the cylinder is a ring of larger vascular bundles; the central region is occupied by a parenchymatous pith, sometimes breaking down to form a central cavity.

This simple form might be thought of as "primitive" but lacking conclusive evidence for this we must of course observe that it could equally well be the simplified product of secondary reduction. For the purpose of my arguments I shall regard this as the basic type, since it shows no vestiges of cells or tissues that suggest it to be secondarily reduced. In the closely related family Anarthriaceae, reduction of a basic type was evident in traces of certain tissue systems in the simplified forms (Cutler and Shaw, 1965).

POSSIBLE COURSE OF EVOLUTION WITHIN THE FAMILY

Since there is a lack of macro-fossil record and records of fossil pollen are sparse and of little help for this family (Couper, 1960), broad speculations are the best we can allow in attempting to depict the stages of evolution within the family. The times at which these stages might have occurred are less clear than the stages themselves (Fig. 2).

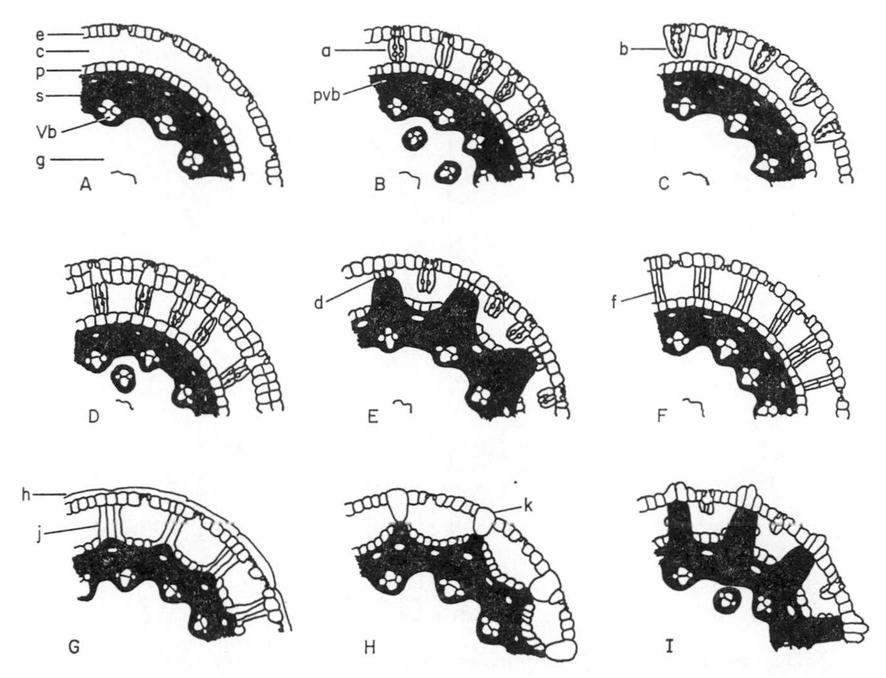

FIG. 1. Diagrams of sectors of culm transverse sections in Restionaceae. Details of chlorenchymatous cells omitted for clarity. A, simple or primitive; e, epidermis; c, chlorenchyma; p, parenchyma sheath; s, sclerenchyma; Vb, larger vascular bundle; g, parenchymatous ground tissue. B, basic type with protective cells, a; pvb, small peripheral vascular bundle. C, with inwardly elongated epidermal cells, b. D, protective cells in two layers, double epidermis. E, protective cells; girders from sclerenchyma cylinder extending to thickened chlorenchyma cells, d. F, thickened chlorenchyma cells, f, similar to protective cells but not lining substomatal cavities. G, pillar cells, j, derived from parenchyma sheath, and plate-like hairs, h. H, *Dielsia* with rows of enlarged epidermal cells, k. I, *Phyllocomos* with all sclerenchyma girders reaching the epidermis. For further explanation, see text.

Restionaceae may have begun their evolution as a family when the climate in the southern hemisphere land mass was less extreme than in present-day Cape Province or Australia, probably in the late Cretaceous.

As the family increased its territory, and while S. Africa and Australia were still within range of the means of dispersal available to the proto-restionaceae (perhaps by way of the antarctic land mass), one can envisage the relatively unspecialized form spreading into suitable semi-aquatic habitats. Africa and Australia gradually separated, and the climate became drier. New stresses would have been imposed on the family. Specimens evolved which were able to cope with seasonally dry conditions. In

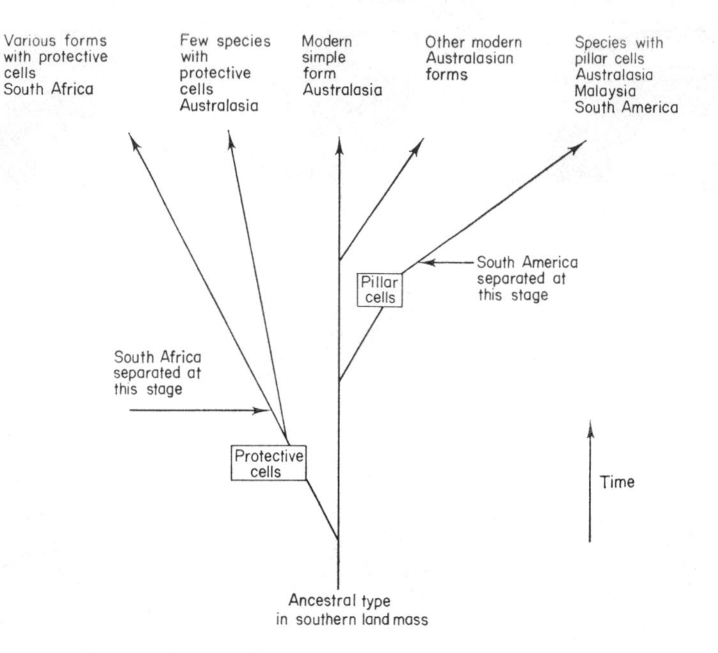

FIG. 2. Diagram showing the possible course of evolution within the Restionaceae.

S. Africa these xeromorphic forms were successful and their ancestors did not survive. In Australia some of the simple types persist today in the wetter habitats; others evolved and took positions in more arid situations.

At a much later stage than the separation of S. Africa and Australia there was a territorial expansion of *Leptocarpus sensu stricto*, already highly specialized xeromorphs. This group extended from Australia to N. Zealand, Malaysia, S. Vietnam and S. America. *Leptocarpus* had either a dispersal mechanism far superior to those of its close relatives, or a much stronger constitution and now represents the relic of a richer restiad flora in the outposts of its present-day range. Alternatively it could have extended its range while S. America and Australia were still in land connection with or at least close to the antarctic land mass. Its dispersal via the island chains in Malaysia follows a distribution pattern of many other plants.

It is interesting to note in passing that there are no Restionaceae in India. This agrees well with geological evidence that India had become disconnected from the southern land mass before S. Africa separated (Tarling and Tarling, 1971), probably before the time of evolution of the protorestionaceae.

APPLICATION OF ANATOMICAL DATA: DISCUSSION AND CONCLUSIONS

How does the anatomical evidence support the foregoing hypothesis about the origin of the present-day distribution of Restionaceae, and the probable sequence of events? Quite simply, when the different groups of restiads in S. Africa and Australia responded to water stress gradually imposed by climatic deterioration, modification occurred in different tissues in the two groups.

It appears that prior to the isolation of S. Africa from Australia a modification to some cells of the chlorenchyma had occurred in certain members of the family. The modified cells are called "protective-cells" (Cutler, 1964). They surround each substomatal cavity (Fig. 1B). The cells are usually thick-walled and lignified but may be poorly differentiated in some species. Most species with protective cells have 2–3 layers of palisade-like chlorenchyma and the protective cells may extend part or all of the way between the epidermis and the parenchyma sheath in 1, 2 or 3 layers (in the radial direction). The lining of the substomatal cavity is only one protective cell thick and has cuticle on the surfaces away from the chlorenchyma. There are various arrangements permitting gas exchange between the substomatal cavity and the chlorenchyma. The most frequent is one in which the protective cells have small gaps between adjacent cells towards the inner end of the substomatal cavity.

South African species all have protective cells. Certain genera also have some additional modifications not found in Australian species. For example a double or multiple epidermis occurs in some *Elegia* and *Chondropetalum* species (Fig. 1D). Girders of sclerenchyma extending from the sclerenchymatous cylinder to the epidermis are a feature of some *Restio* species and *Phyllocomos insignis* Mast. (Fig. 1I). Girders of sclerenchyma are also prominent in a number of *Hypodiscus* and *Willdenowia* species, but here they rarely reach the epidermis and either end in the chlorenchyma or are separated from the epidermis by a layer of modified, thick-walled chlorenchyma cells (Fig. 1E). The girders in these species alternate with the peripheral vascular bundles. The single species from Madagascar, *Restio madagascariensis* H. Chermez., has protective cells and looks very much like many S. African *Restio* species, except that it has a short papilla on each epidermal cell.

In Australia and New Zealand a very small proportion of the species has protective cells. *Calorophus elongatus* Labill. has them and so does

Sporadanthus traversii F.v. Muell. ex J. Buchanan and the species of *Lepyrodia* which should properly be referred to *Sporadanthus* (Cutler, 1969). Thickened chlorenchyma cells also occur in *Lyginia barbata* R.Br., but these are arranged in a network in the meshes of which are areas of substomatal chlorenchyma cells, so the thickened cells do not closely surround the substomatal cavities and are not strictly protective cells (Fig. 1F).

Discontinuity among S. African and Australian genera is shown more strongly in the bulk of Australian species which are either of the "simple" type described earlier or show other modifications. Firstly there is the group of species with modified epidermal cells. In this group epidermal cells surrounding the stomata are elongated to a greater or lesser degree into the chlorenchyma. In the best developed form they line the substomatal cavity and have apertures between them similar to those between protective cells (Fig. 1C). The cuticle lines the inner side of the cavity, away from the chlorenchyma. This is a most remarkable instance of parallel evolution where cells from different tissue systems have developed to produce structures very similar in appearance and function. The cells are most elongated in *Harperia lateriflora* W. V. Fitzgerald; they are also present in all species of *Loxocarya* examined except *L. pubescens*, in *Coleocarya gracilis* S. T. Blake, some *Lepidobolus* species, *Onychosepalum laxiflorum* Steud. and *Restio confertospicatus* Steud. Also classified among epidermal modifications is the unusual structure found in *Dielsia cygnorum* Gilg (Fig. 1H), which has longitudinal files of much enlarged epidermal cells separated by up to 10–12 files of smaller cells and stomata. Girders of sclerenchyma from the sclerenchyma cylinder occur opposite and extend to each file of enlarged cells. A small peripheral vascular bundle is usually found at the inner end of each girder.

The second main developmental trend in Australian species is quite significant. In this group cells from the parenchyma sheath have become elongated radially at more or less regular intervals round the culm, and they extend in files of about 2–3 from slight, domed ridges on the sclerenchyma cylinder to the epidermis, thus dividing the chlorenchyma into a series of separate channels (Fig. 1G). The radially elongated cells have thickened lignified walls and are termed "pillar-cells". A small peripheral vascular bundle is found in each sclerenchyma ridge. (Note that in those S. African species of *Hypodiscus* and *Willdenowia* which have sclerenchyma girders, the small vascular bundles alternate with them—another difference.) Pillar cells are found in *Leptocarpus* species *sensu stricto*, *Meeboldina denmarkica* Suess., *Loxocarya pubescens* Benth., *Chaetanthus leptocarpoides* R. Br. and certain "*Restio*" species from Australia.

The division of the chlorenchyma into self-contained longitudinal channels may have survival significance for the plant. I have seen specimens in which fungal infection through stomata has been confined to one of these channels. The mechanics of such a structure also make it very rigid, and drying damage must be lessened in species with this type of construction.

In certain S. African *Hypodiscus* and *Willdenowia* species the cells between sclerenchyma girders and the epidermis resemble pillar cells, but they are derivatives of the chlorenchyma and not of the parenchyma sheath. They are not, therefore, true pillar cells.

Since true pillar cells do not occur in S. African species it seems probable that this cell type evolved after it was possible for interchange of plants between S. Africa and Australia to take place. This observation is of importance because those members of the Restionaceae which are found in S. America, the Malay Peninsula and S. Vietnam all have pillar cells. The plants having them are obviously successful and would be expected to thrive in S. Africa had it been accessible to them. The occurrence of species with pillar cells in Chile and Patagonia would suggest that some form of connection between S. America and Australia remained after S. Africa and Australia had separated. Members with pillar cells having this wide dispersal are all from the genus *Leptocarpus sensu stricto*. It could be that they are either relics of a larger representation of the family, or more probably, that *Leptocarpus sensu stricto* has particular qualities which makes it more successful as an invader than others (but it has no special seed character or dispersal mechanism). The invader theory seems the more probable, for it is a *Leptocarpus* species, *L. tenax* (Labill.) R.Br., which is the sole member of the family known to me to occur in both east and west Australia. It is also found in Tasmania.

However the present-day distribution has come about, it is within this group of *Leptocarpus* species that I consider true vicariads can be demonstrated. In addition to pillar cells, *L. aristatus* R.Br., *L. brownii* Hk.f., *L. canus* Nees, *L. chilensis* (Steud.) Mast., *L. coangustatus* Nees, *L. disjunctus* Mast., *L. erianthus* Benth., *L. similis* Edgar, and *L. tenax* (Labill.) R.Br. all have a most unusual hair form. The hairs are multicellular, sub-rhombic plates, closely fitting at their margins with one another and attached to the culm at their proximal end by a very short somewhat sunken stalk. The constituent cells are very thick walled and the hairs are held very close to the epidermis. When they were initially observed in section it was only with difficulty that they were recognized to be hairs and not the outer cells of a double epidermis. Such hairs do not occur in any other genus in the family. *Meeboldina* has plate-like hairs, but these are of a more complex structure (Cutler, 1969). *Leptocarpus ramosus* R.Br. has pillar cells, but hairs were not seen on this plant and *L. spathaceus* R.Br. has pillar cells and multicellular, simple uniseriate unbranched hairs with thick-walled basal cells and thin-walled filament cells. Among the species with plate-like hairs, species identification from transverse sections of the culm can be difficult. The position of the stomata relative to the culm surface may be diagnostic. Here again the very close interrelationship of these plants is clearly shown.

CONCLUSIONS

Vicarious species are defined by Cain (1944) as ''closely related allopatric species which have descended from a common ancestral population and attained at least spatial isolation''. Since I have demonstrated that no genus of the Restionaceae occurs both in S. Africa and Australia, if we wish to find an intercontinental distribution of vicarious species we must look to *Leptocarpus sensu stricto* in Australasia, S. America, Malaysia and S. Vietnam. Within the continental groups it is quite possible that other vicarious species exist, but until the taxonomic treatment of the family is brought thoroughly up to date, their identity will remain unknown.

SUMMARY

The current taxonomic treatment of the Restionaceae is most unsatisfactory. Anatomical evidence has demonstrated that no one genus is really represented in both South Africa and Australia, and that at present truly vicarious species can only be discerned with any degree of certainty among species of the genus *Leptocarpus* occurring in Australia, New Zealand, South America, Malaysia and South Vietnam.

ACKNOWLEDGEMENTS

I am grateful to my colleagues at the Jodrell Laboratory for their comments on the manuscript and to Dr. P. Barnard, Department of Botany, University of Reading, for his guidance on the palaeobotanical aspects of this paper.

REFERENCES

CAIN, S. A. (1944). "Foundations of Plant Geography". Harper Bros., New York.

COUPER, R. A. (1960). *N.Z. Geol. Surv. Paleont. Bull.* 32.

CUTLER, D. F. (1964). *Notes Jodrell Lab.* **1**, 11–13.

CUTLER, D. F. (1969). "Anatomy of the Monocotyledons, Vol. IV Juncales" (C. R. Metcalfe, ed.), Clarendon Press, Oxford.

CUTLER, D. F. and SHAW, H. K. A. (1965). *Kew Bull.* **19**, 489–499.

GILG, E. (1890). *Bot. Jb.* **13**, 541–606.

GILG-BENEDICT, C. (1930). Restionaceae *In* "Die natürlichen Pflanzenfamilien" (Engler and Prantl, eds), 2nd Edition, Vol. 15a, pp. 8–27.

HUTCHINSON, J. (1959). "Families of Flowering Plants. II Monocotyledons", 2nd Edition, pp. 700–702. Clarendon Press, Oxford.

PILLANS, N. S. (1928). *Trans. R. Soc. S. Afr.* **16**, 207–440.

PILLANS, N. S. (1950). Restionaceae *In* "Flora of the Cape Peninsula" (R. S. Adamson and T. M. Salters, eds). Cape Town and Johannesburg.

TARLING, D. H. and TARLING, M. P. (1971). "Continental Drift". Bell and Sons, London.

6 | New Approaches to the Study of Disjunctions with Special Emphasis on the American Amphitropical Desert Disjunctions

OTTO T. SOLBRIG

Department of Biology and Gray Herbarium, Harvard University, Cambridge, Massachusetts, U.S.A.

INTRODUCTION

Disjunctions in the distribution range of plant species have attracted the attention of botanists for a long time. This is particularly true in the case of distributional gaps involving long distances and/or geographical barriers such as mountains or oceans. The approach to the investigation of species with such distributions has usually involved a detailed morphological–geographical study of the species with the aim of explaining the means by which the gap in the distribution of the species under study may have arisen. In this paper I would like to discuss a different approach to the study of disjunctions, which emphasizes the problem of establishment of a species and the various facets of convergent evolution.

In a strict sense, all species of plants have disjunct distributions, since there are always some discontinuities between populations. More important than the existence of range discontinuities is the distance separating the populations and the way large discontinuities are established and maintained.

Whenever populations become separated from each other, they become subjected to different selection pressures. Our general model (Stebbins, 1950; Mayr, 1963; Solbrig, 1970; Grant, 1971) states that this is the first step towards differentiation and possible eventual speciation. Generally accepted theory (Wright, 1969) also equates distance with isolation, and degree of isolation with differentiation given enough time, since some environmental differences between localities are always assumed to exist. Consequently, when range disjunctions involve thousands of miles and big natural barriers, and there is evidence that the disjunction is sufficiently ancient, the situation presents a problem. Occasional exceptions to any biological theory can be absorbed intellectually by *ad hoc* explanations, such as assuming low rates of differentiation or parallel evolution.

However, when important segments of a flora diverge from the accepted pattern of differentiation, a very special problem is created.

Intercontinental disjunctions present further problems, since they must either involve long-range dispersal or alternatively must be old enough to go back to a time when the continents were together. Five major intercontinental disjunctions are known: (1) Amphiatlantic (Löve and Löve, 1963), (2) Eastern Asia–Eastern North America (Gray, 1859; Wood, 1971), (3) Africa–South America (Mayr, 1952), (4) South America–Australasia (Pantin, 1960; Darlington, 1965), and (5) Temperate South America–Temperate North America (Johnston, 1940; Raven, 1963).

The first four of these disjunctions have become less puzzling since a viable, independently arrived at theory of continental drift has been proposed (Dietz, 1961; Dietz and Holden, 1970). The temperate South America–temperate North America floristic disjunctions cannot be explained that way. In this paper, I will address myself to the problem of the disjunctions involving the desert floras of temperate South America and temperate North America. More specifically, I will emphasize the disjunction between the phytogeographical province of the "Monte" in Argentina (Hauman, 1947; Cabrera, 1953; Morello, 1958) and the phytogeographical province of the "Sonoran Desert" in Mexico and the United States (Shreve, 1951; Shreve and Wiggins, 1964).

DESCRIPTION OF THE AREAS

If the broad outlines of the vegetation of the American continent are mapped (Fig. 1), it can be seen that there is a band of tropical forest around the equator flanked north and south by belts of scrubland, savanna and desert. These areas on both sides of the tropical belt are roughly equivalent, although the transition between tropical vegetation and drylands is more abrupt in the Northern Hemisphere than in the Southern. In addition, the continent has a backbone of high mountains that stretches almost uninterrupted from the Arctic to the sub-Antarctic. These mountains share the same kind of high mountain vegetation, but the middle and lower elevation flora varies considerably according to the climate of the area and the exposure of the mountains. Finally, there are relatively dry areas of both savanna and scrubland in the geographical tropics, such as the Brazilian "cerrados" and "caatinga," the Venezuelan and Colombian "llanos," more or less linking those of the northern and southern temperate areas.

The areas normally referred to as "deserts" in the United States and

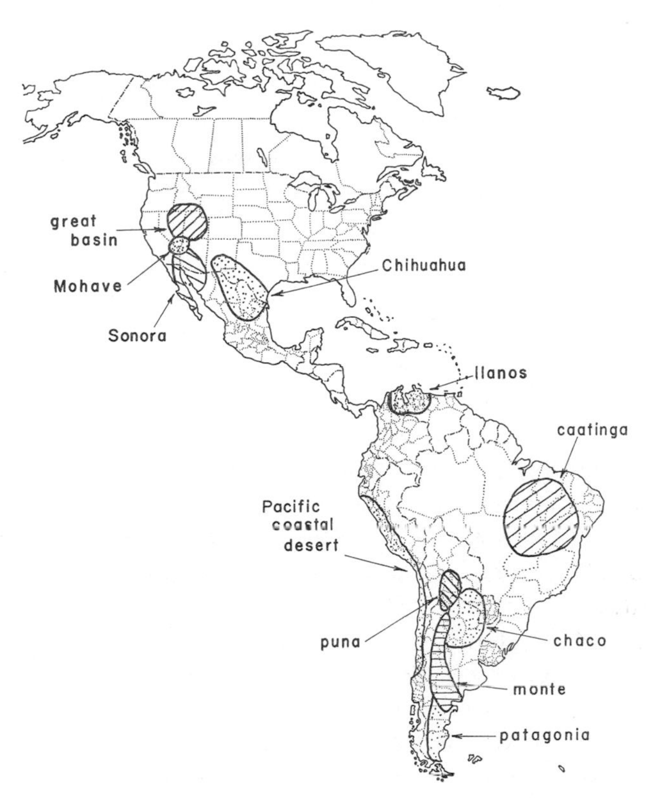

FIG. 1. Map showing the approximate area of the major semi-desert areas of North and South America.

Mexico are ecologically semideserts. They are the Great Basin Desert, the Mohave Desert, the Sonoran Desert, and the Chihuahuan Desert. The Great Basin is a cold semidesert with a tendency toward summer droughts and winter rains; the Mohave Desert is characterized by winter rains and higher temperatures than those of the Great Basin; the Sonoran and Chihuahuan Deserts have summer rains and warm temperatures, the latter experiencing more rainfall than the former.

In Argentina, we find three formations that are semi-deserts: "Patagonia," which is a semi-desert with cold climate, largely winter rains and with a seasonal variation in excess of the daily variations; "Puna," which is a high mountain semi-desert with cold, dry climate, summer rains and daily temperature fluctuations in excess of seasonal variations; and the "Monte," which is a semi-desert with moderate temperature, summer rains, and seasonal variations in excess of daily temperature.

The Patagonian Desert is in broad features comparable to the Great Basin, the Monte to the Sonoran and Chihuahuan Deserts, but there are no counterparts to the Mohave Desert in the south or to the Puna in the north. Of the amphitropical formations under consideration, the Monte and the Sonoran Desert are by far the most similar in climate, topography, and vegetation.

The Monte and the Sonoran Desert are almost unique in their great similarity in climate and in vegetational structure. Although each of these communities is composed in great part of different species with different evolutionary histories, they have come to look remarkably alike and presumably to function alike. There are, however, a number of species that are present on both sides, some fifty in number. These have drawn the attention of botanists familiar with both areas since the time of Asa Gray and will be the focus of the present paper.

The Monte (Lorentz, 1876; Hauman, 1947; Morello, 1958; Cabrera, 1953) is a phytogeographical province that extends from 24° 35′ lat. S to 44° 20′ lat. S, and from 62° 54′ long. W on the Atlantic coast to 69° 50′ long. W. Rains in the region average less than 200 mm a year in most localities and never exceed 600 mm; evaporation exceeds rainfall throughout the region. The rain falls in spring and summer. The area is bordered on the west by the Cordillera de los Andes, which varies in height between 5000 and 7000 m in this area and thus provides a barrier to winds (and moisture) from the west. On the north, the area is bordered by the high Bolivian Plateau (3000–5000 m high) and on the east by a series of mountain chains (Sierras Pampeanas) that vary in height from 3000 to 5000 m in the north (Aconquija, Famatina and Velazco) to less than 1000 m (Sierra de Hauca Mahuida) in the south. Physiographically, the northern part is formed by a continuous phase of high mountains which become less important further south as well as lower in height. The Monte vegetation occupies the valleys between these mountains, as a discontinuous phase in the northern region, and as a more or less continuous phase from approximately 32° S, south.

The temperature regime is warm-temperate with mild winters (mean minimum temperatures average 0°C to 5°C and mean maximum temperatures 13°C to 21°C) and warm summers (mean minimum temperatures 12°C to 19°C and warm maximum temperatures 25°C to 34°C) (Hauman, 1947).

Physiognomically, the vegetation can be described as a typical subdesert scrub with marked xeromorphic adaptations (Morello, 1955,

1958; Böcher *et al.*, 1963) such as small leaves with thick cuticles and sunken stomata; and presence of hairs, resins, or wax on the leaf and stem surfaces. Another very conspicuous characteristic is the great number of aphyllous shrubs. Floristically, the most conspicuous elements are shrubs of the family Zygophyllaceae, particularly species of the genus *Larrea*, but also *Bulnesia*; Leguminosae, especially *Gourliea spinosa* Skeels, *Zuccagnia punctata* Cav., *Cassia aphylla* Cav., and *Prosopis* spp.; *Condalia microphylla* Cav. in the family Rhamnaceae; the capparidaceous shrub *Atamisquea emarginata* Miers; *Bougainvillea spinosa* Heimerl in the Nyctaginaceae; several species of *Lycium* in the Solanaceae; many Compositae, particularly species of *Baccharis* and *Senecio*; a great number of Cactaceae, most notably species of the genera *Cereus*, *Trichocereus*, *Opuntia*, *Echinocactus*, and *Echinopsis*; and last but not least, particularly in the northern sectors, a number of terrestrial Bromeliaceae of the genera *Dyckia* and *Deuterocohnia*.

The Sonoran Desert (Shreve, 1951; Shreve and Wiggins, 1964) is the region that immediately surrounds the head of the Gulf of California. It extends from approximately 35° lat. N, north of Needles on the California–Arizona border in the United States, to 23° lat. N at San Jose del Cabo in the Peninsula of Baja California, Mexico, and from approximately 109° 30′ long. W in Sonora, Mexico, to 116° 40′ long. W at Indio, California, covering an area of approximately 310,362 km² (Shreve, 1951). The surface of the Sonoran Desert is essentially a series of slightly inclined planes on which are superimposed innumerable hills and mountain ranges, most of which are less than 1000 m in height. Basically, the "desert" area is the continuous phase with the mountains forming a discontinuous phase. In this general topography, it differs from the Monte, particularly from the northern part of the Monte. The Sonoran Desert is limited by the Pacific Ocean and the Gulf of California for a great part of its western boundary; to the north and east it is limited by a series of mountain ranges, none of which reaches the heights of the Andes and the Sierras Pampeanas that border the Monte.

Climatically, a relatively uniform regime prevails throughout the Sonoran Desert. In spite of the nearness of the Pacific Ocean and the intrusion of the Gulf of California, the climate is distinctly of the continental type, more so than the Monte, particularly in its northern part, which lies in the lee of high mountain ranges and is not influenced by the narrow waters of the Gulf of California. The difference in continentality of the climate between Monte and Sonora is due also in part to the distribution of land and water masses in the northern and southern hemisphere.

Some of the highest temperatures in North America have been recorded in the Sonoran Desert. In general, summer mean maximum temperatures range from 29°C to 40°C and summer mean minima from 24°C to 30°C, while winter mean maxima range from 16°C to 19°C and winter mean minima go from 1°C to 16°C (Shreve, 1951). Average rainfall decreases from east to west, going from approximately 375 mm on the Sonoran–Chihuahuan border to less than 100 mm in parts of Baja California. Another important feature is the rainfall distribution, which is almost entirely confined to late winter and early spring in the western areas of Baja California and California; it is a mixture of winter and summer rains in central Arizona and is largely summer rains in southern Sonora (Shreve, 1951).

The vegetation of the Sonoran Desert is varied, exceeding the other three North American deserts in the number and variety of its life forms and of its plant communities. We find a wealth of succulent and semisucculent forms, large numbers of microphyllous shrubs and some small trees.

Floristically among the evergreen shrubs the most conspicuous element is *Larrea divaricata* Cav., followed by *Simmondsia chinensis* (Link) Schneider, *Condalia spathulata* A. Gray, *Celtis spinosa* var. *pallida* (Torr.) M. C. Johnst., *Atamisquea emarginata* and *Cordia sonorae* Rose. Among the trees are three species of *Cercidium*, *C. microphyllum* (Torr.) Rose & Johnston, *C. sonorae* Rose & Johnston, and *C. floridum* Benth. as well as *Olneya tesota* A. Gray, two species of *Jatropha*, three of *Bursera*, *B. microphylla* A. Gray, *B. hindsiana* Engl., and *B. filicifolia* Brandg., *Ipomoea arborescens* Sweet, and among the riparian element, *Prosopis glandulosa* var. *torreyana* (L. Benson) M. C. Johnst. and *P. velutina* Wooton, *Populus fremontii* S. Wats., and *Celtis reticulata* Torr. Other important elements are species of Leguminosae such as *Acacia greggii* A. Gray, *A. farnesiana* (L.) Willd., *Caesalpinia pumila* (Britton & Rose) Hermann, *Parkinsonia aculeata* L., and *Mimosa laxiflora* Benth. One of the most characteristic elements is the succulents, particularly species of Cactaceae of such genera as *Pachycereus*, *Cereus*, *Opuntia*, *Echinocereus*, *Mammillaria*, *Ferocactus*; species of Agavaceae, of genera such as *Yucca* and *Agave*; and two species of Fouquieriaceae, *Fouquieria splendens* Engelm. and *Idria columnaris* Kell.

THE FLORISTIC SIMILARITIES

Although there are some differences in the physical set-up, in the physiognomy of the vegetation, and in the floristic composition, there are also some notable similarities. Most striking is the sharing of certain species with disjunct ranges that are dominant elements of the vegetation. Among them is the zygophyllaceous shrub *Larrea divaricata*, which is the dominant element, albeit often the only shrub, distributed over the greater part of the Monte and Sonoran Desert; various species of *Prosopis*, which constitute the principal riparian element (although no common species are shared); closely related species of *Cercidium*, another conspicuous element (*C. sonorae*, *C. microphyllum* and *C. floridum* in the north; *C. australe* I. M. Johnst. and *C. andicola* Griesb. in the south); and the capparidaceous shrub *Atamisquea emarginata*. Other genera with common species or pairs of closely related species which are not necessarily as conspicuous as the preceding are: *Celtis* (*C. spinosa* Spreng. in the Monte, *C. pallida* and *C. laevigata* Willd. in Sonora), *Tribulus terrestris* L., *Jatropha*, *Maytenus*, *Condalia* (*C. microphylla* Cav. in the Monte, *C. globosa* I. M. Johnst. and *C. brandegeei* I. M. Johnst. in Sonora), *Ziziphus* (*Z. mistol* Griesb. in the south, *Z. sonorensis* S. Wats. in the north), *Ayenia*, *Mentzelia*, *Frankenia*, *Cereus*, *Opuntia*, *Vallesia glabra* (Cav.) Link, *Gilia sinuata* Dougl., *Aloysia*, *Glandularia*, *Verbena*, *Lycium*, *Mimosa strigosa* Willd., *Acacia*, *Caesalpinia gilliesii* (Hook.) Benth., *Cassia*, *Hoffmanseggia*, *Calandrinia*, *Portulaca*, *Amsinckia*, *Cryptantha*, and *Heliotropium*. Among the Compositae, the two regions share species or pairs of closely related species of *Ambrosia*, *Baccharis*, *Encelia*, *Flourensia*, *Gochnatia*, *Gutierrezia*, *Hymenoxys*, *Pectis*, *Senecio*, *Tessaria*, and *Verbesina*. Finally, a number of species of grasses are common to both areas, among them *Blepharidacne benthamiana* Hitchc., *Andropogon saccharoides* Swartz, *Cottea pappophoroides* Kunth, *Digitaria californica* Henrard, *Enneapogon desvauxii* Beauv. ex Desv., *Eragrostis megastachya* Link., *Pappophorum mucronulatum* Nees, *Leptochloa uninervia* (Presl) Hitchc. & Chase, *L. dubia* Nees, *L. filiformis* (Lam.) Beauv., *L. fascicularis* (Lam.) A. Gray, *Phragmites communis* Trin., *Setaria geniculata* Beauv., *Stipa speciosa* Trin. & Rupr., and *Trichloris crinata* (Lag.) Parodi. Furthermore several genera such as *Muhlenbergia*, *Bothriochloa* and *Stipa*, have pairs of related species in both areas.

In total, the two regions share approximately 50 species or pairs of very closely related species, among them some of the most conspicuous elements of the flora. Although many of these are herbaceous, most notably the grasses, others are shrubs and even trees.

The existence of identical or closely related plant taxa in the temperate and subtropical zones of North and South America has attracted the attention of botanists since the time of Asa Gray and has led to several interesting studies such as those by Bray (1898), Johnston (1940), Axelrod (1941, 1948), and Campbell (1944).

These workers have been concerned with two major aspects of the problem, namely (1) the taxonomic problem of establishing whether the disjunct taxa are indeed conspecific, and (2) the question of how the disjunctions may have arisen. Three major explanations have been proposed: (1) that the disjunction is the result of long-range dispersal from one zone to the other outside of the tropical areas, (2) that it represents the remains of a once continuous or largely continuous distribution, or (3) that it is not a true disjunction but the result of parallel evolution from common tropical ancestors.

In order to discriminate unmistakably between these three hypotheses, rather complete fossil evidence is required, which unfortunately is largely missing. It also should be added that the explanation for each particular disjunction may be different.

Although the inadequate fossil evidence and the possibility of different explanation for each species pair makes it almost impossible to foresee a complete solution for the problem of the reason for each North–South American disjunct, certain kinds of research with living plants promise to produce explanations that approximate the true historical situation.

A NEW APPROACH

The problem of disjunct ranges of distribution can be approached in two ways. First, there is the systematic species-by-species approach, in which each particular case is investigated in detail and an *ad hoc* explanation consistent with the available evidence is proposed. This has been the system most commonly employed, and it is the only possible one when dealing with rare or unique patterns. The second approach is the systems approach, which involves the investigation of the donor and recipient floras in their totality, aiming to discover major underlying facts that account for whole patterns of distribution. The latter approach is much more demanding in manpower and resources and is more time consuming, but it also gives a greater insight into the complex network of interacting forces that determine the structure and function of ecosystems.

A number of Argentine, Chilean and United States biologists agreed during 1968 and 1969 to establish an Integrated Research Program under the aegis of the International Biological Program to study the nature of biological convergences using the systems approach to the study of ecosystems. The basic question being asked is the following: Given two identical physical environments (topography, geology, climate) and two phylogenetically unrelated floras and faunas, how similar in structure and function will the ensuing ecosystems be? In other words, the question asked is whether convergent evolution will invariably ensue under a given physical environment, or whether the phylogenetical input will determine to a large extent the final structure of the ecosystem.

To answer this question, it was decided to study two pairs of regions with similar physical characteristics. For logistic and other considerations, the areas chosen were: (1) a site in southern California and its counterpart near Santiago, Chile, with a Mediterranean type of climate, and (2) a site in the Sonoran Desert, near Tucson, Arizona and its counterpart in the Monte near Andalgala, Catamarca, Argentina. At each site, a number of detailed studies have been planned (see Figs 2 and 3). They fall into four main categories: (1) studies of the physical parameters, aimed at determining exactly how similar the physical environment at the various sites are, (2) niche studies, aimed at studying the actual convergence (or lack of such) in selected groups of organisms, (3) species interactions, aimed at understanding the competitive and trophic interactions taking place, and (4) studies involving the effect of time on the systems, both in a successional and in a geological sense.

The program, named the "Structure of Ecosystems Program," was approved by the U.S. National Committee for the I.B.P. as one of its official programs, and it is now starting its second year of official life under a grant from the N.S.F.

Although ideally one would like to deal at each of the two points with phylogenetically unrelated species, this is not the case in the two desert sites, Silver Bell Bajada in Tucson and Bajada Coyango in Argentina. Consequently, another dimension is added, the study of species with disjunct ranges. Actually, the existence of species with disjunct ranges permits the two-way comparison of phylogenetically related and ecologically similar species with phylogenetically unrelated and ecologically similar or dissimilar ones.

CONDITIONS FOR THE ESTABLISHMENT OF NEW SPECIES

Three main phases can be identified relative to the establishment of a new population of plants (which always will be technically disjunct, even if it might be just by a few meters). These phases are (1) dispersal, (2)

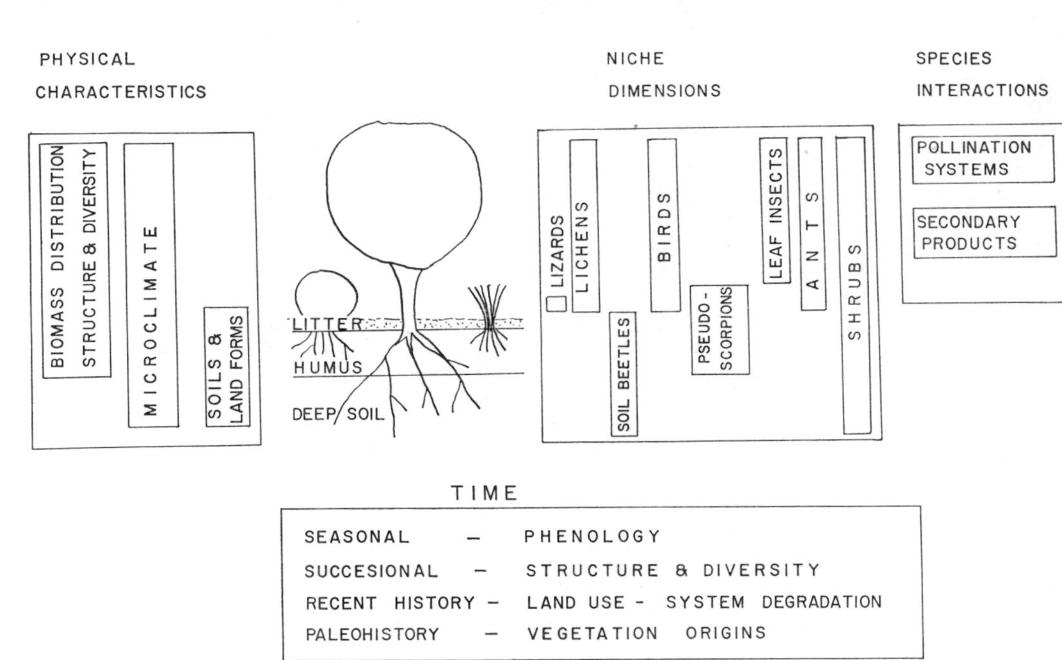

Fig. 2. The main research projects of the Mediterranean Scrub Project of the Structure of Ecosystems Program. Each project is under the direction of one or more individual investigators, the entire project being coordinated by Drs Harold Mooney of Stanford University and Francesco Di Castri of the Universidad Austral de Valdivia.

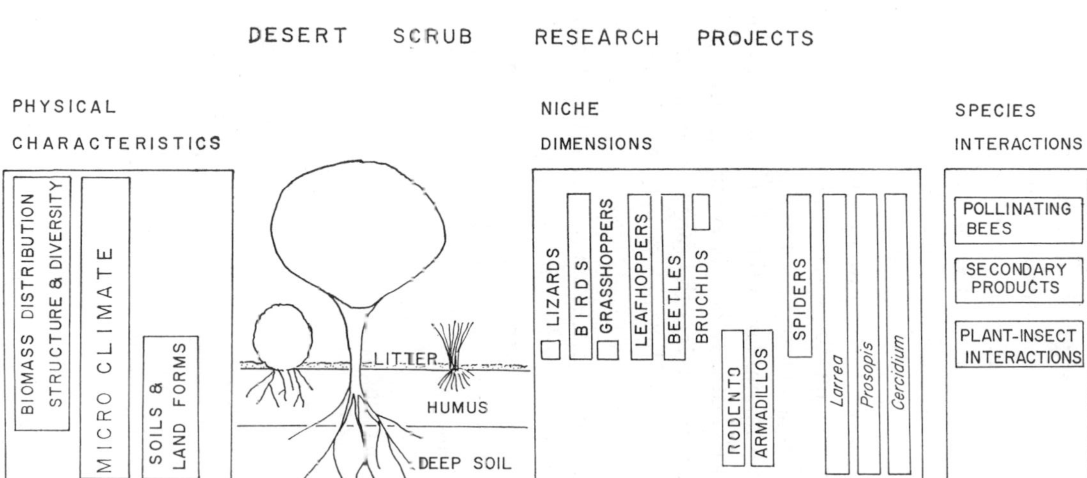

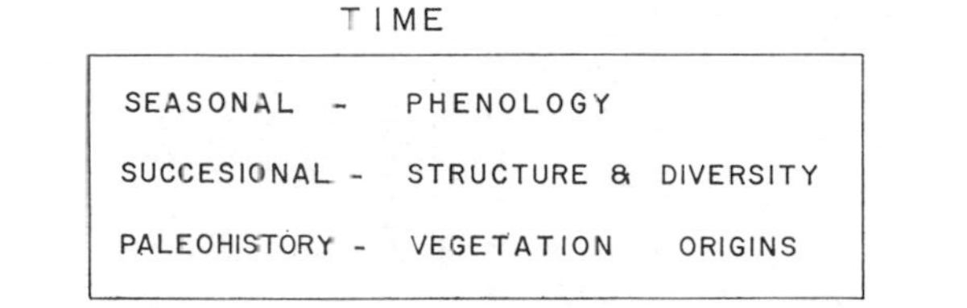

FIG. 3. The main research projects of the Desert Scrub Project of the Structure of Ecosystems Program. Each project is under the direction of one or more individual investigators, the entire project being co-ordinated by Dr O. T. Solbrig of Harvard University and a representative committee of Argentine scientists headed by Drs Jorge Morello from INTA and Juan Hunziker from the University of Buenos Aires.

germination and establishment and (3) survival over time. The dispersal capacity of any propagule is always greater than its capacity to germinate and to become established at the site to which it was dispersed, and likewise, a plant is capable of germinating and growing in more places than those where it is capable of surviving generation after generation.

The first of these two steps, dispersal and establishment, are accomplished with the genetic patrimony that the propagules carry with them from their original population. Survival, at least in the initial stages, is also dependent to a large extent on the original genetic makeup of the plants, although natural selection will start acting on the population after the first generation. Because of this, the probability of establishment and survival of a population is dependent on the environment of the new locality being within the range of tolerance of the arriving genotypes. In other words, the environment must be similar to the environment of the area from which the propagules originate. It is this latter requirement, rather than lack of dispersability, that restricts the ranges of species.

There is one more point that should be mentioned. All favorable environments are overcrowded with plants unless some recent disturbance has occurred (fire, floods, man, etc.). This means that a lot of competition is taking place between and within the species that occupy the area. This competition results in a series of mutual adjustments (character displacements, Brown, 1957; Wilson, 1965) and exclusions of species from microhabitats within an area. These interactions take place between species at the same trophic level and between species at different trophic levels. Consequently, any arriving propagule must cope with the competing local species and must withstand the vagaries of the physical environment. Since competition is partly a numbers game, the invading species with its few propagules is at a greater disadvantage than the local species that produce a great excess of propagules in relation to the number of eventual survivals. To overcome such disadvantages, the invading species must possess a great competitive advantage.

On the other hand, in harsher and more unpredictable environments, such as deserts, or in highly disturbed environments (particularly those that are man-made), the populations of plants and animals are usually below the environmental carrying capacity. Under such circumstances, once a propagule has become established, it can build up a sizable population before it gets into direct competition with local species. This accounts for the relative ease with which species become established in disturbed sites and for the almost world-wide distribution of many such species. In deserts, however, the extremely unfavorable physical environment limits the number of species that are capable of establishment and particularly survival during unfavorable years (Lewis, 1962).

In summary, new populations become established by the dispersal of a propagule to a new locality, by successful germination and establishment, and then by successful reproduction and survival over time, through adjustments to the vagaries of the physical environment and the negative effects of competition of other species for the available resources.

Although the problem of the peripheral population has attracted the interest of geneticists and population biologists (Mayr, 1945, 1963; Carson, 1955; Selander, 1970; Rohlf and Schnell, 1971), long-range disjuncts have attracted much less attention from this angle. In the specific case of the disjunctions between desert temperate South and North America, most work to date has concentrated on describing the problem, on determining the degree of relationship between the taxa, and in postulating possible explanations for the dispersal of propagules from one region to the other. Experimental work is in its infancy and so far has been restricted to herbaceous disjuncts (Raven and Lewis, 1959; Grant, 1971; Solbrig *et al.*, 1968).

As part of the "Structure of Ecosystems Program", an attack on the problem of invasion of new habitats is planned. Three genera with conspicuous disjunctions which play an important role in the ecosystem have been chosen. These genera are: *Larrea*, which has four species in South America, one of which, *L. divaricata*, has a disjunct Monte Sonoran Desert disjunction (Hunziker *et al.*, 1971); *Prosopis*, with five species in North America and about 25 in South America, with an almost continuous distribution (Burkart, 1940); and *Cercidium*, with four species restricted to North America and two to southern South America and a seventh species with an almost continuous distribution from Mexico to Peru (Johnston, 1940). The approaches vary with the genus (see Tables I–III) but basically entail a detailed comparison of the entities in North and South America, as well as a detailed investigation of the role the species play in the ecosystem at the extremes of their distributions. This involves a study of such ecological parameters as soils, species abundance, biomass production, species association, pollinators and seed dispersal. Those studies may reveal whether the structure of the ecosystems in the Monte and the Sonoran Desert are similar, and whether an ecosystem similar to the one from which a propagule comes is a prerequisite for successful establishment.

TABLE I. Research on *Larrea*

FIELD	LABORATORY
Ecological studies: coverture, abundance, diversity, age (Barbour, Diaz, Yang)*	Storage proteins (Hunziker)
Hybridization and chromosomes (Hunziker, Palacios, Poggio, Yang)	Enzymes (Solbrig)
Pollination (Hurd)	Flavonoids (Timmermann, Valessi)
Seedling establishment (Yang)	Water tolerance physiology (Lowe, Morello)
Larrea–insect relationships (Cates, Orians, Schulz)	HERBARIUM
	Morphological variation (Hunziker, Yang)
	Geography (Hunziker, Yang)
	Taxonomy (Hunziker, Yang)

TABLE II. Research on *Prosopis*

FIELD	LABORATORY
Ecological studies: coverture, abundance, diversity, age (Morello, Solbrig, Vuilleumier)	Germination and seedling variation (Solbrig)
Hybridization and chromosomes (Hunziker, Palacios, Poggio)	Storage proteins (Hunziker)
Pollination (Hurd, Vuilleumier)	Serology (Simon)
Seed dispersal and seedling establishment (Solbrig)	Enzymes (Solbrig)
Bruchids and seed predation (Kingsolver, Solbrig, Teran)	Flavonoids (Carman)
	HERBARIUM
	Morphological variation (Vuilleumier)
	Geography (Burkart, Solbrig, Vuilleumier)
	Taxonomy (Burkart)

TABLE III. Research on *Cercidium*

Work on *Cercidium* being conducted jointly by C. Lowe and O. T. Solbrig covers the following aspects:	Germination and seedling variation
Ecological studies: coverture, abundance, diversity, age	Storage proteins
Hybridization and chromosomes	Flavonoids
Pollination	Morphological variation
	Geography
	Taxonomy

* Individual investigators conducting that aspect of the work are indicated in parentheses.

SUMMARY

All species show disjunctions between populations. From the point of view of the evolutionist and phytogeographer what is striking is that some populations unconnected by gene flow over extremely long periods of time remain so similar that the systematist considers them conspecific. It is a problem because we always assume that some environmental differences exist between localities. Obviously more attention has to be paid to the effect of environment and to the way populations diverge in similar and dissimilar environments.

In considering long-range disjunctions, one problem is to ascertain how the seeds got to the disjunct locality. However, each case is an independent event, and a unique event that took place in the past. Only by studying each individual case and proposing *ad hoc* hypotheses can we unravel the problem. Although from the point of view of the systematist and the evolutionist interested in a particular group of plants, the solution of the problem for his species is important, it is unlikely that any theory of general validity regarding long-range dispersal of seeds will emerge in this way.

But disjunctions also entail the successful establishment of a population in an area where it did not grow before. This aspect of disjunctions has been studied much less. By studying the totality of the ecosystem, we hope to ascertain general rules relating to how new populations can become established in an area, what a ''niche'' is and whether it exists, and the relative roles played by ''phylogeny'' and ''environment'' in the evolution of ecosystems and the maintenance of disjunctions.

REFERENCES

Axelrod, D. I. (1941). *Proc. natn. Acad. Sci. U.S.A.* **27**, 545–551.

Axelrod, D. I. (1948). *Evolution* **2**, 127–144.

Böcher, T., Hjerting, J. P. and Rahn, K. (1963). *Dansk bot. Arkiv* **22**, 7–115.

Bray, W. L. (1898). *Bot. Gaz.* **26**.

Brown, W. (1957). *Q. Rev. Biol.* **32**, 247–277.

Burkart, A. (1940). *Darwiniana* **4**, 57–128.

Cabrera, A. L. (1953). Esquema fitogeografico de la Republica Argentina. *Revta Mus. La Plata (bot.)* **8**, 87–168.

Campbell, D. H. (1944). *Proc. Calif. Acad. Sci.*, Ser. 4, 25.

Carson, H. L. (1955). *Cold Spring Harb. Symp. quant. Biol.* **20**, 276–286.

Darlington, R. J. (1965). ''Biogeography of the Southern End of the World''. Harvard University Press, Cambridge, Massachusetts, U.S.A.

Dietz, R. S. (1961). *Nature, Lond.* **190**, 854–857.
Dietz, R. S. and Holden, J. C. (1970). *Scient. Am.* **223**, 30–41.
Grant, V. (1971). "Plant Speciation". Columbia University Press, New York.
Gray, A. (1859). *Mem. Am. Acad. Arts Sci.* II, **6**, 377–452.
Hauman, L. (1947). *GAEA, Geografia de la Republica Argentina* **8**, 208–248.
Hunziker, J., Palacios, R. A., de Valesi, A. G. and Poggio, L. (1971). *In* "Plant Species Disjunctions" (O. T. Solbrig and H. W. Wagner, eds), Harvard University Press, Cambridge, Massachusetts, U.S.A. (in press).
Johnston, I. (1940). *J. Arnold Arbor.* **21**, 356–363.
Lewis, H. (1962). *Evolution* **16**, 257–271.
Lorentz, P. (1876). *In* "La República Argentina" (R. Napp, ed.), pp. 77–136. Buenos Aires.
Löve, A. and Löve, D. (1963). "North Atlantic Biota and Their History". Pergamon Press, Oxford.
Mayr, E. (1945). *Lloydia* **8**, 69–83.
Mayr, E. (1952). *Bull. Am. Mus. nat. Hist.* **99**, 85–258.
Mayr, E. (1963). "Animal Species and Evolution". Belknap Press of Harvard University, Cambridge, Massachusetts, U.S.A.
Morello, J. (1955). *Revta Agron. NE Argent.* **1**, 301–524.
Morello, J. (1958). *Opera Lilloana* **2**, 1–155.
Pantin, C. F. A. (1960). *Proc. R. Soc. (ser. B)* **152**, 431–682.
Raven, P. H. (1963). *Q. Rev. Biol.* **38**, 151–177.
Raven, P. H. and Lewis, H. (1959). *Brittonia* **11**, 193–205.
Rohlf, F. J. and Schnell, G. D. (1971). *Am. Nat.* **105**, 295–324.
Selander, R. K. (1970). *Am. Zool.* **10**, 53–66.
Shreve, F. (1951). "Vegetation of the Sonoran Desert". Carnegie Inst. Wash. Publ. No. 591.
Shreve, F. and Wiggins, I. (1964). "Vegetation and Flora of the Sonoran Desert". Stanford University Press, Stanford, Calif.
Solbrig, O. T. (1970). "Plant Biosystematics". The Macmillan Co., New York.
Solbrig, O. T., Passani, C. and Glass, R. (1968). *Am. J. Bot.* **55**, 1235–1239.
Stebbins, G. L. (1950). "Variation and Evolution in Plants". Columbia University Press, New York.
Wilson, E. (1965). *In* "The Genetics of Colonizing Species" (H. G. Baker and G. L. Stebbins, eds), pp. 8–28. Academic Press, New York and London.
Wood, C. E. Jr. (1971). *In* "The Distributional History of the Biota of the Southern Appalachians. Part II. Flora" (P. C. Holt, ed.), pp. 331–404. Virginia Polytechnic Inst. and State University, Blacksburg, Va., Research Division Monograph 2.
Wright, S. (1969). "Evolution and the Genetics of Populations", Vol. 2, University of Chicago Press, Chicago.

7 | Evolutionary Problems in the Arctic Flora

TYGE W. BÖCHER
Institute of Plant Anatomy and Cytology, University of Copenhagen, Copenhagen, Denmark

INTRODUCTION

While chromosome counts of arctic plants date back to 1927, when Hagerup published his paper on *Empetrum hermaphroditum* Hagerup, it is only very recently that experimental researches on arctic species have been carried out. One of the best examples of such researches is that on *Oxyria digyna* (L.) Hill by Billings *et al.* (1969). Experiments with arctic species are difficult, simply because most species are difficult to cultivate under conditions normally available in botanical gardens or experimental fields. It was therefore a great step forward when the Arctic Greenhouse in the Copenhagen Botanical Garden was built in 1960. Here it is possible to cultivate even typical snow-patch species such as *Phippsia algida* (Sol.) R.Br. and *Ranunculus pygmaeus* Wbg. as well as other true arctic species such as *Ranunculus glacialis* L. and *R. sulphureus* Sol., *Saxifraga cernua* L. and *S. rivularis* L.; and in the last few years, greenhouse cultivations have been combined with experiments in growth chambers. What we are going to deal with here, however, is not only results obtained in the greenhouse and the growth chambers; in fact, a good many species, particularly those from dry habitats, can be grown perfectly well outdoors, as potted plants in frames, at the experimental field outside Copenhagen.

Evolutionary problems in the arctic flora have also been approached on the basis of herbarium material by many taxonomists and plant geographers. The most outstanding personality using such methods only is Dr Eric Hultén; and he has produced a long series of maps which not only provide very important information about the area limits of arctic plants, but also in many cases, by grouping maps of several closely related taxa, tell us a great deal about possible evolutionary trends in the more complex species-groups.

Nowadays the Atlantic Ocean, with its northern extensions, separates the North American and the Eurasian continents so effectively that plant migrations across the ocean are made very difficult. Cases of successful long-distance dispersal from America to Europe or *vice versa* are probably very rare. Plant geographers have distinguished a circumpolar group of such plants which are either west arctic or east arctic. West arctic plants occur in N. America and Greenland but may more exceptionally reach Iceland or Norway; east arctic plants are Eurasiatic but sometimes extend their areas to Greenland and Eastern North America. The two groups which have been able to cross the gap are, from an evolutionary point of view, of particular interest. We may ask here, what happened to them as species during their migration?

One of the problems which we are going to discuss is evolution as a result of biotype depletion during migration; another is gene-pool reductions during possible perglacial survival. Further problems are connected with species formation and the occurrence of endemics in the North Atlantic area.

MIGRATIONS OF ARCTIC-ALPINE AND AMPHI-ATLANTIC SPECIES

Let me begin with some European alpine plants which crossed the North Atlantic area, thus becoming arctic-alpine and amphi-atlantic. *Ranunculus glacialis* reaches East Greenland north of Angmagssalik. It is diploid over its whole range. I have it in culture from Angmagssalik, Iceland, Norway and several places in the Alps (Fig. 1). The alpic plants deviate greatly from the others; they often have 2–4 flowers per stem, and narrow leaf segments of which the larger are stipitate. The northern plants very rarely have more than a single flower per stem; they also have more rounded and non-stipitate leaf segments, and furthermore, they bloom later. The same morphological differences appear from herbarium material, which also show that the alpic populations are richer in different biotypes. In the Alps there is not only the branched type just described; some specimens correspond almost entirely with the Scandinavian ones, while others represent the other extreme, having narrow tapering and acute leaf-segments (Fig. 2).

The conclusion may be drawn that several different biotypes were able

FIG. 1. Cultivated specimens of *Ranunculus glacialis*. From the left: No. 6587 from the Alps, No. 42 from Norway, and No. 10209 from Angmagssalik, East Greenland. The last two specimens have single flowers and broad leaf segments.

to survive glaciation on nunataks in the Alps, while those biotypes which did so in the North Atlantic area (including Scandinavia) were much more uniform. Considering the difference in the range of genetic variation between the Alps and the northern areas, it is unlikely that there was a late-glacial immigration of *Ranunculus glacialis* from the Alps to Scandinavia and further on to the Faeroes, Iceland and East Greenland.

In East Greenland the species has a southern limit near Angmagssalik; it stops almost where the topography changes from sharp crests and peaks to flattened ice-polished mountains. At the present time its southern limit may thus be a result of very slow migration from refugia in the areas with jagged mountains, but we cannot exclude the possibility that it suffered some kind of biotype elimination during the Ice Age. If only biotypes adapted to glacial conditions survived, it is perhaps more understandable that the species does not penetrate further south than Angmagssalik. It grows on the mountain tops in the Faeroes; so it ought to grow at Cape Farewell on the high peaks there, if the populations in the Faeroes and Greenland were ecologically identical.

Some other species with montane European-arctic and amphi-atlantic ranges have been cultivated in order to be able to find other examples of genetic difference between Alpine and North Atlantic populations. Three of them, viz. *Juncus trifidus* L., *Saxifraga paniculata* Mill. and *Arabis alpina* L., seem to be much more uniform in their low-arctic–North Atlantic

Fig. 2. Herbarium specimens of *Ranunculus glacialis* from the Alps showing range of variation in number of flowers per stem and leaf shape. Upper row: both from Tirol; lower row, left: Steiermark, Seewigtale; lower row, right: var. *crithmifolius* Rchb. Mallnitzer Tauern.

area; in the case of *Arabis alpina* and *Saxifraga paniculata*, the low-arctic and montane Scandinavian plants also deviate physiologically from the alpic plants.

In the experiments carried out by Hedberg (1962) with plants of *Arabis alpina* from Kenya and Lappland the afro-alpine plants died, while those from Lappland survived. In the experiments which I have carried out, plants from Greenland, Iceland, Scotland and Finse in Norway, as well as from Kenya, showed a low degree of winter hardiness, while plants from the Alps, the Tatra Mountains and Italy usually survived the winter in Denmark. The peculiar fact that plants from the north are so difficult to grow outdoors during the winter in Denmark is probably connected with a difference in ecology between the northern and the middle- and south-European races. In the south *Arabis alpina* has a wide altitudinal distribution and prefers calcareous gravel. In Greenland the species avoids the high-arctic, dry regions. Here as well as in Norway it shows the greatest affinity to springs and snow beds, as well as winter-snow-covered gravelly screes. It is probably its dependence on snow protection which is decisive. It does not tolerate Danish winters with intermittent frost periods. The middle European and southern montane plants are not killed by black frost. Some introgression from subsp. *caucasica* may have taken place and have resulted in a change in ecology, making the southern strains more adapted to dry conditions, including physiological dryness during frost periods.

Again the comparative cultivation experiments do not support any idea of late-glacial migration. All three species are old, originally European elements and have reached not only Greenland but also eastern North America; *Juncus trifidus*, for example, is very abundant on Mt. Washington (New Hampshire, U.S.A.) in a race closely resembling plants from Greenland and Scandinavia, but differing in many characters from montane strains e.g. from the Pyrenees and the Tatra mountains.

Saxifraga paniculata (earlier *S. aizoon* Jacq.) is very polymorphic in the middle and south European mountains. The Greenland plants are more uniform. In culture they do not deviate essentially from plants from Newfoundland, Iceland and Norway. Certain strains from European mountains south of Scandinavia also resemble Greenland plants, but the European mountains harbour among others low-growing populations producing many reddish flowering stems. It is characteristic of most European plants, that they have long inflorescences, and that they almost every year produce a fairly large number of flowering stems. The Greenland and North Atlantic populations have usually shorter inflorescences and are inclined to produce few or no inflorescences. Some of them have been cultivated for ten years without producing more than 1–3 inflorescences per individual, while alpic plants show an average number of 8 to 20 flowering stems per year. However, some of the Greenland plants approach the European behaviour by producing many flowering stems in a single year or two, after which they return to the normal state of mainly vegetative growth, with very few or no inflorescences. This behaviour of the northern strains resembles that of species which, in the northern parts of their range, are composed of races which have a long rosette—or reinforcement stage before flowering. In any case, the North Atlantic populations of *Saxifraga paniculata* have a distinct genetic composition, probably poorer in biotypes than the middle and south European. A late- or post-glacial crossing of the Atlantic Ocean by this species is hardly possible, especially when we realize that the species in Norway is very rare, its two populations here being widely separated. Nordhagen (1965) and Ryvarden (1966) think that the Rogaland population in S.W. Norway immigrated from the west from a "North Sea continent" and not from the mountains in middle and southern Europe.

A differentiation of a separate western North Atlantic population in an amphi-atlantic species of European origin is more obvious in the case of *Viscaria alpina* (L.) G. Don. Here, however, the dividing line goes between Iceland and Greenland. Iceland *Viscaria* resembles the boreal serpentine races of Scandinavia and the alvar race of Øland in the Baltic Sea. Together these races form a separate subsp. *borealis* Böch. with narrow leaves and a more or less loose and elongate inflorescence. This subspecies does not occur west of the Denmark Strait. In Greenland and Labrador the plants have broad basal leaves and larger flowers in a dense inflorescence. They belong to subsp. *americana* (Fern.) Böch. This low-arctic subspecies approaches the European subsp. *alpina*, but is more robust, with broader leaves than the montane and alpine European plants.

One may speculate here about the evolutionary trends. The species has clearly evolved in Europe, where one finds the greatest diversity of forms and where other related species occur, e.g. *Viscaria vulgaris* Bernh., which can hybridize with *Viscaria alpina*. The latter must have crossed the North Atlantic ocean early enough to allow time for the differentiation of subsp. *americana*. It was probably Norwegian mountain plants which went over and became the ancestors of subsp. *americana*. The boreal biotypes with narrow leaves and more loosely arranged, smaller flowers probably followed the retreating icecaps and are now divided up into three geographically separated varieties, one in Iceland, one on serpentine hills in Scandinavia, and one in shallow soil overlying limestone on Øland. These populations may be regarded as remains of a continuous late-glacial population.

They are more specialized, being adapted to rare soil conditions. But they were not able to cross the Denmark Strait. Thus the Greenland population escaped hybridization with the narrow-leaved, small flowering, more southern biotypes and thereby became able to evolve into subsp. *americana*.

A complete separation of an alpic and a Scandinavian–Greenland species has taken place in the species pair *Braya alpina* Sternb. & Hoppe—*Braya linearis* Rouy. While *B. alpina* is confined to the Alps, *Braya linearis* is found in Norway where it is bicentric, in E. Greenland, and in W. Greenland where it also occupies two separate small areas. These two species resemble one another morphologically, cytologically (having the same chromosome number), and ecologically. In culture they behave similarly, being inclined to bienniality; but they are good species, in spite of the fact that they can cross and that the hybrid is fertile. In the F_2 there is segregation from almost typical *linearis* to almost typical *alpina*; the latter is rarer, suggesting that many *alpina* genes are recessive.

Braya linearis is a soil specialist and tolerates extremely basic and saline soils, e.g. shores of salt lakes in central W. Greenland. It probably had a wider range during late-glacial times, when leaching of the soils had not then taken place. In any case, its present range is striking; if it survived glaciation outside Greenland and Norway, it is now extinct in its refugia. *Braya alpina* is considered, e.g. by Gams, to be a Tertiary relic with south Siberian origin. To the same group also belongs the Asiatic *Braya rosea* (Turcz.) Bunge. This species can also hybridize with *B. linearis*, but in this case the hybrid is sterile. A number of the present day *Braya* species are probably remains of a species complex with 42 chromosomes, a complex which owing to isolation in various refugia evolved into several species. *B. linearis* and *B. alpina* are still able to exchange genes when brought together. They constitute a species pair which probably descends from a common arctic-alpine ancestor. Both species are old; the differentiation into two is not a recent post-glacial event, but may go back to Tertiary or early Quaternary time.

Nowadays when species of *Braya* are sympatric, as for example in N.E. Greenland, they may occasionally hybridize. Th. Sørensen has found and described a decaploid species, *Braya intermedia* Th. Sør., which is endemic to N.E. Greenland. It combines characters from *Braya humilis* (C. A. Mey.) Robins., *B. linearis* and/or *B. purpurascens* (R. Br.) Bge and is perhaps a triple-allodecaploid. Both *B. linearis* and East Greenland *B. humilis* are hexaploids, and an easy way of obtaining a decaploid would be a fertilization between an unreduced gamete of a hexaploid and a normal reduced gamete of an octoploid. The only octoploid *Braya* in N.E. Greenland is in fact *B. purpurascens*. The alloploid evolution in the case of *B. intermedia* is supported by the fact that Jørgensen *et al.* (1968) found exclusively bivalents at meiosis and no or very few multivalents.

N.E. Greenland harbours another very interesting alloploid species, viz. *Saxifraga nathorstii* (Dusén) Hayek with $2n = 52$, or twice as many chromosomes as its parent species, the yellow flowering *Saxifraga aizoides* L. and the purplish flowering *Saxifraga oppositifolia* L., which both have $2n = 26$. While *S. oppositifolia* is circumpolar arctic-alpine, *S. aizoides* is probably of European origin, but it has reached far into the Canadian arctic areas. There are vast areas where the two species are sympatric, but the primary hybrid has so far never been found; yet the alloploid stabilized hybrid is known from a fair number of stations in N.E. Greenland, where it is regarded as a good species. While fertilization of egg cells by male cells from the other species may happen occasionally, it may be impossible for the primary hybrid seeds to germinate; on the other hand, under arctic conditions, we might assume fusion between two unreduced sexual cells, resulting in the fertile hybrid. Another possibility would be hybridization between tetraploid strains of the two species; so far tetraploids have not been found in *S. aizoides*, but a tetraploid has been recorded in *S. oppositifolia*.

Before further discussion of the chromosome evolution in arctic species we may for a while return to the east arctic element. In the very part of N.E. Greenland where the two alloploid endemics have been located, one finds the sole Greenland stations of two Asiatic species, viz. *Potentilla stipularis* L. and *Draba sibirica* (Pall.) Thell. These species are absent from Europe except for a few stations of *Draba sibirica* in Eastern Europe. It is not very likely that these species have recently arrived in East Greenland from Asia, not only because the distance is so great, but also because the East Greenland populations of the two species seem to be distinct from the Asiatic. The *Potentilla stipularis* is described as a separate variety; and *Draba sibirica* (from Jameson Land in Scoresbysund) is very different in culture from plants originating in Eastern Europe or Asia. The *Draba* is diploid in both areas; but the Greenland plants grow vigorously and flower abundantly in the Arctic Greenhouse, which is not the case with Eurasian strains. On the other hand, the Eurasian plants are almost like weeds when grown in gardens or in pots in the experimental field, whereas the Greenland plants look unhealthy. We have no knowledge about the *Draba* populations growing in the Taymir peninsula; they may resemble the East Greenland plants, being low-arctic races of a species which otherwise is mainly sub-arctic-montane. Nor is *Potentilla stipularis* a true arctic species but rather low-arctic—montane; it may have segregated a special arctic Greenlandic race which is able to grow in arctic areas, viz. *P. stipúlaris* var. *groenlaudica* Th. Sør.

We may now introduce a problem concerning the dwarf birch, *Betula nana* L., in Greenland. This species occurs in bogs in northern Europe under temperate climatic conditions, while in Greenland it has a southern limit at about 63° N on the west coast and 65° on the east coast. Here, however, it is bicentric, having one large area in N.E. Greenland and a smaller one at Angmagssalik. It is missing in South Greenland, where its possible habitats are largely occupied by the closely related west arctic species *Betula glandulosa* Michx. The true *Betula nana* is east arctic; it is European radiating into eastern Canada, where it is remarkably rare. In West Greenland there may be a narrow zone where hybrids occur, but the two species mainly keep clear of one another. The southern limit of true *Betula nana* in Greenland is of particular interest. The pair of species *Betula glandulosa–B. nana* behave as two climatic races, which do not overlap because they are adapted to different climatic conditions. *Betula glandulosa* grows in the very maritime climate of South Greenland, and here even in mountain fell-fields, while *Betula nana* is particularly abundant inland and northwards, thus behaving like an arctic-continental species. In Norway *Betula nana* grows as far out as Bergen, in an extreme atlantic climate. This puzzling distributional pattern would be more understandable if we assumed that *Betula nana*, having reached Greenland on its way towards North America, had lost those biotypes which were adapted to maritime and sub-arctic–temperate climates. But such a depletion would hardly happen during a recent migration across the North Atlantic; on the other hand, a considerable gene-pool reduction would very probably take place during a glaciation. Of course these assumptions are based on the distributional pattern and local habitat preferences in Greenland and ought to be substantiated by experimental cultivation in growth chambers; unfortunately *Betula nana* is not well suited to culture experiments.

APOMIXIS AND SPECIES FORMATION

Such reductions in biotype content and alterations of the autecology of a species may be of great importance for a better understanding of its present range, but from an evolutionary point of view they will rarely involve more than a development of ecological races. Species formation is largely due to chromosomal changes and to mutations securing apomictic seed development. This is not the place to make many comments on apomixis, but the importance of this type of propagation is perhaps greater in the arctic regions than elsewhere. Not only do several grasses, *Ranunculus*, *Alchemilla*, *Potentilla*, *Taraxacum*, *Hieracium*, and *Antennaria* contain apomictic taxa, but other genera too, such as *Erigeron*, *Arnica*, and *Arabis*. A great many of the endemic taxa in Greenland are apomictic micro-species. Phytogeographically many of them are of great interest. Among the east arctic taxa we may call to mind *Ranunculus auricomus* L. var. *glabratus* Lynge, a diploid and probably old entity, having very few stations in Scandinavia and East Greenland, where it is found in non-shady habitats. If we compare the behaviour of the east arctic *Hieracium* and *Taraxacum* species a striking difference becomes apparent. While there are swarms of separate endemic apomictic microspecies of *Hieracium* which reach the North Atlantic islands, the Faeroes, Iceland and Greenland, this seems not to be the case in *Taraxacum*. When presenting some collections to a *Taraxacum* specialist for determination, one may be surprised by a statement that the material, collected perhaps in a West Greenland fjord, belongs to a taxon occurring in Norway. Of course long-distance seed dispersal is a possibility, but no records of the taxon between Norway and West Greenland are known, and one may therefore doubt this explanation. Instead we might assume that European *Taraxacum* crossed the North Atlantic with hardly any change in its gene-pool; in Greenland, mutations produced plants which morphologically are almost indistinguishable from certain European Taraxaca and are therefore given the same name.

Among the west arctic taxa, Greenland has received a great number of *Antennaria* microspecies, some of which have very striking distributions. Here again we are faced with the problem of mutations perhaps creating the same morphological type in several places. *Antennaria glabrata* (J. Vahl) Greene is a glabrous plant and a close relative of *Antennaria angustata* Greene. They are also related ecologically, being often found together; but it is striking that they often occur together on geographically far remote and isolated stations. This could be the result of a mechanism by which glabrous plants, such as *A. glabrata*, arose with a definite frequency.

Another west arctic apomict is *Arabis holboellii* Horn. This taxonomically intricate species contains sexual diploids outside Greenland, where it occurs from California and Nevada to Alaska and Canada. But in Greenland some plants are diploid, others triploid, and plants of both ploidy-levels propagate apomictically. In Western North America polyploids occur, which propagate apomictically; some of these are hybrids involving other species of *Arabis*. The *Arabis holboellii* complex has a striking tolerance. It grows up to 800–900 m above sea level in Scoresbysund in East Greenland; it is very abundant almost to the point of being weedy along the highway in dry parts of British Columbia; and it grows in dry pine woods and prairies in Little Valley in Nevada. When cultivated in growth chambers, the southern and arctic strains show different ecological demands; on the other hand, they are all very tolerant with regard to temperature in the growth period. *Arabis holboellii* belongs to a southern life form, being annual or biennial, and is a typical long-shoot herb. In spite of this it is able to tolerate arctic conditions, especially low temperatures, but it droops under moist air-conditions. Thus it is mainly adapted to dry and continental climates; it is not arctic, neither is it subarctic, nor temperate. In Greenland its favoured occurrences are in interior steppe-like habitats.

POLYPLOIDS IN THE ARCTIC FLORA

Since the discovery by Hagerup of the tetraploid *Empetrum hermaphroditum* in the Arctic, discussions have taken place about the importance of polyploids in the arctic flora. It is now assumed that polyploids, because of increased genetic variability, have been capable of occupying newly exposed glaciated surfaces in soils which are more or less wet and unstable due to solifluction. The older diploids seem to prefer stable, more or less dry sites on rocks and sunny slopes. While this may be an attempt to formulate a rule, it becomes clear that there are so many exceptions to it, that one has to be very cautious.

First we can point out cases where the polyploids occupy low- or subarctic areas, while diploids of the same complex occupy a wider range and radiate far north into the arctic regions. This is the case in the *Campanula rotundifolia* L.–complex. Another interesting group is the *Trisetum spicatum* (L.) Richt.–complex, where the widespread arctic species *T. spicatum* is tetraploid, whereas the sub–low-arctic *T. triflorum* (Bigel.) Löve & Löve is hexaploid. The latter is furthermore a clear west arctic entity, reaching Greenland and Iceland but no other points in Europe. The total area of this complex shows a great range extension in

North America towards the south; considering the strictly alpine occurrences in Europe, this seems inexplicable until one realizes that the complex in North America, Greenland and Iceland includes taxa with a higher chromosome number, which ecologically deviate markedly from the widespread arctic tetraploids. While the latter prefer snow-beds and mesic soils, the southern hexaploids can grow in dry habitats, such as rocks and sandy places. The southern *Trisetum* which is hexaploid is probably not the result of a simple chromosomal doubling. How it evolved is unknown, but some kind of alloploid evolution involving the arctic *Trisetum spicatum* and another taxon is probable.

In *Draba* too, the probability of alloploid evolution is great. The *Draba cinerea* Adams group may be taken as a good example. The typical *Draba cinerea* is a sub–low-arctic continental species and has $2n = 48$. It is a hexaploid with regular meiosis without multivalents. In many characters it resembles the continental *Draba lanceolata* Royle, which is tetraploid with $2n = 32$. In other characters it resembles the arctic *D. arctica* J. Vahl subsp. *groenlandica* (Ekm.) Böch., which is octoploid with $2n = 64$. These two taxa may be the parents of *Draba cinerea*. *Draba arctica* subsp. *arctica* is a high-arctic taxon and is decaploid with $2n = 80$; it has long been confused with the hexaploid *D. cinerea* with which it is closely related, but morphologically it is closer to the octoploid *D. arctica* subsp. *groenlandica*. If an unreduced sexual cell in *D. cinerea* with 48 chromosomes was fertilized by a reduced one from *D. arctica* subsp. *groenlandica* with 32, the result would probably be a decaploid. In the *Draba cinerea* group hybridization between closely related species on different ploidy levels has probably taken place; but the sterility which usually follows such crossings is replaced by fertility due to chromosome doubling. Now, in this case the high ploidy levels, octoploids and decaploids, occur in the real high-arctic regions; but it is probably not the genetic diversity due to polyploidy which is advantageous, but rather the hybrid origin and the possibility of the production of several parallel alloploids from different crosses between closely related strains.

CONCLUSION

To sum up: evolution in the arctic zone does not deviate much from that in other climatic regimes. Perhaps one trend is more pronounced in arctic areas, viz. irregularities during meiosis which on the female side lead to formation of unreduced egg cells; these may develop apomictically or, by fertilization with reduced or unreduced pollen grains, result in plants of a higher level of polyploidy, in many cases alloploids, which become stabilized new taxa.

Geographical isolation also seems to have been effective and is surely still effective in arctic areas, where land masses or smaller nunataks are so often isolated by icesheets. However, isolated populations also often arise in the ice-free areas. A species may happen to migrate to a suitable locality which represents the only possible habitat within a huge area. The chance of such a pocket receiving newcomers of the same species from outside is in many cases minimal; the population remains isolated and small, and genetic drift may well occur.

REFERENCES

BILLINGS, W. D., GODFREY, P. J., CHABOT, B. F. and BOURQUE, D. P. (1969). XI International Botanical Congress, Abstracts, p. 15. Seattle, Washington.

BÖCHER, T. W. (1941). *Meddr. Grønland* **131**, No. 2.

BÖCHER, T. W. (1951). *Biol. Skr.* **6**, No. 7.

BÖCHER, T. W. (1959). *Bot. Tidsskr.* **55**, 23–29.

BÖCHER, T. W. (1963). *Biol. Skr.* **11**, No. 6.

BÖCHER, T. W. (1966). *Biol. Skr.* **14**, No. 7.

HEDBERG, O. (1962). *Caryologia* **15**, 253–260.

HULTÉN, E. (1958). *Kungl. svenska Vetensk-Akad. Handl.* 4 ser. **7**, No. 1.

JØRGENSEN, C. A., SØRENSEN, TH. and WESTERGAARD, M. (1958). The flowering plants of Greenland. *Biol. Skr.* **9**, No. 4.

NORDHAGEN, R. (1965). *Blyttia* **23**, 145–162.

RYVARDEN, L. (1966). *Blyttia* **24**, 323–330.

8 | Connections Between Cool Temperate Floras, with Particular Reference to Southern South America

D. M. MOORE

Department of Botany, University of Reading, Reading, England

INTRODUCTION

1. Origins and Early History of Temperate Floras

The general latitudinal differentiation of the world's climates, and consequently of the major vegetation zones which they determine, appears to be long-established. The separation in the Northern and Southern Hemispheres of regions with a temperate climate, therefore, has a long history and is one of the factors in the perennial interest in the development and distribution of plants from such regions. The temperate climate, of course, comprises a mosaic of warm and cool, oceanic and continental conditions which, together with local edaphic and climatic factors, determines the physiognomy of the vegetation and the distribution of temperate taxa, but a more or less prolonged cold season is characteristic. The relationship between altitude and temperature, which can result in quasi-temperate conditions on mountain massifs at lower latitudes, is of course important when considering possible connections between the temperate zones of the two hemispheres and there is still a great need for data on such features in relation to the early history of the world's floras.

The extratropical floras of the Northern and Southern Hemispheres probably had a common origin, as evidenced by the many affinities at family level, whether by transtropical migration via upland regions of pre-Cretaceous fold-belts (e.g. Du Rietz, 1940) or by migration to temperate lowlands of both hemispheres from their centres of origin in Triassic to middle Jurassic tropical uplands (e.g. Axelrod, 1970). Whatever the details of their earlier separation, however, the northern and southern temperate floras appear to have been differentiated by middle Cretaceous times (Berry, 1937; Axelrod, 1952; Oliver, 1955; Takhtajan, 1969).

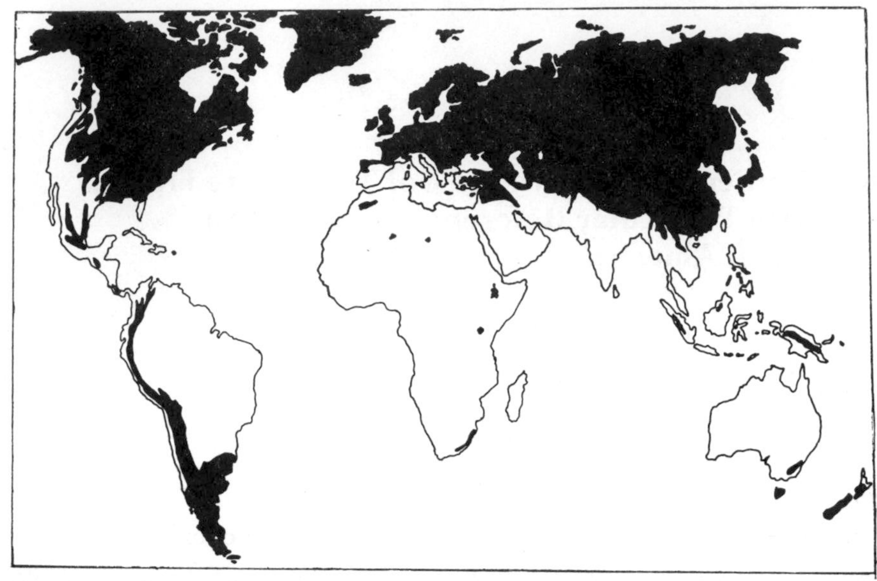

FIG. 1. Distribution of temperate (including high montane) regions of the world. Areas with a "Mediterranean" climate are not shown.

Within each hemisphere the differentiation of regional floras was strongly influenced by the rifting of the modern continents from the Paleozoic land masses of Laurasia and Gondwanaland. In the late Cretaceous the boreal flora, much more homogeneous than it is now, showed the development of regional differences in partial response to the opening of the Atlantic Ocean and, later, by the cessation of major migration between E. Asia and Europe in apparent response to the spread of the arid Asiatic regions when the Tethys Sea disappeared. This Arcto-Tertiary temperate flora then gradually extended south to approach the present boundary of the tropical zone (Takhtajan, 1969). Whether or not Gondwanaland played a central role in the origin of the angiosperms (e.g. Hawkes and Smith, 1965) it is clear that a temperate Antarcto–Cretaceous flora can be discerned (Cranwell, 1964) in Antarctica, which appears to have remained in polar latitudes from Permian times to the present (Tarling, 1971), and in New Zealand. Regional differentiation undoubtedly accompanied the Cretaceous–early Tertiary fragmentation of Gondwanaland and there has been a northward spread of this flora which has drawn attention to the central role occupied by Antarctica (e.g. Cranwell, 1964). In both hemispheres the structure of the temperate floras was considerably modified by the Pleistocene glaciations.

2. Floristic Connections between Temperate Regions

It has long been known that connections exist at all taxonomic levels between the temperate floras of the world. Because they span the tropics or the oceans these connections have excited considerable interest and they have been variously explained by parallel evolution, long-distance dispersal, or shorter dispersal over land connections of one sort or another.

Despite numerous intergradations, warm and cool temperate regions have modally distinct floras and their environmental and biological histories are probably sufficiently distinct to justify considering them separately. Warm temperate floras with a "Mediterranean" climate have been dealt with by Raven (1963, 1971), while Axelrod (1970) reviewed desert and related warm temperate floras, a topic considered by Solbrig (1972).

This paper deals solely with cool temperate floras. It will not consider such connections in the Northern Hemisphere, where they are particularly strong, because the basic geographical situation is simpler. Other contributions (e.g. Hara, 1972; Kornaś, 1972) have already dealt with the region, and the major problem posed by the Atlantic Ocean was the subject of a recent review (Löve and Löve, 1963), which provides references to the abundant earlier work. The South Temperate and North–South connections are in many ways parallel, both in the types of distribution and in focusing on the problems of major disjunctions over inhospitable terrain, and they are both directly relevant to current interests in the flora of southern South America.

NORTH–SOUTH COOL TEMPERATE CONNECTIONS

These connections, first noted early in the nineteenth century by Willdenow, Humboldt and Schouw, have been extensively documented by such workers as Hooker (1853), von Hofsten (1916), Steffen (1939), Du Rietz (1940), Good (1964) and Raven (1963). As might be expected, the connections are greatest at the family level, with *ca* 80–85% of families in New Zealand and in South America south of *ca* 40° S occurring in the Northern Hemisphere, as do *ca* 75% of those in Tasmania. Connections are less at the generic level and there is a difference between the Old and New Worlds. Thus, whilst New Zealand shares *ca* 10% of its genera with the Northern Hemisphere alone, Tierra del Fuego shares 35%; in each case a further 20% and 27% respectively occur in New

Zealand, South America and the Northern Hemisphere. The dissimilarity between the Old and New Worlds becomes even more striking at the species level.

1. Amphitropical Cool Temperate Genera

Such genera fall into a series of groupings between which there are many intermediates but which represent generally different patterns of variation related to differing evolutionary, and probably migrational, histories. In genera such as *Caltha* and *Coriaria* the Northern and Southern Hemisphere representatives belong to quite different sections or subgenera, while the strongly derived, very high polyploid Australasian species of *Asperula* are quite different from the Eurasiatic species (Ehrendorfer, 1971), and they may be considered to approach the situation where related genera are confined to different hemispheres.

In *Geum* and *Plantago*, for example, a further pattern is superimposed upon this basic situation. This can be illustrated by *Geum* (Gajewski, 1957), which also exemplifies the frequently noted generalization that most amphitropical genera appear to have migrated from the Northern Hemisphere. All but three of its subgenera are in the Northern Hemisphere, together with other genera of the tribe. The subgenus *Oncostylus*, present in both South America and Australasia, is very distinct and may be of generic status. It shows the full evolutionary sequence, from species with straight styles to those with conspicuously hooked stigmas adapted to epizoochory, as is shown in the Northern Hemisphere subgenera. This pattern, comparable to those mentioned above, certainly results from a more ancient disjunction than that represented by the five South American species of subg. *Geum*, which is richly developed in North Temperate regions. *Viola*, well-developed in the temperate regions of both hemispheres, provides a further series of overlapping patterns. Thus, in South America the very distinct sections *Andinium*, *Rubellium* and *Tridens* are present, and in Australasia the rather distinctive section *Erpetion*. These presumably represent older disjunctions than do the yellow-flowered species of the South American sect. *Chilenium* which appears close to sect. *Chamaemelanium* in North America (Clausen, 1929), and the possibly still more recent sect. *Viola* (*Nomimium* Ging.), subsect. *Adnatae*, which ranges from Australasia through Malaysia to northern Eurasia and includes the widespread forms of *Viola betonicifolia* Sm.

These older disjunctions may also be exemplified by *Hypochoeris*, in which Stebbins (1971) has shown the South American species to be chromosomally derived from the less numerous species in Eurasia, where generic relatives occur. A similar distribution is shown, for example, by *Briza* in which there is indication of striking phytochemical differences between European and South American species (Williams and Murray, 1972), and the absence from North America presumably results from extinction there after migration to the Southern Hemisphere. The aquatic *Littorella*, absent from North America, is effectively an amphitropical vicariad and certainly results from a much more recent migration.

The younger distributions shown by parts of *Geum* and *Viola*, for example, intergrade via such genera as *Gentianella* and *Saxifraga*, which are best developed in the Northern Hemisphere but which have several austral species, to those like *Primula*, *Potentilla* and *Antennaria* in which the Southern Hemisphere representatives constitute a rather limited series of self-compatible, often autogamous biotypes obviously derived from the major northern centres. *Littorella*, *Adenocaulon* and *Empetrum*, which effectively constitute amphitropical vicariads, may at least in part be considered along with this last group, although they are a biologically heterogeneous group of open and woodland habitats containing both autogamous and strongly outcrossing representatives.

At first sight the occurrence in the Northern Hemisphere of such genera as *Drusa* and *Naufraga*, with their strong South American affinities, and of representatives of the typically austral perennial species of *Euphrasia* appears to indicate northward migration. Yeo (1968) has pointed out that the plants of other genera most closely resembling the prevailing type of flower in *Euphrasia* are Australasian, like some of the more primitive *Euphrasia* species (see also van Steenis, 1962). However, it is also likely that the perennial Euphrasias diverged in the north and south from common ancestors and that they were largely eliminated in North Temperate regions by the late Tertiary or Quaternary climatic changes which permitted the derived annual species to spread and hybridize there. This would accord with the possibility that the Southern Hemisphere can serve as a reservoir of old types as suggested by Ehrendorfer (1971) for *Hebe* and perhaps *Asperula*. Interestingly, the highest chromosome number ($n = 44$) in *Euphrasia* is for the only austral species known cytologically (Moore, 1967), a comparable situation to that noted by Ehrendorfer (1971) for *Asperula* and the *Veronica chamaedrys* group. That migration has taken place from the south is shown by *Acaena*, largely developed in South Temperate regions, which has obviously crossed the tropics to reach Central America and California with three species (Fig. 2).

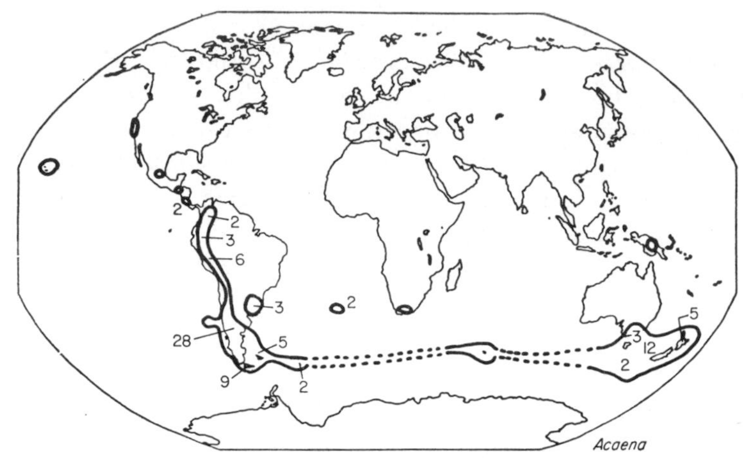

FIG. 2. Distribution of *Acaena* showing the approximate number of species in different parts of its range. Regions without a figure have one species.

2. *Amphitropical Cool Temperate Species*

Like the genera, these species can occur in the Northern Hemisphere and in either or both the South American and Australasian regions. There is a further similarity in the much closer affinities in the New World—neglecting species present in all three areas, southern South America shares more than 5 times as many species with the Northern Hemisphere than does New Zealand. Southern Africa is not involved in such distributions, although a number of North Temperate species penetrate to or just across the equator on the East African mountains (Hedberg, 1961).

In his excellent review of these species in the Americas Raven (1963) summarized much of the available information, which I do not intend to duplicate here. Subsequent studies have tended to confirm his general conclusions that in most instances migration was from north to south [*Colobanthus quitensis* (Kunth) Bartl. being a clear exception (Moore, 1970)], that the southern plants are usually self-compatible and often at least facultatively autogamous, that they comprise a disharmonious selection of the North Temperate flora and that most occur in relatively open habitats. There is, however, a clear need to know more about these species. Not only will further exploration clarify the *facts* of distribution—the very wide disjunction of, for example, *Koenigia islandica* L. compared to *Carex macloviana* D'Urv. which penetrates along the cordilleras to much lower latitudes,—but the various tools of modern taxonomy must be further employed to examine the structure, relationships and capabilities of the populations involved.

Like the genera, amphitropical species can show every degree of relationship between the Northern and Southern Hemisphere populations, even intergrading with the generic patterns. Thus, with increasing study, many of the cool-temperate amphitropical taxa mentioned by Hooker (1853) have been shown to be specifically distinct in the two hemispheres. Lovis (1959), for example, mentions several cases amongst the ferns in which New Zealand populations formerly attributed to European species such as *Hymenophyllum unilaterale* Bory and *H. tunbridgense* (L.) Sm. have been shown to belong to chromosomally distinct species. Chromosomal divergence between Northern and Southern Hemisphere populations currently assigned to one species appears to be infrequent. A rare example is *Asplenium trichomanes* L., which is hexaploid ($n = 108$) or sometimes tetraploid ($n = 72$) in New Zealand and diploid or tetraploid in Britain (Lovis, 1959). All the other species for which there is information have the same chromosome number in both Northern and Southern Hemispheres. Thus, New Zealand populations of *Cystopteris fragilis* (L.) Bernh. (Lovis, 1959), *Deschampsia caespitosa* (L.) P. Beauv. (Hair, 1966; D. M. Moore, unpublished) and *Trisetum spicatum* (L.) Richt. (D. M. Moore, unpublished) have the same chromosome number as in North Temperate regions, while the same is true for South American representatives of *Anemone multifida* Poir., *Festuca megalura* Nutt. (Raven, 1963), *Fragaria chiloensis* (L.) Duchn. (Staudt, 1962), *Microsteris gracilis* (Hook.) Greene, *Polemonium micranthum* Benth. (Grant, 1959), *Koenigia islandica* L., *Chenopodium macrospermum* Hook.f., *Spergularia marina* (L.) Griseb., *Hippuris vulgaris* L., *Armeria maritima* L., *Plantago maritima* L., *Triglochin concinna* Davy, *T. palustre* L., *Montia fontana* L., *Phleum commutatum* Gaud. (Moore, 1967, unpub.), *Carex curta* Gooden, *C. magellanica* Lam. and *C. microglochin* Wahlenb. (Moore, 1967; Moore and Chater, 1971).

Other characters similarly differ in the extent of differentiation. Thus, Northern and Southern Hemisphere populations may have virtually identical morphological ranges, as in *Koenigia islandica* and *Hippuris vulgaris* (D. M. Moore, unpublished), or be clearly distinguishable and even subspecifically distinct as in *Armeria maritima*. This can even occur within one genus and, for example, Moore and Chater (1971) showed that while Northern and Southern Hemisphere populations of *Carex capitata* L. and *C. maritima* Gunn. could not be distinguished morphologically, in *C. microglochin* and *C. magellanica* they were clearly different; *C. macloviana*

D'Urv. and *C. curta* showed intermediate degrees of differentiation. This variation could result from different times of migration or different evolutionary rates but the need for caution in interpreting such data is emphasized when other results are included. Thus, preliminary studies (T. T. Aye, unpublished) show that Northern and Southern Hemisphere *C. microglochin* and *C. macloviana* have clear differences in the occurrence of several flavonoids, including luteolin 7-glucoside and tricin 5-glucoside, while *C. magellanica* and *C. curta* do not differ chemically. On the other hand in *Empetrum* the chemical data support the modal morphological differentiation of the Northern and Southern Hemisphere diploids, suggesting that they are subspecifically distinct (Moore *et al.*, 1970), while in *Phleum commutatum* there is much less chemical and morphological amphitropical differentiation (D. M. Moore and J. B. Harborne, unpublished).

Detailed studies of such species as *Deschampsia caespitosa*, *Trisetum spicatum* and *Carex curta* are particularly interesting here since they show both the cool temperate disjunctions dealt with. In *T. spicatum* Hultén (1959) has shown that the circumboreal group of subspecies having short, wide glumes have crossed the tropics to reach Bolivia, while those with longer, narrower glumes have reached both southern South America and Australasia. This might suggest two waves of transtropical migration within this species, the first in both New and Old Worlds, the second only in the New World. Indeed, the very close relationship noted between Alaskan and Patagonian plants could indicate a third, more recent migration or result from parallel selection. Southern Hemisphere plants of *Carex curta*, from both Old and New Worlds, modally differ from those in the Northern Hemisphere (Moore and Chater, 1971). This would suggest transtropical migration, presumably in the Americas, followed by circum-Antarctic dispersal, except that Malaysian plants are intermediate between Australian and European forms (Nelmes, 1951), which might suggest southward migration in Old and New Worlds with subsequent selection of the Southern Hemisphere populations. This draws attention to the possibility of striking differences between the northern and southern Cool Temperate environments about which we as yet know little and which are perhaps underemphasized in such discussions.

As noted, for example, by Raven (1963), maritime species such as *Plantago maritima*, *Honckenya peploides* (L.) Ehrh. and *Lathyrus japonicus* Willd. form a rather special group in that their amphitropical distributions cannot be related to migration via low-latitude cordilleras. Detailed studies of South American *Plantago maritima* (D. M. Moore *et al.*, unpublished) show that it readily gives fertile hybrids, having normal chromosome pairing, in crosses with European plants, thus according with Gregor's (1939) data for hybrids between North American and European material. It can be distinguished from the more variable Northern Hemisphere material on several morphological and some chemical characters, showing closest affinity with North American populations, which are also self-compatible. Furthermore, they are more similar to eastern than western North American material although in a few characters, such as the constant absence of annuli on the pollen-pores, they are like European plants. Clearly, migration took place along the Atlantic coast, in contrast to the predominantly western routes of most amphitropical species (Raven, 1963). Since the Atlantic currents do not appear favourable for such transport, and the seeds are scarcely suited for wind dispersal, the distribution must result from birds migrating between the North and South Atlantic Oceans to which the seeds, glutinous when moist, could be attached.

3. *Age and Mode of Migrations*

The distinctive austral groups of amphitropical genera such as *Geum*, *Plantago* and *Caltha* seem likely to reflect the middle Cretaceous differentiation of the northern and southern temperate floras, of which the typically austral genera like *Nothofagus*, *Acaena*, *Gunnera* and *Astelia* are an older indication. Since the closest relatives of, for example, *Acaena*, *Colobanthus* and *Nothofagus* are Eurasiatic, this points to the importance of the Old World tropics for their common origin and it is doubtless significant that Takhtajan (1969) and others have pointed to S.E. Asia in considering early angiosperm diversification.

The migrations of genera well-developed in the North Temperate zone and with comparable or relatively restricted development in the south must have occurred at a later stage. The temperate genera which occur fairly continuously through the montane neo- and palaeotropics, such as *Anemone*, *Bromus*, *Clematis*, *Deschampsia*, *Eleocharis*, *Juncus*, *Luzula*, *Rubus* and *Ranunculus*, or those which only do so in the Americas, like *Alnus*, *Draba*, *Vicia* and *Ribes*, suggest the routes by which these migrations could have taken place for taxa with wide photoperiodic capacities. Their absence from lower latitude mountains, as in *Saxifraga* and *Agoseris*, is perhaps due to subsequent extinction there. These routes are also clearly indicated for other groups of plants, such as the Jungermanniae (Schuster, 1969). The great importance of the New World in transtropical generic migrations, noted earlier, must be related to the links between North and South America, which pulled apart in the late Triassic and appear only to have rejoined by the middle Eocene (Freeland and Dietz, 1971). Unless, therefore, one concedes an early Mesozoic age to the distributions of such genera they must have migrated via transtropical Tertiary uplands, for which there is little evidence, or by dispersal across tropical lowlands much wider than they are now. The route afforded by the American Cordilleras, which attained their maximum development during the Pliocene, would become available during the Miocene, although the Central American gaps would be present and at least as great as they are now. One can, of course, take the view that all such genera must have crossed such gaps by distance dispersal, but the dioecious *Empetrum*, the woodland *Adenocaulon*, *Ribes* and *Alnus*, for example, would seem unsuitable candidates for this. Early to Middle Tertiary migration would also seem suitable for such genera as *Saxifraga*, *Vicia* and the yellow-flowered *Viola* species which are well-differentiated in South America and which neither seem particularly suited to long distance dispersal nor do their distributions elsewhere in the world suggest it has been employed. More information is needed on such genera as *Gentianella*, *Gentiana* and *Littorella* (present in Europe and South America) but they may be later Tertiary migrants, as seems to be probable for genera like *Armeria*, *Primula*, *Cardamine* and *Antennaria*, which are self-compatible, often autogamous, of open habitats and well capable of long distance dispersal. The northward migration of *Acaena* also clearly falls into this category.

These genera probably grade into the amphitropical, cool temperate species whose most likely time of migration was the Pliocene–Pleistocene. Whether the differing degrees of morphological and chemical differentiation between Northern and Southern Hemisphere populations point to dissimilar evolutionary rates or different times of migration, perhaps including the post-Pleistocene, still needs much more investigation. In all events there seems no biological reason why such species could not establish populations in open habitats following chance long distance dispersal of isolated propagules. *Colobanthus quitensis* is a comparable example of a "natural weed" which has migrated from south to north. In view of the major influx of Northern Hemisphere species into the Southern Cool Temperate zone following European colonization (e.g. Holdgate and Wace, 1961), man may be a post-Pleistocene agent responsible for some, but certainly not many, of the apparently natural disjunctions under discussion.

SOUTHERN COOL TEMPERATE CONNECTIONS

Lacking the extensive continental interiors of the Northern Hemisphere, the cool temperate lands of the Southern Hemisphere mainly have oceanic climates, apart from the east Patagonian dry steppe in the rain shadow of the Andes. Connections between their floras have been documented by Hooker (1853), Skottsberg (1915, 1960), Good (1964), van Steenis (1962) and others, and because they span the broad oceans of the Southern Hemisphere such distributions, particularly of genera and species, are very conspicuous and of great interest. The largest cool temperate area in the Southern Hemisphere is in southern South America and much the closest floristic affinities at generic and specific levels are shown between it and New Zealand, while Tasmania and montane S.E. Australia are involved to lesser extents. The very small regions in the mountains of Southern Africa share no species with the southern Cool Temperate Zone and are linked only by *Acaena*, *Gunnera* and *Tetragonia*, as well as the more widespread temperate genera like *Viola*, *Geum*, *Triglochin* and *Myosotis*. Since the Pliocene Antarctica has been almost completely removed from consideration, but the surrounding islands have taken part in the present southern distribution patterns since at least the Pleistocene.

As in the amphitropical distributions, connections between eastern and western parts of the Southern Hemisphere occur at all taxonomic levels; and it is likewise convenient to consider generic and specific affinities separately. Information from a variety of sources is available on the degree of differentiation, and therefore their evolutionary age and perhaps also the time of disjunction, and on the direction of migration.

1. *Amphi-Antarctic Genera*

Between 50 and 60 genera or distinctive parts of more widespread genera occur principally in southern South America and Australasia, about half of them in New Zealand alone and, as Good (1964) has pointed out, most of them are rather small and equally diverse in the two regions. Of the genera which are well-differentiated in the Southern Cool Temperate regions some, such as *Nothofagus*, *Acaena* and *Uncinia* Pers., have their major taxonomic groupings in both the Old and New Worlds, as is the case, on chromosomal evidence, for the Loranthaceae (Barlow and Wiens,

1971). These distributions have been important in emphasizing the importance of Antarctica in the early history of austral distributions (e.g. Cranwell, 1964), particularly since it houses Cretaceous fossils of such genera (Couper, 1960; Cranwell, 1964).

Most of the other amphi-Antarctic genera appear to have migrated from east to west. Thus, South American representatives of *Lebetanthus*, *Astelia*, *Donatia*, *Phyllachne*, *Drimvs* and *Coprosma* seem to be derived from the Old World, where related genera occur. *Lagenophora* also probably belongs here, but affinities with the C. American *Laestidia*, a partly shrubby relative, are not clear (Cabrera, 1966). *Jovellana*, closely related to *Calceolaria*, may be a rare case of New World origins on these criteria.

Whilst remembering that secondary centres of speciation can obscure migrational trends it can be noted that again, there seems to be a preponderance of cases in which the major diversity is found in the Old World—e.g. *Pseudopanax*, *Ourisia*, *Drapetes*, *Pratia*, *Oreobolus*, *Drosera*, *Geum* subg. *Oncostylus* (only sect. *Neo-Oncostylus* reaches South America) and *Abrotanella*. On these criteria New World centres may be postulated for a few genera such as *Marsippospermum* and *Discaria*. Although *Colobanthus* Bartl. has about 4 times as many species in Australasia as in South America (Fig. 3), the endemic *C. kerguelensis* Hook.f., a relict of the Antarcto–Tertiary flora, is a reminder of the distorted view we have of austral floras since the Pleistocene glaciations.

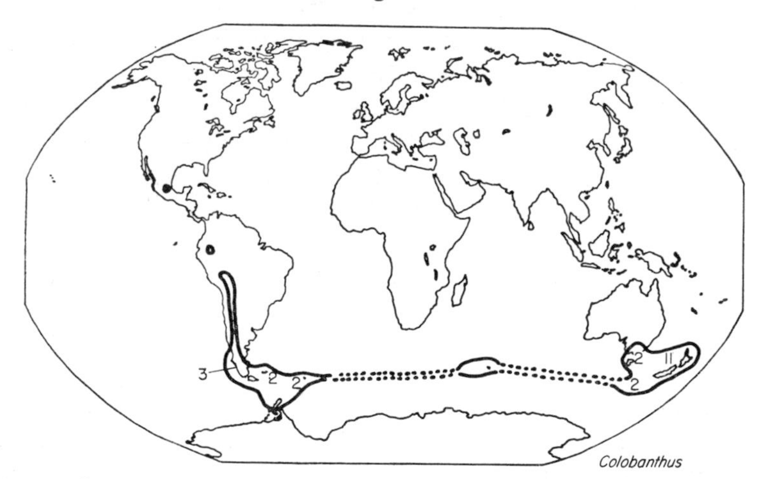

FIG. 3. Distribution of *Colobanthus* showing the number of species in different parts of its range. Regions without a figure have one species. The area delimited along the central and north Andes, and in Mexico, is occupied by one species, *C. quitensis* (Kunth) Bartl., which continues south to the Antarctic Peninsula.

The apparent ability of numerical and structural chromosome changes to increase the capacity of plants to expand into new areas (e.g. Darlington, 1963; Stebbins, 1971) is also of value in such cases, supporting the distributional data in, for example, *Pratia*, in which the New Zealand species have $n = 7$, 21 or 35 (Beuzenberg and Hair, 1959) but the highest number ($n = 42$) is known from South America (Moore, 1967). *Fuchsia* and *Gaultheria-Pernettya* have their centres of diversity in the New World, where polyploids (and diploids in the former) occur, but diploids alone are found in Australasia. In contrast, *Schizeilema* has 10 species in New Zealand with $n = 16$ or higher (J. B. Hair, unpublished), 1 species in S.E. Australia and a single diploid ($n = 8$) in South America (Moore, 1967), to which the closely related *Azorella* is virtually restricted.

Morpho-geographical data do not indicate the direction of migration in such genera as *Carpha*, *Tetrachondra* and *Gaimardia*, effectively constituting amphi-Antarctic vicariads, as well as, for example, *Caltha*, *Plantago* and *Eucryphia*. However, other data can help with the two latter. Thus, although South American and New Zealand species of the circum-Antarctic *Plantago* sect. *Oliganthos* have $n = 24$ (Moore, 1967; Rahn, 1957; Rattenbury, 1957), $n = 36$ has only been found in the South American *P. barbata* Forst. f. (Rahn, 1957) and *P. uniglumis* Wallr. & Walpers (D. M. Moore, unpublished), while $n = 6$ and $n = 18$ are known from *P. muelleri* Pilger of montane S.E. Australia (D. M. Moore, unpublished) and $n = 12$ from the New Zealand *P. triantha* Spreng. (Rattenbury, 1957). All of which accords with migration from East to West. In *Eucryphia*, Bate-Smith *et al.* (1967) showed that the Australian species lacked the 5-*O*-methylated and 3·5-di-*O*-methylated flavonols present in the South American species. Since the 3- and 3·5-*O*-methylation of flavonoids appear to be primitive angiosperm characters, this would suggest that the genus, like *Schizeilema* has evolved in an easterly direction from South America.

Of particular interest in demonstrating the potential value of chromosome studies in austral distributions is *Cotula*, whose major subgeneric groupings accord with the regional separations. The circum-Antarctic subg. *Leptinella* has a basic number of $x = 13$, with tetraploids and higher polyploids in Australasia (Hair, 1962) and southern South America (*C. scariosa* (Cass.) Franchet $2n = 262$; D. M. Moore, unpublished). It has been suggested by Hair (1962) that either *Leptinella* is an amphidiploid combination of the mainly Australian subg. *Strongylosperma* ($x = 9$) and the mainly South African subg. *Cotula* ($x = 5$), with subsequent reduction from $x = 14$ to $x = 13$, or that *Strongylosperma* ($x = 18$) is the amphidiploid. On either hypothesis he favours Australasia as the centre of origin.

2. *Amphi-Antarctic Species*

Some 32 species† occur in both the Australasian and South American cool temperate regions, excluding such cosmopolitan species as *Montia fontana*, *Cystopteris fragilis*, *Trisetum spicatum* and *Deschampsia caespitosa*. Eleven species do not occur in the intervening area, but the remainder are found on one or more of the sub-Antarctic islands and some, such as *Blechnum penna-marina*, *Callitriche antarctica* and *Ranunculus biternatus*, are present on most of them. A further 17 species penetrate to the sub-Antarctic islands from one or other of the continental areas (see Greene and Greene, 1963), two of them [*Deschampsia antarctica* Desv. and *Colobanthus quitensis* (Kunth) Bartl.] reaching the Antarctic peninsula.

Although there is still a great need for further information on the structure of the species' populations and their affinities, it is clear that there has been migration in both easterly and westerly directions. Thus, for example, *Hebe elliptica*, *H. salicifolia* and *Ranunculus acaulis* have obviously dispersed from New Zealand, where their closest relatives occur, while *Azorella selago* has migrated eastwards from the South American centre of the genus. Where cytological information is available it shows no amphi-Antarctic differences in chromosomes number (footnote), but *Selliera radicans* ($x = 8$), with diploids in Chile (Moore, 1963) and New Zealand (Hair and Beuzenberg, 1960), has diploids and hexaploids in Tasmania (Jackson, 1958). Since this species is the only South American representative of the Australasian Goodeniaceae, migration was obviously

† The following species occur in the Australasian and South American sectors of the Southern Cool Temperate Zone. Those marked by an asterisk have the same chromosome number in both areas. *Hymenophyllum ferrugineum* Colla, *Polystichum mohrioides* (Bory) Presl, *Blechnum penna-marina* (Poir.) Kuhn, *Schizaea fistulosa* Labill., *Grammitis armstrongii* Armstr., *Stellaria decipiens* Hook.f., *Ranunculus acaulis* Banks & Sol. ex DC.*, *R. biternatus* Sm.*, *Crassula moschata* Forst.f., *Tetragonia tetragonioides* (Pallas) Kuntze, *Acaena magellanica* (Lam.) Vahl*, *Geum parviflorum* Sm., *Sophora microphylla* Aiton*, *Oxalis magellanica* Forst. f.*, *Azorella selago* Hook.f., *Apium australe* Thouars*, *Nertera depressa* Banks & Sol. ex Gaertn., *Callitriche antarctica* Engelm. ex Hegelm.*, *Hebe elliptica* (Forst.f.) Pennell*, *H. rosmarinifolia* (Forst.f.) Pennell, *Selliera radicans* Cav.*, *Taraxacum gilliesii* Hook. & Arn.*, *Juncus planifolius* R.Br., *J. pusillus* Buchenau, *J. scheuzerioides* Gaudich.*, *Rostkovia magellanica* (Lam.) Hook.f., *Festuca contracta* T. Kirk, *Agrostis magellanica* Lam., *Hierochloë redolens* (Sol. ex Vahl) Roem. & Schult., *Scirpus inundatus* (R.Br.) Poir., *Uncinia macrolepis* Decne, *Carex trifida* Cav.

to the west either before polyploidization took place or, if later, only involving diploid biotypes.

The only other circum-Antarctic species showing chromosomal variation is *Acaena magellanica* (Lam.) Vahl (Moore and Walton, 1970), which occurs from South America to Macquarie Island (Fig. 4). Of the 20 or more species of *Acaena* for which there are chromosome counts all but three have $2n = 42$; one New Zealand and one Chilean species have $2n = 84$ and *A. californica* Bitter has $2n = ca\ 126$ (D. M. Moore, 1964, and unpublished). *A. magellanica* is "diploid" on the sub-Antarctic islands and in the climatically similar regions of easternmost Tierra del Fuego, while only "tetraploid" populations are found in the forested western parts of Fuegia, in South Chile and in the Falkland Islands, which are floristically part of the Andean summer forest zone (Skottsberg, 1913; Moore, 1968). This differs from the more usual situation (e.g. *Biscutella*

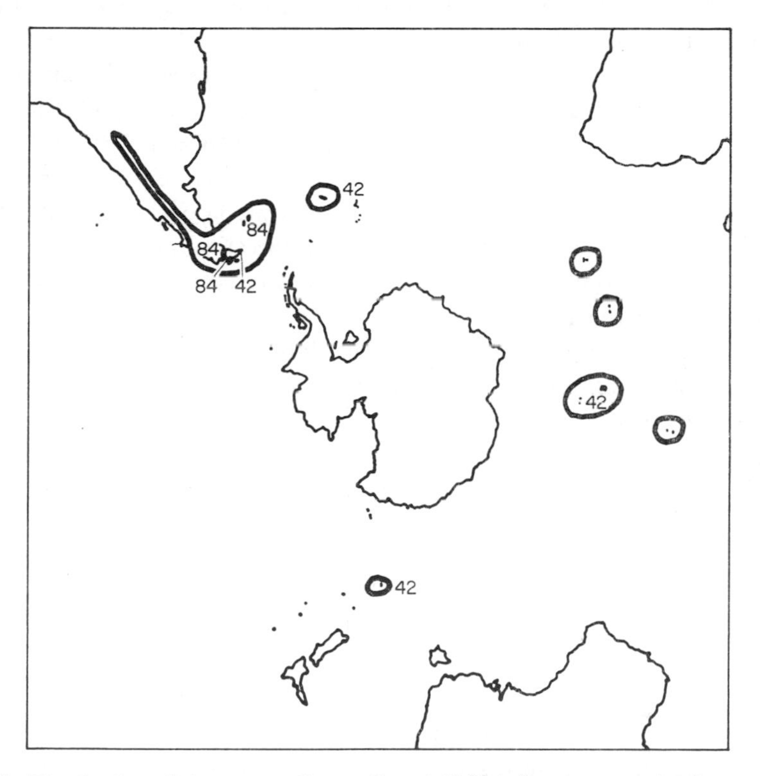

FIG. 4. Distribution of *Acaena magellanica* (Lam.) Vahl indicating somatic chromosome numbers ($2n$) known in different parts of its range (Data from Moore, 1963, 1964, 1967, unpublished; Moore and Walton, 1970).

laevigata, Manton, 1934) in which the polyploids have occupied the glaciated regions from the ice-free refugia of the diploids. The 42-chromosome race of *A. magellanica* either survived the Pleistocene glaciations beyond the icefront in southern South America or in at least partly unglaciated islands like Kerguelen and then spread elsewhere subsequent to the retreat of the ice, colonizing the southward-spreading forest zone as the 84-chromosome form. *Acaena* seems to be basically a 42-chromosome genus of open habitats, such as screes, grasslands and herbfields, in which forms with higher chromosome numbers and/or of forest habitats are derived.

3. *Age and Mode of Migration*

Central to any consideration of southern cool temperate migrations are the extent to which Antarctica was joined to the other components of Gondwanaland, the time when the links were sundered by continental drift and the extent to which subsequent connections were possible. Although there is still uncertainty about the precise details it seems clear that Africa split off first, probably in the early to middle Cretaceous (Freeland and Dietz, 1971), with New Zealand separating from Antarctica about the late Cretaceous (Heirtzler *et al.*, 1968) and Australia rather later, probably by the early Tertiary (Tarling, 1971). Subsequent connections between East Antarctica and Australasia are very difficult to support on the geological evidence (Adie, 1963), although major elevation of the intervening submarine ridges would provide some "stepping stones". With regard to the geologically younger West Antarctica, not only do present maps suggest a closer relationship between South America and the Antarctic Peninsula but current evidence on the evolution of the Scotia Arc suggests that they were connected until at least the early Tertiary (Dalziel and Elliott, 1971). Antarctica, unglaciated during the Cretaceous and much of the Tertiary, appears to have borne a cool temperate flora from at least the late Cretaceous (Couper, 1960; Cranwell, 1964) until its extinction by the Pliocene–Pleistocene glaciations. Traces of this flora remain in the relict endemics *Pringlea*, *Lyallia* and probably *Colobanthus kerguelensis*, which persisted in unglaciated parts of the Iles de Kerguelen.

As Fleming (1963) has pointed out, the interpretation of amphi-Antarctic connections requires the recognition, or attempted recognition, of a Neo-austral element dispersed under present geographical conditions and a Palaeo-austral element dispersed under earlier, probably more continuous, land conditions. Undoubted species probably belong to the former while well-established groups, diversified on both sides of Antarctica, presumably belong to the latter element.

(a) *Palaeo-austral Connections* It seems inescapable that at least some of the Palaeo-austral distributions date from the Cretaceous, particularly for groups such as *Nothofagus* and members of the Loranthaceae and Myrtaceae, which are known as Cretaceous fossils in Antarctica (Cranwell, 1964). The oldest fossils of *Nothofagus*, for example, occur in the Cretaceous of Antarctica and New Zealand; it appeared more recently in Australia, later still in South America and only reached New Guinea in the Pliocene (Cranwell, 1964). As Fleming (1963) notes, "New Zealand botanists and foresters who know living *Nothofagus* are almost unanimous in maintaining that it requires land for dispersal", and one is forced to the conclusion that *Nothofagus* achieved migration between New Zealand and Antarctica before their late Cretaceous separation, at a time when its major groups were already differentiated, crossing to Australia and South America while connections were still available during the late Cretaceous–early Tertiary and subsequently moving northwards in the east and west.* The chromosomal evidence (Hair, 1963) for *Podocarpus*, in which the most derived arrangements are near its northern limits in the S.E. Asia–W. Pacific area and in northern South America, accord with this picture. Similarly, the $x = 11$ and 12 forms of Loranthaceae (Barlow and Wiens, 1971) could have moved across temperate Gondwanaland connections, establishing $x = 9$ and $x = 8$ in the Old and New Worlds respectively following their separation and subsequently showing massive secondary radiation in the tropics. Direct evidence of these ancient austral migrations in other genera must await further fossil studies, particularly in Antarctica, but further information on the biological features of such genera is also clearly needed. It can be considered, for example, that dioecious members of *Gunnera* (which also occurs in the South African late Cretaceous; Cranwell, 1964) and *Coprosma* are less suited to long-distance dispersal than the self-compatible, self-pollinating members of *Plantago*, *Colobanthus* and *Oreomyrrhis*. Similarly more or less continuous land connections seem necessary for genera such as *Caltha*, *Discaria* and *Drimys*, for which the fruit biology gives little support for long-distance dispersal between New Zealand and Antarctica. *Acaena* has its major groups present in Australasia and America and, like *Nothofagus*, it might

*See also Van Steenis (1972).

have differentiated in the Cretaceous. However, its modern species, like those in *Colobanthus*, for example, are demonstrably capable of long-distance dispersal and the generic distributions may result from this process in Tertiary times. In this context it is interesting to note the preponderance of genera which seem to have migrated from Australasia to South America, in accordance with the prevailing westerly direction of the winds and ocean currents.

It seems clear that throughout the Tertiary period Antarctica, with cool temperate forests (Cranwell, 1959) in at least its oceanic parts, was a route by which distance dispersal from New Zealand and South America could take place. It seems likely that vicarious species-pairs such as *Drosera uniflora* Willd.,–*D. stenopetala* Hook.f., *Gaimardia australis* Gaudich.,–*G. setacea* Hook.f., *Daucus glochidiatus* (Labill.) Fisch., Meyer & Lallem.,–*D. australis* Poepp. ex DC., *Colobanthus quitensis*–*C. affinis* (Hook.) Hook.f., and *Potentilla anserina* L.–*P. anserinoides* Raoul, result from this phase. The latter two examples at least are capable of distance dispersal but, in view of the genetic barriers (Seddon *in* Moore, 1970; Rousi, 1965) and the morphological differences, this seems unlikely to have occurred after the Pleistocene, as it does in the next group to be considered.

(b) *Neo-austral Connections* The presence of many species on sub-Antarctic islands apparently completely glaciated during the Quaternary ice-ages, and the close morphological and cytological relationships within them, strongly suggests that they attained their present distribution in the post-Pleistocene. During this period there is no doubt that the distribution of land and sea in the Southern Hemisphere was as it is now and one is forced to the conclusion that these species attained much of their present distribution by long-distance dispersal over oceanic gaps of up to 6000 km. Possible mechanisms are, of course, by birds, ocean currents or winds. Many birds, such as giant petrels, shearwaters and albatrosses (Falla, 1960), migrate widely in the Southern Temperate Zone and Taylor (1954) suggested them to be responsible for the colonization of Macquarie Island. However, there is still no direct evidence of birds migrating with propagules, such as the hooked fruits of *Acaena*, *Uncinia* or *Galium*, attached to them or with the fleshy fruits of *Nertera*, *Coprosma* or *Gunnera* in their stomachs. Ocean currents are demonstrably able to carry material between the southern continents (Barber *et al.*, 1959) at about 13 km a day (Deacon, 1960) and have been invoked by several workers since the classic researches of Guppy (1906) to account for transoceanic dispersal.

Sykes and Godley (1968) have produced compelling evidence that this mechanism permitted *Sophora microphylla* Aiton to migrate from New Zealand to South Chile and Gough Island, and such species as *Rostkovia magellanica* (Lam.) Hook.f., *Juncus scheuzerioides* Gaudich. and perhaps *Hebe elliptica* and *H. salicifolia*, which seem to have potentially buoyant fruits and coastal habitats, should be further investigated in this context.

The plumed achenes of *Taraxacum gilliesii* Hook. & Arn. and, more particularly, the light spores of ferns would seem suited to long-distance dispersal by the high winds of the "roaring forties". However, Lovis (1959) has pointed out that, although from this one might expect a wider diffusion of Pteridophytes than flowering plants, this is not so; ferns in the Southern Temperate Zone, for example, recognize the same barriers, fall into the same phytogeographical groupings and are not noticeably more widely dispersed than the flowering plants. Nevertheless, Wace and Dickson (1965) point to the unusually high proportion of cryptogams in the flora of Tristan da Cunha, apparently isolated in the middle of the South Atlantic since its origin in the late Tertiary, whose colonization must be ascribed to long-distance dispersal and with which, for example, the information on the phytochemistry and breeding systems of *Empetrum rubrum* (Moore *et al.*, 1970) is in complete accord.

SUMMARY AND CONCLUSIONS

Following their origin in Triassic to middle Jurassic tropical uplands the temperate floras migrated to lower latitudes, their strong family affinities reflecting this common origin. The northern and southern cool temperate floras have subsequently remained substantially separate and had differentiated by the early to middle Cretaceous. During this period the differentiation of such families as the Myrtaceae, Loranthaceae and Restionaceae, and genera like *Nothofagus*, *Gunnera*, *Drimys* and *Astelia*, occurred in temperate parts of Gondwanaland, and comparable events took place in Laurasia. The development of the distinctive palaeo-austral groups of *Caltha*, *Plantago* and *Geum* could have followed the same pattern or their precursors could have entered temperate Gondwanaland somewhat later over fold belts resulting from the Jurassic to early Cretaceous orogenesis in Malaysia (Takhtajan, 1969); perhaps even the perennial Euphrasias also date from this period. The primitive forms of these groups, as in *Geum* for example, do not seem adapted to distance dispersal. During this period North and South America were separated, the closest relatives of many of these genera and groups are in Eurasia and, where there is evidence, they have moved in a westerly direction, so that the importance of the Palaeo-tropics is emphasized, which links with the postulated origin of angiosperms in S.E. Asia. Although the early to middle Cretaceous removal of Africa from Gondwanaland might explain its smaller involvement in circum-Antarctic distributions, this is much more likely due to its northward movement reducing the available cool-temperate areas. *Gunnera*, and presumably the precursors of *Cotula* subg. *Cotula*, were there before its separation. By the late Cretaceous separation of New Zealand, these palaeo-Antarctic genera had spread between it and Antarctica, except perhaps for genera like *Acaena* where subsequent dispersal over an increasing oceanic gap was feasible. Overland migration from Antarctica to Australia was possible until the early Tertiary and probably later between Antarctica and South America. Following the rifting of Gondwanaland there was the morphological, chemical and chromosomal differentiation in many groups which reflects the present disposition of the southern continents, and a northward spread in both Old and New Worlds, usually within the temperate regions available.

Throughout the Tertiary a cool-temperate Antarctica acted as a route by which distance dispersal between New Zealand and South America could take place. The vicarious species–pairs present in these two areas probably result from this phase.

During the early Tertiary North and South America were united and this timing would appear about correct for the migration of the amphitropical temperate genera which are so pronounced in the New World. However, the lack of suitably continuous transtropical uplands suggests that distance dispersal was involved, although several of these genera seem quite unsuited for it. In the absence of late Cretaceous or Tertiary temperate connections between North and South America, one is inevitably forced to consider parallel evolution or distance dispersal, neither of which seems appropriate. Genera well developed in the Northern Hemisphere and poorly represented in South America, like *Primula*, or *vice versa*, like *Acaena*, have all the biological signs of being able to cross the tropical gaps during the Miocene–Pliocene; and the amphitropical species would be able to do so during the Pliocene–Pleistocene, when the North and South Temperate zones approached their closest points. These same Quaternary glaciations which greatly modified the North Temperate floras, virtually eliminated the Antarcto–Tertiary flora, only a few remnants of which survive, and subsequent recolonization of many sub-Antarctic islands was by post-Pleistocene long-distance dispersal, a method responsible for the colonization of other isolated islands like Tristan da Cunha.

The importance of distance dispersal during at least the Tertiary and Quaternary periods is attested by the general clockwise affinities of the recent floras around the Southern Hemisphere, which accord with the general west wind drift of currents and bird migration. All species involved are in generally open habitats, many are coastal and all oceanic, and most appear to be self-compatible and at least facultatively autogamous. In the absence of the direct demonstration of migrating propagules, the evidence is, of course, circumstantial; but it seems overwhelmingly preferable to the alternative of post-Cretaceous land bridges between New Zealand and Antarctica and to many of the oceanic islands, or to such species having evolutionary rates so slow that they have not diverged, while their congeneric relatives have developed new species and sections since direct land connections were geologically feasible.

Further detailed information is needed on the taxonomy and evolution of the genera and species involved in these major temperate disjunctions. Only then will it be possible to narrow down the range of botanical possibilities open for the solution of the many as yet unresolved difficulties.

ACKNOWLEDGEMENTS

I am grateful to my colleagues—Miss T. T. Aye, Dr J. B. Harborne, Mr B. Murray and Miss C. A. Williams—for permission to cite their unpublished data. Financial support from the Natural Environmental Research Council and the Royal Society of London is gratefully acknowledged.

REFERENCES

ADIE, R. J. (1963). *In* "Pacific Basin Biogeography" (J. L. Gressit, ed.), pp. 455–463. Bishop Museum Press, Honolulu, Hawaii.

AXELROD, D. I. (1952). *Evolution* **6**, 29–60.

AXELROD, D. I. (1970). *Bot. Rev.* **36**, 277–319.

BARBER, H. N., DADSWELL, H. E. and INGLE, H. D. (1959). *Nature, Lond.* **184**, 203–204.

BARLOW, B. A. and WIENS, D. (1971). *Taxon* **20**, 291–312.

BATE-SMITH, E. C., DAVENPORT, S. M. and HARBORNE, J. B. (1967). *Phytochemistry* **6**, 1407–1413.

BERRY, E. W. (1937). *Bot. Rev.* **3**, 31–46.

Beuzenberg, E. J. and Hair, J. B. (1959). *N.Z. Jl Bot.* **1**, 53–67.
Cabrera, A. L. (1966). *Blumea* **14**, 285–308.
Clausen, J. (1929). *Ann. Bot.* **43**, 741–764.
Couper, R. A. (1960). *Proc. R. Soc.*, Ser. B **152**, 491–500.
Cranwell, L. M. (1959). *Nature, Lond.* **184**, 1782–1785.
Cranwell, L. M. (1964). *In* "Ancient Pacific Floras", pp. 87–93. University of Hawaii Press, Honolulu, Hawaii.
Dalziel, I. W. D. and Elliott, D. H. (1971). *Nature, Lond.* **233**, 246–252.
Darlington, C. D. (1963). "Chromosome Botany", 2nd Edition. Allen and Unwin, London.
Deacon, G. E. R. (1960). *Proc. R. Soc.*, Ser. B **152**, 441–447.
Du Rietz, G. E. (1940). *Acta phytogeogr. suec.* **13**, 215–282.
Greene, S. W. and Greene, D. M. (1963). *Polar Rec.* **11**, 411–418.
Ehrendorfer, F. (1971). Discussion *In* "Plant Life of South-West Asia" (P. H. Davis, P. C. Harper and I. C. Hedge, eds), pp. 150–151. Bot. Soc. Edinburgh.
Falla, R. A. (1960). *Proc. R. Soc.*, Ser. B **152**, 655–659.
Fleming, C. A. (1963). *In* "Pacific Basin Biogeography" (J. L. Gressitt, ed.), pp. 455–463. Bishop Museum Press, Honolulu, Hawaii.
Freeland, G. L. and Dietz, R. S. (1971). *Nature, Lond.* **232**, 20–23.
Gajewski, W. (1957). *Monogr. Bot.* **4**, 1–416.
Good, R. D'O. (1964). "The Geography of the Flowering Plants", 3rd Edition. Longmans, London.
Grant, V. (1959). "Natural History of the Phlox Family", Vol. 1. Nijhoff, The Hague.
Gregor, J. W. (1939). *New Phytol.* **38**, 293–322.
Guppy, H. B. (1906). "Observations of a Naturalist in the Pacific between 1896 and 1899", Vol. 2. Macmillan, London.
Hair, J. B. (1962). *Chromosome Information Service* No. **3**, 41–42.
Hair, J. B. (1963). *In* "Pacific Basin Biogeography" (J. L. Gressitt, ed.), pp. 401–414. Bishop Museum Press, Honolulu, Hawaii.
Hair, J. B. (1966). *N.Z. Jl Bot.* **4**, 559–595.
Hair, J. B. and Beuzenberg, E. J. (1960). *N.Z. Jl Sci. Technol.* **3**, 432–440.
Hara, H. (1972). *In* "Taxonomy, Phytogeography and Evolution" (W. H. Valentine, ed.) pp. 61-72, Academic Press, New York and London.
Hawkes, J. G. and Smith, P. (1965). *Nature, Lond.* **207**, 48–50.
Hedberg, O. (1961). *Rec. Adv. Bot.* **1**, 914–919.
Heirtzler, J. R., Dickson, G. O., Herron, E. M., Pitman, W. C. and Le Pichon, X. (1968). *J. Geophys. Res.* **73**, 2119–2136.
Hofsten, N. von (1916). *Zool. Annln.* **7**, 197–353.
Holdgate, M. W. and Wace, N. M. (1961). *Polar Rec.* **10**, 475–493.
Hooker, J. D. (1853). "Flora Novae-Zelandiae", Vol. 1. "Flowering Plants". Reeve, London.

HULTÉN, E. (1959). *Svensk. bot. Tidskr.* **53**, 203–228.
JACKSON, W. D. (1958). *Pap. Proc. R. Soc. Tasm.* **92**, 161–163.
KORNAŚ, J. (1972). *In* "Taxonomy, Phytogeography and Evolution" (W. H. Valentine, ed.), pp. 37–59. Academic Press, London and New York.
LÖVE, Á. and LÖVE, D. (eds) (1963). "North Atlantic Biota and their History". Pergamon, Oxford.
LOVIS, J. D. (1959). *Br. Fern. Gaz.* **10**, 1–8.
MANTON, I. (1934). *Z. indukt. Abstamm.-u. Vererb-Lehre* **67**, 41–57.
MOORE, D. M. (1963). *Madroño* **17**, 52–53.
MOORE, D. M. (1964). *In* "Biologie Antarctique" (R. Carrick, M. Holdgate and J. Prevost, eds), pp. 195–202. Hermann, Paris.
MOORE, D. M. (1967). *Br. Antarct. Surv. Bull.* **14**, 69–82.
MOORE, D. M. (1968). "The Vascular Flora of the Falkland Islands", *Scient. Rep. Br. Antarct. Surv.* **60**, 1–202.
MOORE, D. M. (1970). *Br. Antarct. Surv. Bull.* **23**, 63–80.
MOORE, D. M. and CHATER, A. O. (1971). *Bot. Not.* **124**, 317–334.
MOORE, D. M. and WALTON, D. W. H. (1970). *Br. Antarct. Surv. Bull.* **23**, 101–103.
MOORE, D. M., HARBORNE, J. B. and WILLIAMS, C. A. (1970). *Bot. J. Linn. Soc.* **63**, 277–293.
NELMES, E. (1951). *Reinwardtia* **1**, 221–450.
OLIVER, W. R. B. (1955). *Svensk bot. Tidskr.* **49**, 9–18.
RAHN, K. (1957). *Bot. Tidsskr.* **53**, 369–378.
RATTENBURY, G. G. (1957). *Trans. R. Soc. N.Z.* **84**, 936–938.
RAVEN, P. H. (1963). *Q. Rev. Biol.* **38**, 151–177.
RAVEN, P. H. (1971). *In* "Plant Life of South-West Asia" (P. H. Davis, P. C. Harper and I. C. Hedge, eds), pp. 119–134. Bot. Soc. Edinburgh.
ROUSI, A. (1965). *Ann. Bot. Fenn.* **2**, 47–112.
SCHUSTER, R. M. (1969). *Taxon* **18**, 46–91.
SKOTTSBERG, C. (1913). *Kungl. svenska Vetensk.-Akad. Handl.* **51**, 1–128.
SKOTTSBERG, C. (1915). *Pl. Wld.* **18**, 129–142.
SKOTTSBERG, C. (1960). *Proc. R. Soc.*, Ser. B **152**, 447–456.
SOLBRIG, O. T. (1972). *In* "Taxonomy, Phytogeography and Evolution" (W. H. Valentine, ed.), pp. 85–100. Academic Press, London and New York.
STAUDT, G. (1962). *Can. J. Bot.* **40**, 869–886.
STEBBINS, G. L. (1971). "Chromosomal Evolution in Higher Plants". Arnold, London.
STEFFEN, H. (1939). *Beih. bot. Zbl.* **59b**, 531–560.
SYKES, W. R. and GODLEY, E. J. (1968). *Nature, Lond.* **218**, 495–496.
TAKHTAJAN, A. (1969). "Flowering Plants: Origin and Dispersal". Oliver and Boyd, Edinburgh.
TARLING, D. H. (1971). *Nature, Lond.* **229**, 17–21.

TAYLOR, B. W. (1954). *Ecology* **35**, 569–572.

VAN STEENIS, C. G. G. J. (1972). *In* "Taxonomy, Phytogeography and Evolution" (W. H. Valentine, ed.), pp. 275–288. Academic Press, London and New York.

WACE, N. M. and DICKSON, J. H. (1965). *Phil. Trans. R. Soc.* Ser. B **249**, 273–360.

WILLIAMS, C. A. and MURRAY, B. G. (1972). *Phytochemistry* **11**, 2507–2512.

YEO, P. F. (1968). *Evolution* **22**, 736–747.

Section III

ENDEMISM

9 | Endemism in the Flora of the Canary Islands

D. BRAMWELL

Department of Botany, University of Reading, Reading, England

INTRODUCTION

The Canary Islands form part of the phytogeographical region known as Macaronesia which encompasses several small groups of islands in the northern tropical and subtropical regions of the eastern Atlantic Ocean. From north to south these are the Azores, Madeira, Salvage Islands, Canaries and Cape Verde Islands. Floristically the Canaries are by far the richest group with about 1700 species of flowering plants (probably about 1000 native), of which some 470 constitute the endemic flora.

As was clearly pointed out by Engler (1882) there are two main types of endemics, the relicts (variously known as palaeoendemics, conservative endemics or epibiotics) which are taxa which have survived in a limited portion of their past territory, and the so-called new, autochthonous endemics (known as secondary, progressive or neoendemics) which have arisen by means of differential evolution in the area concerned. Wulff (1950) discusses this point and considers that there are few difficulties involved in distinguishing the types, as palaeoendemics rarely have any close relatives in the same or adjacent regions whereas neoendemics tend to have numerous, often close bonds with other species in the same or nearby regions. Wulff, however, notes that certain problems arise in some areas where, because of the effect of phenomena such as adaptive radiation and vicariance, instances occur of endemic genera, which would generally be considered to be palaeoendemic but which comprise what appear to be more recently evolved groups of species which might possibly be classified as neoendemics. Rikli (see Wulff, 1950) has termed such genera "active epibiotics" and there are a number of examples of this type of endemism in the Canary Islands.

AGE OF THE ENDEMIC FLORA

There is very little doubt that the endemic flora of the Canaries is basically a very old one with the vast majority of taxa either palaeoendemics or their derivatives, relicts of a series of much older floras that probably occupied Southern Europe and North Africa in the Tertiary Epoch (Takhtajan, 1969). There are several strong lines of evidence to support this view; (1) the predominance of woody life-forms in the endemic flora; (2) the taxonomically isolated position of many of the endemic taxa; (3) major disjunctions in the distribution of many endemics; (4) the fossil evidence provided by the Pliocene deposits of the Mediterranean region and (5) the extremely low level of polyploidy in the endemic flora. Each of these lines of evidence is considered in more detail below.

1. Woody Life-forms

The predominance of woody life-forms in the Canaries in genera such as *Echium*, *Sonchus*, *Sideritis*, *Argyranthemum*, *Senecio*, *Pterocephalus*, *Convolvulus*, *Crambe*, *Descurainia*, *Erysimum* and *Plantago* has been considered by several authors (Meusel, 1952; Lems, 1960a) to be a characteristic of relict plants. Meusel derives many modern Mediterranean growth-forms from basic forms similar to those now found only in Macaronesia. Lems based his study of adaptive radiation in the genus *Aeonium* on the premise that the most woody growth-form was the ancestral one and he cites the disjunct Macaronesian/East African distribution of this growth form as evidence of its relict nature.

In most cases of widespread distribution within the Macaronesian region woody life-forms are generally predominant in, for example, *Echium*, *Sonchus* and *Isoplexis*. Further, many of the Mediterranean or East African/Macaronesian disjunct taxa are also frutescent in their extra-Macaronesian outposts, as for example, *Argyranthemum*, *Euphorbia* sect. *Pachycladae*, *Aeonium*, *Plantago* and *Convolvulus*.

If one considers the history of the Mediterranean flora in general, it is evident that Macaronesia, particularly the Canaries where the eastern islands probably formed early Tertiary land-bridges, seems likely to have been originally colonized as a group of "continental islands" by a series of predominantly woody floras originating from the forest floras of the Tethyan–Tertiary region. The fossil evidence which is outlined below supports this view (Takhtajan, 1969).

In view of this it appears improbable that there would be similar evolutionary opportunities to those afforded by oceanic islands such as

Hawaii. Such islands were colonized by small fragments of floras following long-distance dispersal (Carlquist, 1965), and predominantly herbaceous plants appear to have become secondarily woody in order to fill unoccupied ecological niches suitable for frutescent or arborescent plants. It is more probable that in the Canarian situation, where a woody flora seems to have been present from a very early stage, unoccupied niches for frutescent plants would have been filled from the already woody families and genera present, rather than by the adoption of a frutescent or arborescent habit by herbaceous plants. It is likely, therefore, that the Macaronesian shrubs were already woody in the Tethyan–Tertiary region before they ever reached Macaronesia, and that, as Meusel (1952) points out, the herbaceous members of these genera now found in the modern Mediterranean flora were probably derived from woody ancestors similar to the surviving Macaronesian relicts.

2. *Taxonomically Isolated Position of Endemics*

The taxonomically isolated position of many of the endemics can be demonstrated by considering the number of endemic genera or sections of genera. In all about 17 genera are endemic to the Canary Islands themselves and another 12 are found only in Macaronesia. The Canarian endemics are *Allagopappus*, *Gonospermum*, *Lactucosonchus*, *Heywoodiella*, *Schizogyne*, *Sventenia* and *Vieraea* (Compositae), *Ixanthus* (Gentianaceae), *Legendraea* (Convolvulaceae), *Dendriopoterium* and *Bencomia* (Rosaceae), *Dicheranthus* (Caryophyllaceae), *Gesnouinia* (Urticaceae), *Greenovia* (Crassulaceae), *Parolina* (Cruciferae), *Spartocytisus* (Leguminosae), *Todaroa* (Umbelliferae).

The Macaronesian genera tend to have their greatest concentration of species in the Canaries. These are *Aichryson* and *Monanthes* (Crassulaceae), *Marcetella* (Rosaceae), *Phyllis* and *Plocama* (Rubiaceae), *Drusa* and *Melanoselinum* (Umbelliferae), *Sinapodendron* (Cruciferae), *Visnea* (Ternstromiaceae), *Picconia* (Oleaceae), *Isoplexis* (Scrophulariaceae), *Semele* (Liliaceae). In the majority of cases these are woody plants.

In addition there are many examples of Mediterranean or sub-Mediterranean genera which have endemic sections in Macaronesia. The most important of these are *Sideritis* sect. *Leucophae*, *Echium* with several frutescent sections, *Crambe* sect. *Dendrocrambe*, *Sonchus* sect. *Dendrosonchus*, sect. *Atalanthus*, *Senecio* sect. *Bethancouria*, sect. *Pericallis*, *Aeonium* with several sections, and *Convolvulus* sect. *Frutescentes*, sect. *Floridi*. The names of several of these sections reflect their woody habit.

Many of these Macaronesian genera and sections are mono- or oligotypic and their taxonomic isolation is probably a reflection of their great age, as they have been separated from their nearest relatives perhaps early in the Tertiary period.

3. Disjunct Distributions

Further evidence of the relict nature of many Canarian endemics is afforded by consideration of their distribution and that of their nearest relatives. If one takes examples such as *Aeonium*, *Canarina*, *Jasminum odoratissimum* L., *Adiantum reniforme* L. and *Osyris quadripartita* Decne, it can be seen that the areas occupied by these genera and species are decidedly discontinuous with a wide gap across North Africa and the Sahara, suggesting a relict distribution. *Canarina* (Campanulaceae) has a single species in the Canaries and two in the mountains of East Africa (Ethiopia, Uganda, Kenya, Sudan). Its distribution is shown in Fig. 1 and Hedberg (1961) suggests that the disjunction in distribution dates from the Tertiary.

Aeonium (Crassulaceae) has a similar distribution pattern but as a result of adaptive radiation the genus is represented in Macaronesia by about 30 species. The only East African member of the genus is *A. leucoblepharum* Webb from Ethiopia and Somalia.

The fern *Adiantum reniforme* L. is found only in Macaronesia and on the Indian Ocean islands of Madagascar and Reunion (Fig. 2). This species also appears to be a Tertiary relict (Tardieu-Blot, 1946) and may date from the early part of the Tertiary period.

At the species level *Jasminum odoratissimum* L. also provides evidence of an old disjunction. This species occurs in Macaronesia, on several of the island groups, and also in Rhodesia. In both areas it appears to be a species of fairly moist local climates and its distribution may be a remnant of pre-Saharan times.

A further important example of such a disjunct distribution is that of *Osyris quadripartita* Decne (Santalaceae). This is a Canarian/Mediterranean species which extends into Asia and Southern Africa (Fig. 3). The Mediterranean and North African forms of this species have small leaves and a single female flower in the leaf-axils of female plants, whereas the East African form has much larger leaves and two or three female flowers in each inflorescence. The Canarian material is more or less identical with the East African and, therefore, perhaps provides one of the most notable examples of this kind of disjunction.

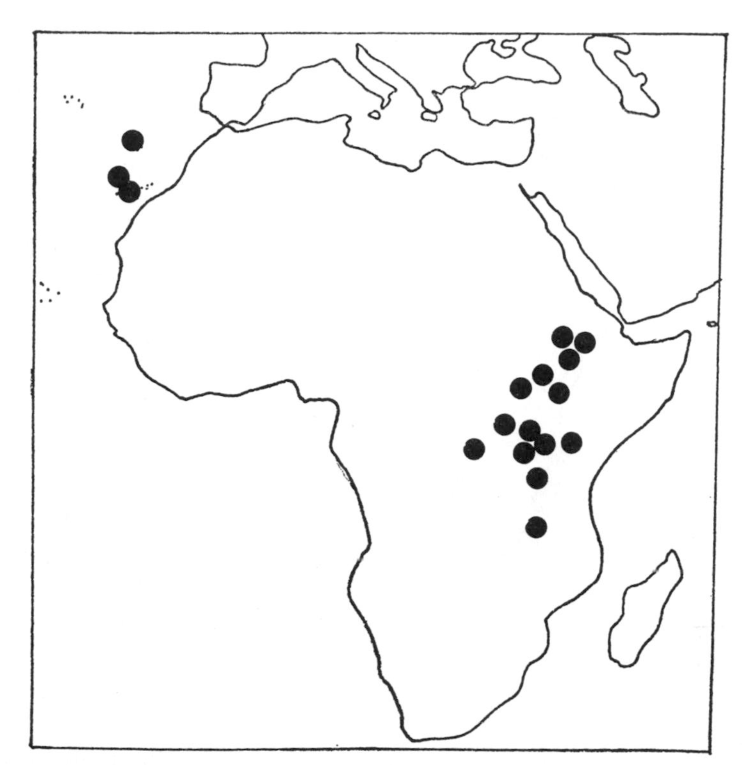

Fig. 1. Distribution of the genus *Canarina* L. showing East African/Macaronesian Disjunction.

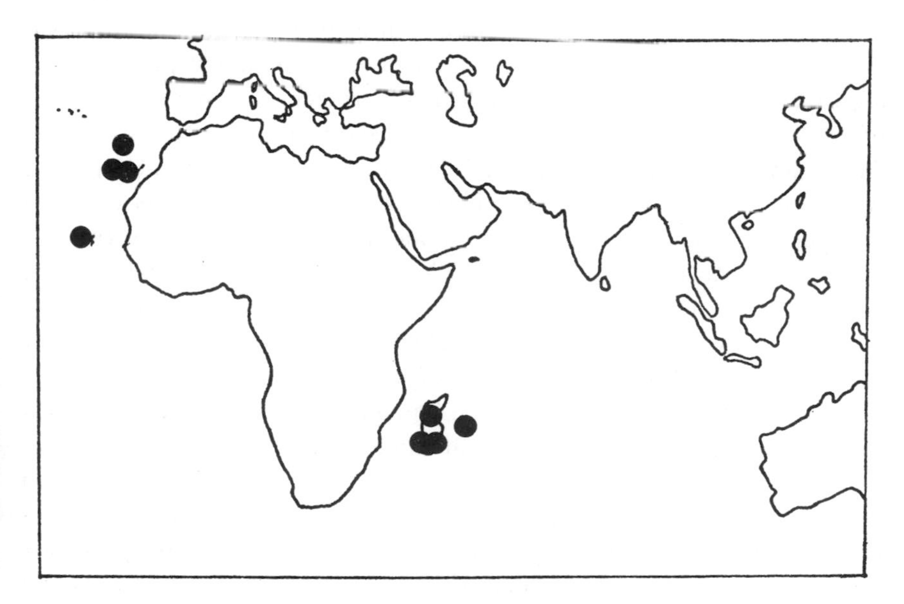

Fig. 2. Distribution of *Adiantum reniforme* L. showing Madagascar/Macaronesia Disjunction.

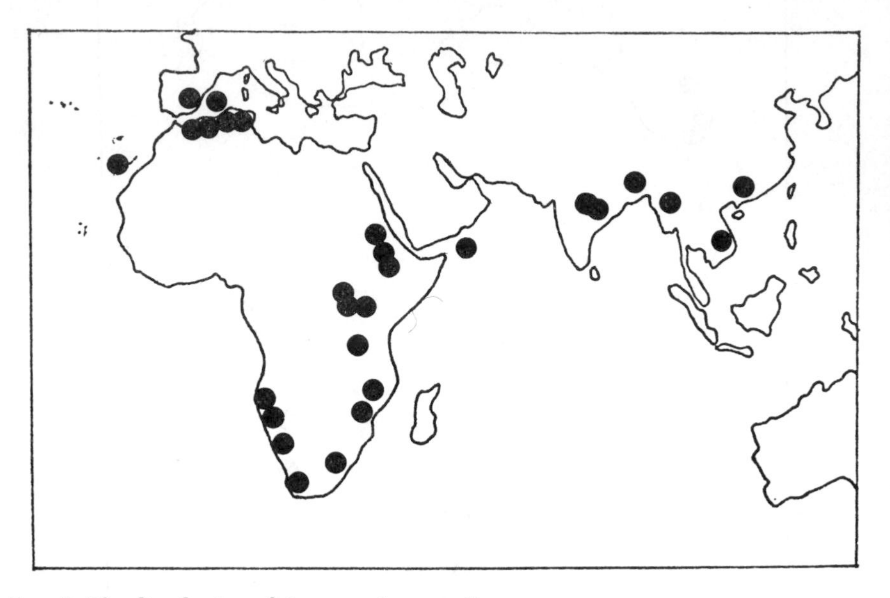

FIG. 3. The distribution of *Osyris quadripartita* Decne.

In addition to Macaronesian/East African disjunctions there is also a second important type. This consists of a number of genera which have their nearest relatives in central and southern America. These plants are often considered to make up an American element in the Macaronesian flora and are sometimes used as evidence of continental drift, so that they are of considerable phytogeographical importance. The genus *Clethra* is often cited as an example (Fig. 4); it occurs in eastern Asia and Central America with a single species, *C. arborea* Aiton in Macaronesia. Hu (1960) writes of this species "It is very clear that *C. arborea* Aiton is an introduction from the East Indies to the Madeira Islands in historical time. It was probably accidentally introduced in connection with the spice trade, and was firmly established in the late Eighteenth Century when it first caught the attention of British naturalists". Unfortunately there is very little evidence to support this view. *Clethra arborea* Aiton is known only from Macaronesia and has never been found in the East Indies. It is known from early Quaternary lignite deposits on Madeira laid down long before there was any spice trade and it is also known as a Tertiary fossil from several Mediterranean localities. The species, as pointed out by Tavares (1965) is one of deep ravines in mountain forests, an unlikely habitat for the establishment of an introduced species. Taxonomically *C. arborea*

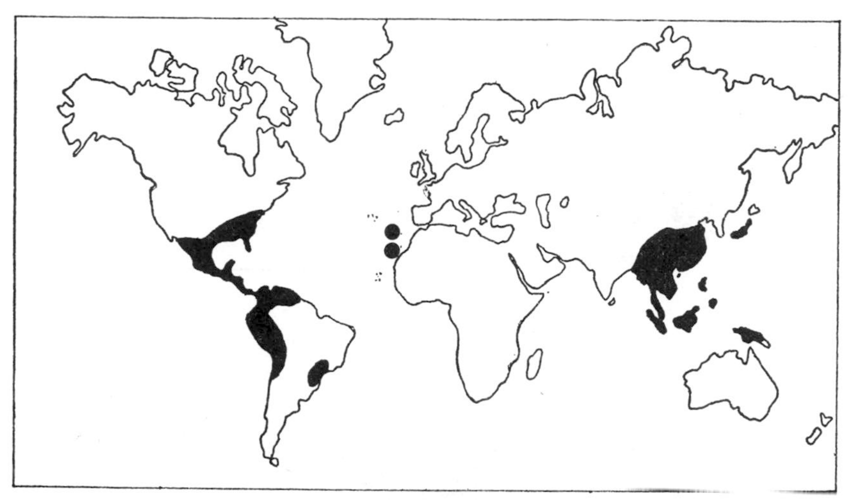

FIG. 4. Worldwide distribution of the Genus *Clethra* L.

Aiton occupies an isolated position, but in its seed and capsule characters it resembles the American group of species rather than the Asiatic.

The genus *Drusa* (Umbelliferae, Hydrocotyloideae) is endemic to Macaronesia and its nearest relatives, members of the genera *Bowlesia* and *Homalocarpus*, are found on the American continent. *Bystropogon* (Labiatae) has an endemic Macaronesian section and a further section in western South America. The distribution of both these genera and their relatives is shown in Figs 5 and 6.

Dansereau (1968) without giving any reasons, does not consider the American element in the Canarian flora to be a valid one and Lems (1960b) suggests that its presence can be explained by long-distance dispersal from America by water. In view of the level of separation between the disjunct groups (generally generic or sectional) it seems more probable that this type of disjunction can be explained by "continental drift" and that this small element is a group of survivors of a pre-drift landmass flora perhaps from the late Cretaceous period. This explanation is suggested by Good (1964) for a number of discontinuously distributed taxa of the New and Old World. The fact that none of the Macaronesian–American disjunct groups seem to be particularly adapted for such long distance dispersal seems also to lend weight to this view.

A third group of taxa exhibit East Mediterranean or Iberian and Macaronesian disjunctions both at generic and specific levels. For example, *Erysimum* sect. *Cheiranthus* has one centre of distribution in the Aegean and

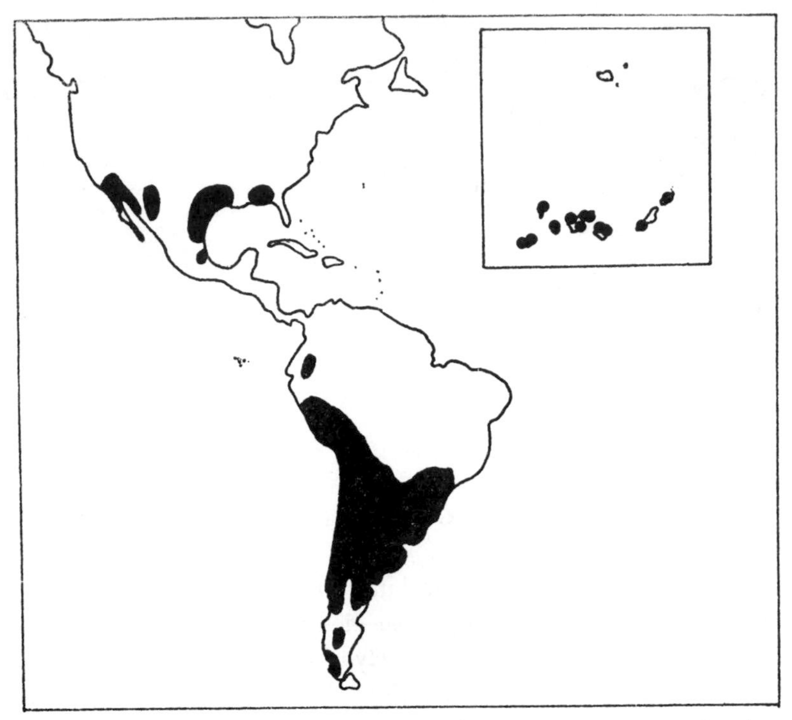

FIG. 5. Distribution of the genera *Bowlesia* Ruiz & Pavon and *Homalocarpus* Hook & Arn. on the American Continent with the Genus *Drusa* DC. (inset).

Balkan regions and a second in Macaronesia (Madeira and Canary Islands). The Mediterranean taxa of these groups tend to be classical chasmophytic relicts as is the case in *Sideritis* sect. *Empedoclea* in the Balkans. In this particular group the Canarian species are also confined to Tertiary basalt cliff habitats on the geologically ancient island of La Gomera and appear to comprise a very old group.

As a further example the Iberian endemic *Digitalis obscura* L. might well be cited. This species occupies a very isolated position taxonomically. It differs from most other *Digitalis* species in its woody habit, its coriaceous leaves and its bright orange flowers. In these characters, however, it closely resembles the Macaronesian genus *Isoplexis* and the species seems to be a Tertiary relict of a Pliocene forest flora now extinct in Europe but still surviving in a modified form in Macaronesia (Rivas Goday, 1946). Several other species such as *Umbilicus heylandianus* Webb & Berth. and *Leuzea cyanaroides* (Link) Font Quer are also found only in a few localities in Spain and in the Canaries (Font Quer, 1957).

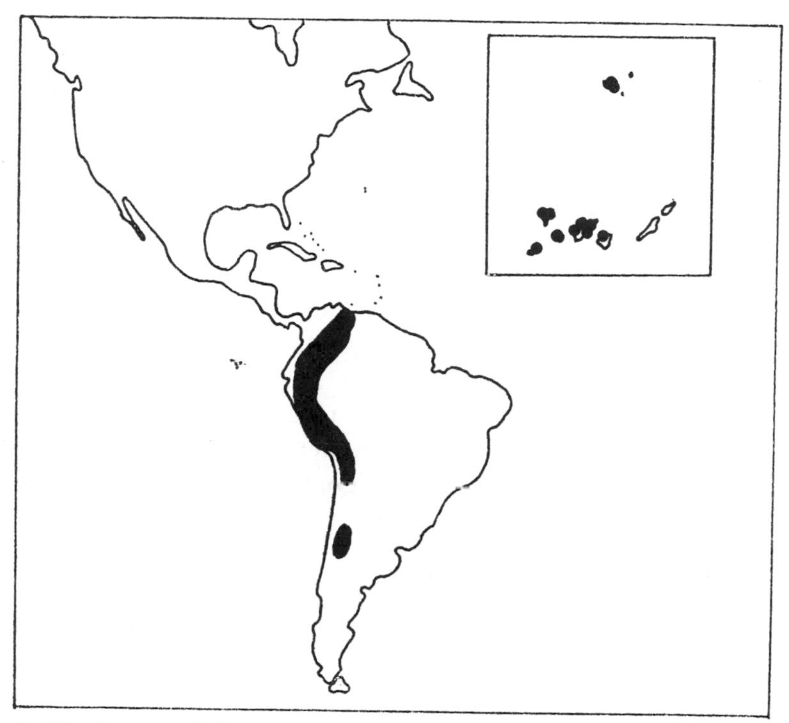

FIG. 6. Distribution of the genus *Bystropogon* L'Hér. in South America and (inset) Macaronesia.

4. *Pliocene Fossil Floras*

The Tertiary flora of southern Europe is relatively well known and has been considered by a number of authors (Schenck, 1907; Depape, 1922; Takhtajan, 1969). The Miocene and Pliocene floras were tropical or subtropical forest ones which were widely distributed from Spain, southern France, Italy and North Africa eastwards to South Georgia. At the end of the Tertiary this flora was displaced by the changes in climatic zones and the southern migration of a cooler, probably drier climate. The subtropical flora was replaced by a temperate one but survived in a modified state in Macaronesia and to a small extent in a few favourable places in the south-west of the Iberian peninsula. The following species were reported by Depape (1922) from Pliocene deposits in the Rhône Valley and near Barcelona and all of these still survive in Macaronesia: *Laurus azorica* (Seub.) Franco, *Persea indica* Spreng., *Ocotea foetens* (Aiton) Benth., *Catha cassinoides* Webb & Berth. (as *Celastrus gardonensis* Saporta), *Ilex canariensis* Poiret, *Viburnum rugosum* Pers., *Picconia excelsa* DC., *Phillyrea angustifolia*

L., *Smilax canariensis* Willd., *Pinus canariensis* Spreng., *Woodwardia radicans* (L) Sm., *Visnea mocanera* L.fil. Several other species are still found both in Macaronesia and in Spain or Portugal: for example, *Prunus lusitanica* L., *Myrica faya* Aiton, *Culcita macrocarpa* C. Presl. and *Davallia canariensis* (L) Sm.

It must be noted that a number of important subtropical elements found as Pliocene fossils are, however, missing from the modern Macaronesian flora, for example *Cinnamomum*, *Sapindus* and *Buettneria*, so it is rather difficult to agree entirely with the views expressed by some authors that the present-day Canarian forests represent actual communities that occurred in Tertiary Europe (Ciferri, 1962; Oberdorfer, 1965).

Many of the present-day Canarian forest plants probably had their origin in the Tertiary floras of the Mediterranean and their present distribution perhaps only represents a small fragment of their former range so that they are good examples of palaeoendemics.

5. *Polyploidy Level in the Endemic Flora*

About 200 endemic species have been studied cytologically and less than 25% of these shown any degree of polyploidy. As pointed out by Larsen (1960) and Borgen (1969), this is one of the lowest levels of polyploidy for any known flora. If one applies the system of classification of endemics proposed by Favarger and Contandriopoulos (1961) and Favarger (1964), the high proportion of diploid species should be considered as being palaeoendemics in the cytological sense as well as the classical. Some of the large genera such as *Sonchus*, *Echium* and *Argyranthemum* have been studied and are entirely diploid in the Canaries though especially in the case of *Echium* and *Sonchus* their continental relatives tend to be polyploid. Several other genera are polyploid in Macaronesia, for example *Bystropogon*, *Bencomia*, *Isoplexis* and *Senecio* sect. *Pericallis*. These tend not to have closely related diploids in nearby regions and seem to fit Favarger's (1967) concept of palaeopolyploids as they generally have a polyploidy level lower than their closest relatives.

There is, therefore, considerable weight of evidence to support the view that the Canarian flora is a very old one and that, as suggested by Meusel (1952), the woody habit predominant in the endemic groups is a relict characteristic. This view is opposed by Carlquist (1965) who suggests that the woody habit in Canarian plants is a derived condition resulting from insular isolation and an equable climate. Carlquist's concepts were largely worked out for the Hawaiian flora which, as pointed out above, probably originated and responded to distinct habitat conditions in a different way from the Canarian flora; and the weight of palaeobotanical and cytological evidence as well as the distribution patterns and taxonomic isolation of the Canarian endemic flora seem to support strongly Meusel's hypothesis.

EVOLUTION IN THE CANARIAN ENDEMIC FLORA

Many of the apparently relict endemic genera and sections of genera are represented in the Canaries by numerous species which seem to have evolved since the arrival in the islands of the parent types. These relicts seem to have had a second lease of life in the Canaries and appear to be examples of Rikli's active epibiotics.

The Canaries are volcanic islands; the highest, Tenerife, rises to over 3700 m. A considerable amount of evidence has recently been presented which suggests that the archipelago is made up of two distinct groups of islands (Rothe and Schmincke, 1968). The eastern group, Lanzarote, Fuerteventura, Graciosa and Lobos, were once probably connected to the African mainland, whereas the western islands, Gran Canaria, Tenerife, Hierro, La Palma and Gomera, are probably purely volcanic in origin, that is, true oceanic islands in the geological sense.

The eastern islands, being close to the continent, tend to be influenced by Saharan climatic conditions. The western group, however, are situated between 200 and 360 km from the coast and have a much more equable, oceanic climate. This wide range of climatic conditions coupled with the altitude and rugged topography of the islands means that there is a good deal of scope for the establishment of widely differing vegetation types and these are shown in Table I. Within each major zone there are numerous individual habitats and it is this wide range of potential niches which seems to have led to the evolution of so many species in the relict groups. This evolution has been further stimulated by the replication of similar habitats on a series of islands each differing from the other in topographic and, for the western group at least, minor climatic features.

Evolution in the palaeoendemic flora seems to have proceeded mainly by the processes of adaptive radiation and vicariance. The two processes differ in that adaptive radiation, the diversification of form in response to the pressures of different ecological habitats, is a positive process where genetical response to the stimulus of the environment is the main factor; but vicariance, that is divergent evolution in which geographical isolation

Table I. Principal zones of natural vegetation in the Canary Islands

Vegetation	Altitude m	Climate	Life-forms	Floristic affinity
Subalpine Violetum	2600	Subcontinental cold winter, hot dry summer	perennial herbs	North African–Mediterranean, Afro-montane
Montane scrub	1900–2500		microphyll shrubs	
Pine savanna	800–1900	Dry Mediterranean	needle-leaved trees	South European Tertiary
Tree heath and evergreen forest	400–1300	Wet Mediterranean	broad-leaved trees, shrubby heaths	
Juniper scrub	S. slopes 200–600	Mediterranean	microphyll shrubs	Mediterranean, North African
Semi-desert succulent scrub	0–800	Hot, dry Mediterranean	Candelabra shrubs and spiny microphylls, therophytes	Macaronesian facies of Indo–Saharan semi-desert scrub

has been a very important factor, is perhaps a more passive process where the interaction of genetic drift and weak selection result in the establishment of distinctive characters in populations which occupy essentially similar ecological habitats to their parents. A high degree of vicariant evolution is to be expected in an archipelago such as the Canaries where, as previously stated, sets of ecological conditions are replicated from east to west on a series of islands with similar vegetation zones.

1. *Adaptive Radiation*

There is one well documented case of adaptive radiation in the plants of the Canaries and this is in the genus *Aeonium* (Crassulaceae). This genus was studied by Lems (1960a) and he demonstrated that a wide range of different growth-forms evolved from woody shrubs, through many stages of reduction of the woody axis, to perennial and biennial herbs in response to the many habitats available. Habit changes in *Aeonium* are also correlated with modifications of leaf-succulence, inflorescence structure and branching patterns, all features which are strongly influenced by environmental factors.

Several other examples of adaptive radiation in Canarian plants can be equally well demonstrated. A particularly striking example is the genus *Sonchus*, in which the initial growth-form can be taken as being similar to that of *S. pinnatus* Aiton, a tall (4 m), arborescent species found in the Canaries on several of the islands and also on Madeira; a very similar species, *S. daltonii* Webb also occurs in the Cape Verde group. This widespread distribution pattern along with the woody habit and large capitula may be considered as primitive characteristics and it is possible to derive the other major forms of *Sonchus* found in the Canaries from it (Fig. 7).

From the basic form several evolutionary trends which are closely correlated with ecological conditions emerge and each results in an adapted form fitted to a particular habitat. Reduction in thickness of the stems and branches, capitulum-size and leaf-area in response to dry conditions produces form B (Fig. 7) which is based on *S. regis-jubae* Pitard from dry areas on the north coast of Gomera. This trend culminates in form C, the *S. leptocephalus* Cass. form with very small capitula and linear leaf-segments. Species such as *S. leptocephalus* Cass. occur in very dry areas of the western islands usually below 700 m. In forest habitats where species occupy more mesic conditions the size of the plant rarely exceeds two metres and the leaf-area is increased, probably in response to shade conditions. This type is shown in Fig. 7D which is based on *S. jacquini* DC.

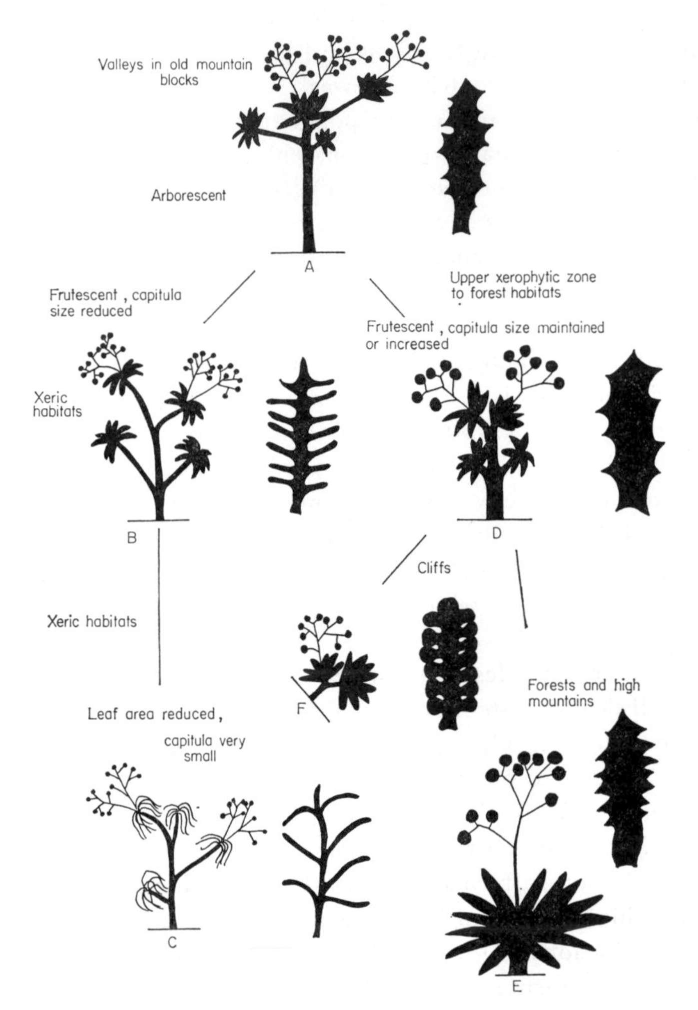

Fig. 7. Adaptive radiation in growth-form, leaf-shape and capitulum-size in the Canarian members of the genus *Sonchus*.

and *S. abbreviatus* Link, both found in the laurel forests of Tenerife and Gran Canaria. This trend culminates in forms such as *S. acaulis* Dum-Cours. and *S. platylepis* Webb & Berth. which have a very short woody stem of only a few decimetres, a rosette of very large leaves and extremely large capitula. These species (Fig. 7E) are found in the high mountain forests of Tenerife and Gran Canaria and even in the montane regions above the forests.

In cliff areas of the basal zone the least woody species are found. These resemble *S. radicatus* Aiton (Fig. 7F) and have a very short woody stem of about 10–15 cm and a dense rosette of small leaves with rounded, often overlapping lobes. The capitula are usually large but few in each inflorescence. In very exposed conditions the stem is usually very short and the rosette closely appressed to the rock-surface.

A number of other genera show similar types of adaptive radiation into widely differing habitats. These include *Echium*, *Sideritis* and *Argyranthemum* which have different species occupying many habitats from the coastal region to the subalpine zones of the high mountains.

2. *Vicariance*

In addition to the adaptive evolution of form in these genera a considerable amount of vicariant evolution has also taken place. In *Sonchus* this has occurred in several of the main growth-form groups; for example, six species of type F, the chasmophytic group, are known. These all occupy similar niches on basalt or phonolite cliffs and each species is effectively isolated from the other members of the group by geographical or topographical barriers. Being very small populations indeed, the individual colonies have evolved into separate easily distinguishable species with genetic drift probably having played an important role in their formation. Each of the individual areas occupied by these species is a mountain block or outcrop of ancient Tertiary rock, probably the original base-rock of the islands, and is separated by areas of more recent volcanic activity (Fig. 8).

A similar situation is found in the xeromorphic, narrow-leaved group of species (*S. leptocephalus* group) where five vicariant species are known.

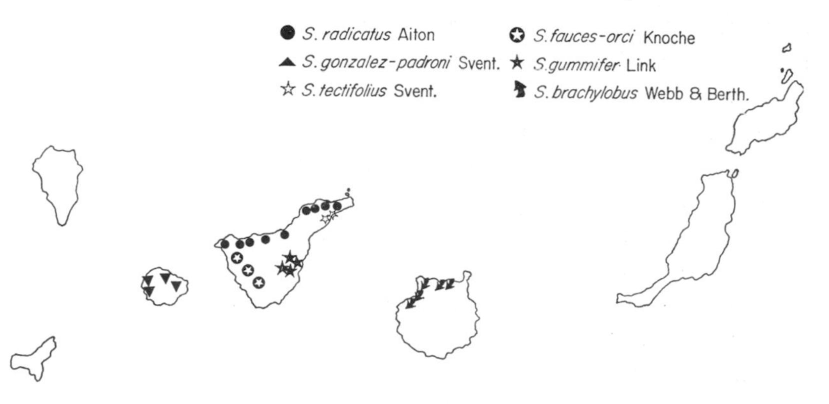

FIG. 8. Distribution of the vicariant species of the *Sonchus radicatus* group in the Canary Islands.

The most widespread is *S. leptocephalus* Cass. itself which occurs on Tenerife and Gran Canaria especially along the north coasts. *S. capillaris* Svent. occurs in the south-west of Tenerife and *S. microcarpus* Boulos on cliffs at Ladera de Guimar in the south of the island. *S. filifolius* Svent. is confined to the south-west of Gomera and *S. fatagae* (*sp. nov. ined.*) is known only from a single valley in the south of Gran Canaria (Fig. 9).

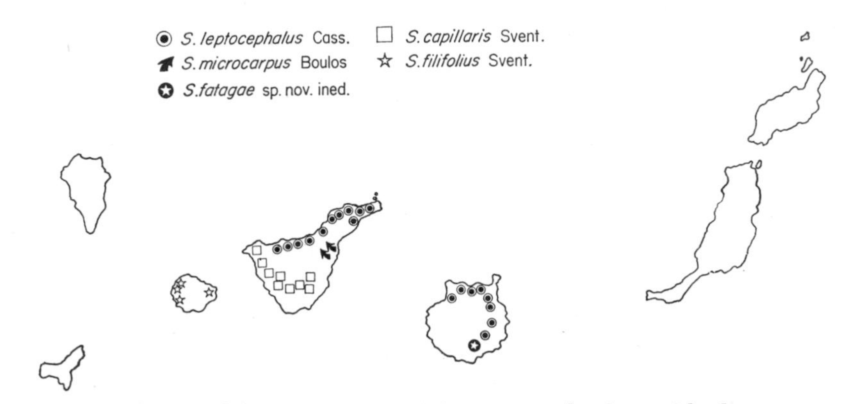

FIG. 9. Distribution of the *Sonchus leptocephalus* group in the Canary Islands.

Of the forest species, *S. jacquini* DC. and *S. abbreviatus* Link occur on Tenerife. The vicariants *S. hierrense* (Pitard) Boulos found on Hierro (with a distinct subspecies *benehoavensis* Svent. on La Palma), and *S. ortunoi* Svent. of Gomera, are known from similar ecological situations on other islands.

In some genera vicariance seems to have taken place without the stimulus of adaptive radiation. In most of the large genera geographical counterparts of particular habit-forms are found as distinct species on each of several of the islands. The yellow-flowered group (subsect. *Flaviflorae*) of *Centaurea* section *Cheirolophus* consists of eight very similar species all occurring on outcrops of Tertiary basalt (Fig. 10). Each of these species exists as a very small population, in most cases of less than 30 individuals, and in *C. junoniana* Svent. and *C. tagananensis* Svent. probably no more than 10 to 15. Species formation in this case has probably taken place by fragmentation of a once widespread parent species, as the old basalt regions of the islands have been isolated from each other by more recent volcanic activity.

A similar situation is found in *Crambe* section *Dendrocrambe* with eight species, four of which are confined to small areas of Tenerife (Fig. 11) and in *Parolinia* (Bramwell, 1970).

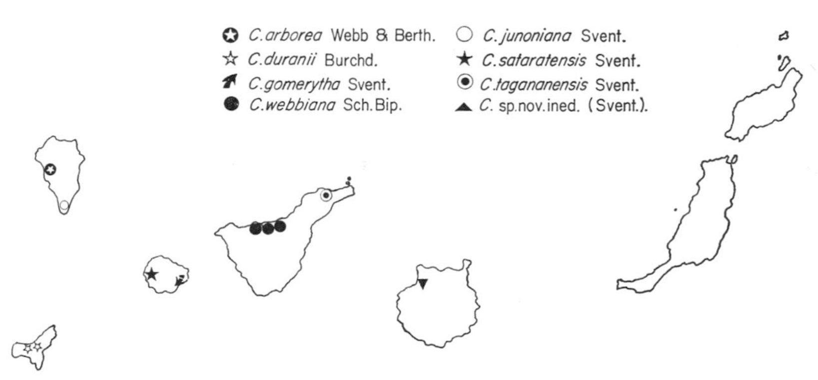

FIG. 10. Distribution of the vicariant species of *Centaurea* sect. *Cheirolophus* subsect. *Flaviflorae* in the Canary Islands.

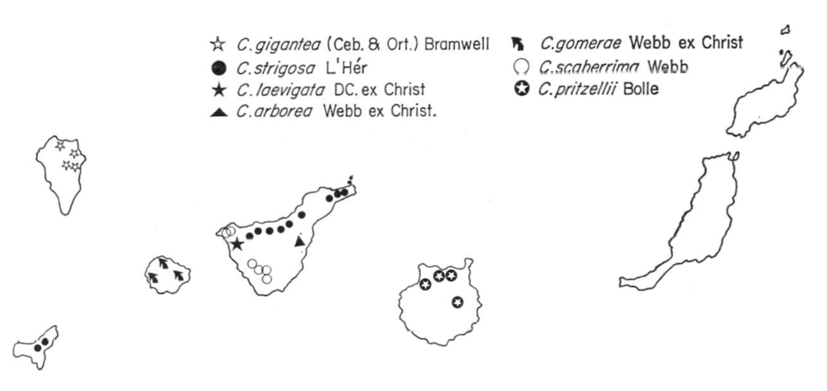

FIG. 11. Distribution of the vicariant species of *Crambe* sect. *Dendrocrambe* in the Canary Islands.

The genera illustrated above are only a few examples of adaptive and vicariant evolution in the Canaries and appear to demonstrate Cain's principle (Cain, 1944) that the restriction of numbers of biotypes in small populations of relicts is in no way an indication of limited evolutionary potential, but is a reflection of the low range of habitat types occupied by species of limited distribution.

3. *Neoendemism in the Canaries*

A number of taxa, mostly of infraspecific rank, for example *Viola odorata* var. *maderensis* Webb, *Frankenia laevis* subsp. *capitata* Webb & Berth., *Rumex bucephalophorus* subsp. *canariensis* Rech. fil., *Orchis patens* subsp. *canariensis* Lindl., *Urginea maritima* var. *hesperia* (Webb) Svent., *Romulea columnae* var. *grandiscapa* Gay, *Erica scoparia* subsp. *platycodon* Webb & Berth. and many others, have very close relatives amongst the species of the modern Mediterranean flora. In many cases the chromosome numbers have not yet been determined but where they have, they are usually the same as their relatives and these are considered to be recent schizoendemics (Favarger and Contandriopoulos, 1961).

In several other species where the chromosome numbers are known there is evidence of cryptic polyploidy. *Silene vulgaris* (Moench) Garcke which is normally diploid in Europe is tetraploid in the Canaries; *Asphodelus fistulosus* L. is tetraploid in Europe and octoploid in the Canaries and *A. microcarpus* Viv. is octoploid in Europe but dodecaploid on Gran Canaria. These taxa must be considered as perhaps the first stages in the formation of polyploid neoendemics.

SUMMARY

In this paper several aspects of endemism in the Canarian flora are considered. Three types of endemics are distinguished, neoendemics, palaeoendemics and active epibiotics. The endemic flora of the Canary Islands is, for the most part, considered to be a very old one and evidence for this, drawn from the fields of taxonomy, cytology, phytogeography, palaeobotany and habit and growth-form studies is reviewed.

The major processes of evolution in the Canaries are shown to be adaptive radiation and vicariance, and several examples of these processes are described. Adaptive radiation, previously known from the genus *Aeonium*, is also demonstrated in *Sonchus*; while vicariant evolution is found in the genera *Centaurea*, *Crambe* and *Sonchus*. These genera are considered to be active epibiotics.

Some examples of neoendemics in the Canarian flora are also given.

REFERENCES

Borgen, L. (1969). *Nytt. Mag. Bot.* **16**, 81–121.

Bramwell, D. (1970). *Bot. Not.* **123**, 394–400.

Cain, S. A. (1944). "Foundations of Plant Geography". New York.

Carlquist, S. (1965). "Island Life: a Natural History of Islands of the World". New York.

Ciferri, R. (1962). *Ric. Sci.* **32**, 111–134.

Dansereau, P. (1968). *Coll. bot.* **71**, 227–280.

Depape, G. (1922). *Annls Sci. Nat.* (*a*) *Bot.* 10e, sér. **4**, 73–265.

Engler, A. (1882). "Versuch einer Entwicklungsgeschichte der Pflanzenwelt", Vol. 2. Leipzig.

FAVARGER, C. (1964). *C.r. somm. Séanc. Soc. biogéogr.* **357**, 23–44.
FAVARGER, C. (1967). *Biol. Rev.* **42**, 163–206.
FAVARGER, C. and CONTANDRIOPOULOS, J. (1961). *Bull. Soc. bot. Suisse* **71**, 384–408.
FONT QUER, P. (1957). *Anuario de Estudios Atlanticos* **3**, 40–50.
GOOD, R. (1964). "The Geography of the Flowering Plants". Ed. 3. London.
HEDBERG, O. (1961). *Svensk Bot. Tidskr.* **55**, 17–62.
HU, S. Y. (1960). *J. Arnold Arbor.* **41**, 164–190.
LARSEN, K. (1960). *Biol. Skr.* **11**, **3**, 1–60.
LEMS, K. (1960a). *Ecology* **41**, 1–17.
LEMS, K. (1960b). *Sarracenia* **5**, 1–94.
MEUSEL, H. (1952). *Flora, Jena* **139**, 333–393.
OBERDORFER, E. (1965). *Beitr. Naturk. Forsch. Sudw. Dtl.* **24**, 47–104.
RIVAS GODAY, S. (1946). *Anales del Instituto Jose Celestino Mutis de Farmacognosia* **V** (9), 123–154.
ROTHE, P. and SCHMINCKE, H. U. (1968). *Nature, Lond.* 281, 1152–1154.
SCHENCK, H. (1907). "Beiträge zur Kenntniss der Vegetation der Kanarischen Inseln". Wissensch. der Deutschen Tiefsee-Expedition auf dem Dampfer Valdivia Bd. 2, 1 Teil, 2 Leiferung: 225–406 and 12 pl.
TAKHTAJAN, A. (1969). "Flowering Plants: Origin and Dispersal". Edinburgh.
TARDIEU-BLOT, M. (1946). *In* "Contribution a L'Etude du Peuplement des Iles Atlantides". *Mem. Soc. Biogéogr.* **8**, 325–348.
TAVARES, C. N. (1965). *Revta Fac. Cienc. Univ. Lisboa.* sér. 2, **13**, 51–174.
WULFF, E. V. (1950). "An Introduction to Historical Plant Geography". Massachusetts, U.S.A.

10 | The Relict Element of the Flora of Crete and Its Evolutionary Significance

WERNER GREUTER

Conservatoire botanique, Genève, Switzerland

INTRODUCTION

Ever since Darwin's well-known studies on the Galapagos fauna, islands have constituted a challenge to the evolutionary biologist. The fact that they house finite, quite independent biological systems makes them particularly suitable for a study of evolution and speciation under relatively simple, almost experimental conditions. In several cases, it has been confirmed that the island condition may be extremely favourable to evolutionary change. The peculiar ecological conditions of islands, and the multitude of ecological niches not yet occupied by the rare, lucky immigrants, cause adaptive shifts and adaptive radiation to take place, thus creating a wealth of new, endemic taxa. Remote islands or island groups constitute, in a way, most effective evolutionary laboratories.

This is not true, however, in every case. Several islands, especially among the larger ones which were once connected with continental areas, house what one might call "subcontinental biosystems", not true "island biosystems". The faunas and floras of such islands, which were formerly part of a continent, have evolved to their actual state from a complete, balanced, continental fauna and flora, through a gradual impoverishment by the extinction of species, counterbalanced to some extent by the establishment of occasional newcomers.

The flora of Crete is of this subcontinental type. Our present knowledge of the geological history of this area, although greatly improved during the last few years, is still rather fragmentary. There is no doubt, however, about the fact that Crete was part of a continental area in mid-Tertiary times; the mountains of present-day Crete are the remnants of an old mountain system connecting the Balkans with southern Anatolia. Nevertheless, it is reasonably certain that Crete remained isolated from the continents during the whole of the Pliocene and Pleistocene. Recent investigation of the neogene sediments (Meulenkamp, 1971) makes it probable that this isolation began in the mid-Tortonian (upper Miocene) and has lasted ever since, i.e., according to the most recent attempts in absolute dating, for about ten million years. Even in macroevolutionary terms this is an appreciable span of time (it can be evaluated as representing 5–10% of the life span of the whole of the angiosperms). One should thus expect that the Cretan flora has evolved considerably since it was isolated.

In fact, according to present knowledge, 132 species of vascular plants, corresponding to some 9% of the wild flora, are Cretan endemics (Greuter, 1971b). If the infraspecific taxa are also considered, the percentage of endemism is even higher; but it is impossible to give reliable figures, since the knowledge of infraspecific variation in Aegean phanerogams is still extremely scanty and uneven. However, endemism is not necessarily to be equated with evolution. Some of the present endemics might be remnants of the old continental flora which have not evolved appreciably since mid-Tortonian times. In the terminology of Favarger and Contandriopoulos (1961), they would correspond to the palaeo-endemic element (if they became extinct outside Crete), or to the patro-endemic element (if they evolved elsewhere and remained more or less unchanged in Crete).

ECOLOGICAL DIFFERENTIATION

The most obvious cases of evolution within the Cretan flora are those where two ecologically vicarious taxa coexist on the island. Most of these cases arise from the differentiation of mountain ecotypes from lowland species. This phenomenon has been discussed in some detail by Rechinger (1947). A list of examples, which may not be quite exhaustive, is given in Table I. In several cases the mountain ecotypes have been described as different species; but some may represent mere modifications due to the environment, as suggested by Rechinger (1943b) for *Crepis mungieri*, and most of them deserve, at any rate, only subspecific or varietal rank. There are also several Cretan mountain ecotypes that have not yet been recognized taxonomically, although they differ to some extent from the lowland populations.

A series of other species, however, goes through from sea-level right to the mountain tops without being modified appreciably. A list of such species, with maximum altitudinal records, is given in Table II. (Since

TABLE I. Vicarious lowland and mountain taxa in the Cretan flora
(Binomials are used wherever they exist, irrespective of the appropriate status of the taxa)

Lowland taxa	Mountain taxa	References
Paronychia macrosepala Boiss.	*P. macrosepala* var.	Greuter (1965)
Hypericum empetrifolium Willd.	*H. empetrifolium* var. *tortuosum* Rech. f.	Rechinger (1943b)
Linum arboreum L.	*L. caespitosum* Sm.	
Sedum creticum Boiss. & Heldr.	*S. hierapetrae* Rech. f.	
Trifolium uniflorum L.	*T. uniflorum* var. *breviflorum* Boiss.	
Erica manipuliflora Salisb.	*E. manipuliflora* var.	
Teucrium gossypinum Rech. f.	*T. alpestre* Sm.	Strid (1965)
Scutellaria sieberi Bentham	*S. hirta* Sm.	
Phlomis lanata Willd.	*Ph. lanata* var. *biflora* Hal.	
Satureja thymbra L.	*S. biroi* Jav.	Rechinger (1943b)
Coridothymus capitatus (L.) Reichenb. f.	*C. capitatus* var. *albospinosus* (Bald.) Rech. f.	
Asperula incana Sm.	*A. incana* var.	
Galium heldreichii Hal.	*G. samothracicum* Rech. f.	
Galium amorginum Hal.	*G. incurvum* Sm.	
Bellis sylvestris Cyr.	*B. longifolia* Boiss. & Heldr.	
Tragopogon sinuatus Avé-Lall.	*T. lassithicus* Rech. f.	Rechinger (1943b)
Taraxacum sect. *Scariosa* Hand.-Mazz. (2 taxa)	*T.* sect. *Scariosa* (2 taxa)	Richards (personal communication)
Lactuca alpestris var.	*L. alpestris* (Gand.) Rech. f.	
Crepis fraasii Schulte Bip.	*C. mungieri* Boiss. & Heldr.	Rechinger (1943a, b)
Colchicum pusillum Sieber	*C. cretense* Greuter	Greuter (1967)
"*Muscari cousturieri*" Gand., "*M. creticum*" Vierh.	*Leopoldia spreitzenhoferi* Heldr. (= *M. amoenocomum* Rech. f.)	Bentzer (ined.)
Crocus cartwrightianus Herb.	*C. oreocreticus* Burtt	Rechinger (1949)
Dactylis hispanica Roth	*D. rigida* Boiss. & Heldr.	Borrill (1961)

TABLE II. Examples of lowland species with wide altitudinal ranges. Maximum altitudinal records (in metres) are indicated in brackets

Species	Altitude	Species	Altitude
Euphorbia acanthothamnos Boiss.	(2200)	*Centaurea raphanina* Sm.	(2150)
Andrachne telephioides L.	(2300)	*Centaurea idaea* Boiss. & Heldr.	(2150)
Onosma erecta Sm.	(2000)	*Poa bulbosa* L.	(2100)
Verbascum spinosum L.	(2100)	*Arum creticum* Boiss. & Heldr.	(2100)
Asperula rigida Sm.	(1800)		

altitudinal records for Cretan plants are still scarce, one can assume that the maxima indicated are, in most cases, understatements.) Besides, a surprisingly large number of lowland species reach the altitudinal belt between 1000 and 1500 m, where they meet mountain species descending into the deforested areas.

The conditions under which the colonization of the Cretan mountains by lowland plants took place can be roughly reconstructed as follows.

(a) Forested areas severely restricted by drought. (Without the influence of man, the present climax vegetation of Crete, from sea level to timberline, would be forest, and the light-loving species of our lists would be restricted, in the lowland, to occasional small open areas.)

(b) The presence of mountains considerably exceeding the timberline.

(c) An old mountain flora greatly impoverished by the previous extinction of many species.

Effectively, the Cretan mountain flora contains a rather small number of old relict species, mainly endemic or with a very disjunct distribution on the eastern Mediterranean mountains. The remainder consists of the derivatives of lowland species (Tables I and II) and of taxa which, being identical with those of the nearby continental Greek mountains, may have arrived by long-distance dispersal. In comparable continental areas of Greece, the differentiation of mountain ecotypes from lowland taxa is a much less common feature, and the altitudinal ranges of lowland plants are, on the whole, more limited, while the old relict mountain flora is correspondingly richer. It would appear that, if mountains have been a constant feature of Crete throughout the period of isolation, they may not necessarily have always been as high as they are now. It is probable, on the contrary, that the treeless summit areas were at times very much reduced. The present situation arose, according to the geologists, in the course of the Pleistocene.

It is interesting to note that, in the examples of Table I, differentiation is always directed from the lowland to the mountain taxa. The former are either found outside Crete, and are often rather widespread, or, if they are endemic (as in the cases of *Teucrium*, *Scutellaria*, *Phlomis*, *Asperula*, and maybe *Taraxacum* and *Leopoldia*), they are represented elsewhere by vicarious lowland taxa. The mountain ecotypes are all endemic to Crete, with two exceptions (*Satureja biroi* and *Crepis mungieri*) where similar forms have evolved, independently as it seems, on the nearby island of Karpathos. Parallel evolution of apparently identical mountain ecotypes on the different, isolated mountain groups of Crete seems, by the way, to have happened in several instances. This supports the view that evolution from lowland to mountain ecotypes is strongly adaptive, following almost identical lines under given climatic and ecological conditions.

There are some other examples of ecologically vicarious, closely related pairs of taxa on Crete, but they are less numerous and less easy to interpret. Most of them concern littoral ecotypes (Table III). In view of their specialized habitat, which is subject to frequent changes and will often present unoccupied ecological niches, one would expect that the littoral taxa would arise, as a rule, from the corresponding inland taxa. However, matters may not be quite as simple as that. None of the littoral taxa is endemic to Crete; this could partly be explained, of course, by long-distance dispersal, a rather common feature with shore plants. The taxa concerned are at least partly allopatric, and so might be their origin; they might have been formed primarily as a result of geographical, not of ecological differentiation. The correct interpretation is further complicated by the fact that some of the corresponding inland taxa (*Arenaria leptoclados*, *Silene behen*) behave in our area like anthropophytes, so that it is impossible to reconstruct with any degree of certainty their former natural distribution.

In view of the great number of species which are members both of the dwarf scrub communities called "phrygana" and of the chasmophytic communities of calcareous cliffs, it is rather surprising that in practically no cases can a corresponding morphological differentiation be observed. Only two exceptions to this rule are known to me. The chasmophytic Aegean species *Linum arboreum* has given rise on Crete to a rare endemic phrygana taxon inhabiting schist or flysch, which has been described as *Linum doerfleri* Rech.f.; and the widespread phrygana species *Hypericum empetrifolium* corresponds, in eastern Cretan chasmophyte communities, to the endemic *Hypericum amblycalyx* Coust. & Gand. This second case, however, merits further investigation; in spite of appearances, the

TABLE III. Vicarious inland and littoral taxa in the Cretan flora
(Binomials are used wherever they exist, irrespective of the status of the taxa)

Inland taxa	Littoral taxa	References
Paronychia macrosepala Boiss.	*P. insularum* Gand.	Greuter (1965)
Arenaria leptoclados (Reichenb.) Guss.	*A. aegaea* Rech. f.	Runemark (1969)
Silene behen L.	*S. holzmannii* Boiss.	
Scaligeria napiformis (Sprengel) Grande	*S. halophila* (Rech. f.) Rech. f.	Rechinger (1965)
Convolvulus oleifolius Desr.	*C. oleifolius* var. *scopulorum* Rech. f.	Rechinger (1943b)
Filago aegaea Wag. subsp. *aristata* Wag.	*F. aegaea* subsp. *aegaea*	Wagenitz (1970)
Helichrysum barrelieri (Ten.) Greuter	*H. minoum* Gand.	
Anthemis tomentella Greuter	*A. ammanthus* Greuter	Greuter (1968)
Echinops bithynicus Boiss.	*E. viscosus* DC.	Davis (1953)
Allium bourgeaui Rech. f.	*A. bimetrale* Gand.	von Bothmer (1970)
Lolium loliaceum (Bory & Chaub.) Hand.-Mazz.	*L. crassiculme* Rech. f.	Greuter and Rechinger (1967)

chasmophytic taxon might be more than a comparatively young descendant of *Hypericum empetrifolium*.

If we try to summarize the ecological differentiation which took place, within Crete, subsequent to its isolation from the continents, the following picture emerges.

(a) A strong trend towards the differentiation of mountain ecotypes from lowland taxa can be observed. The differentiation is usually weak, and in some cases its genetic fixation remains to be proved. It is paralleled by the conquest of the summit areas by several scarcely differentiated lowland species, and probably, through long-range migration, by taxa from the nearby continental mountain areas. As a reason for this multiple invasion of the Cretan mountains by foreign elements, the great impoverishment, through past vicissitudes, of the old, original mountain flora can be postulated.

(b) Littoral ecotypes have probably also differentiated from inland taxa. Here, however, the picture is less clear. In several instances, the observed differentiation may be primarily geographical and may not have taken place on Crete itself; in some cases, it may be older than the isolation of the area.

(c) There has been very little ecotypic differentiation between plants of different treeless lowland habitats (in particular, between constituents of "phrygana" or "Felstrift" and chasmophytes). The same is true, to an even greater degree, between open habitats and forests.

(d) Cases which can be considered as "adaptive radiation" are exceedingly few. One could mention here *Paronychia macrosepala*, a pioneer on open ground in the lowlands, which has produced weakly differentiated mountain and littoral ecotypes; *Linum arboreum*, a chasmophyte, with two rather clear cut derivatives in the alpine zone and on schist or flysch; and *Hypericum empetrifolium* (if this corresponds indeed to the ancestral form) with parallel taxa on the mountains and on calcareous cliffs. In these three cases, the presumed original taxon, which is not endemic to Crete, is still present on the island.

GEOGRAPHICAL DIFFERENTIATION

Patterns of differentiation linked to geographical distribution are a common feature in the whole Mediterranean area. Rechinger (1947) has discussed them in detail for the Aegean. It is generally thought that this type of differentiation is due to fragmentation of the areas of the original species and that it took place under the conditions of geographical isolation. The evolution of the isolated populations would thus correspond to an adaptation to the particular environmental conditions of each isolated area; alternatively it would consist of a random shift of gene frequencies (especially for neutral, non-adaptive alleles), as might be expected according to mathematical models in finite, isolated populations. These two modes of differentiation combined would make it certain that the isolated populations, after an adequate lapse of time, would show the morphological individuality observed today.

Isolation barriers on continents are usually rather short-lived and may change in space and time. In insular areas, however, where the open sea is the isolating factor, they are more constant and, if they change, this implies modifications of the palaeogeography which are usually detectable. As we have already seen, in the case of Crete the isolation from the continent (and from the other Aegean islands) seems, according to present geological and biogeographical knowledge, to have lasted without interruption since the mid-Tortonian. Moreover, during the whole Pliocene era, i.e. for several million years, Crete itself was subdivided into a number of independent islands, the most important of which coincided roughly with the present highlands. If the above views about geographical differentiation are correct, then it is to be expected that the great majority, if not all, of the plants surviving from pre-isolation times will be represented on Crete by taxa endemic to the island, or even to the smaller territories corresponding to the remnant islands of the Pliocene.

There are of course many species and subspecies endemic to Crete; but the non-endemic, still evidently relict taxa are much more numerous. We saw that of the 23 examples of Table I, which must be fairly old citizens of Crete since they have given rise to special mountain ecotypes, 17 are non-endemic; and so are more than 90% of the taxa of the present wild flora. Some of them are widespread weedy species; but the majority have a limited distribution and fit into patterns of geographical differentiation which look quite natural, except for the fact that the meshes of these patterns do not coincide with the geologically established areas of old isolation.

There are, for instance, 23 species which extend beyond Crete to the east or the west, but which are still endemic to the South Aegean area. Their boundaries thus cross the very old isolation lines between western Crete and Andikithira, or between eastern Crete and the Karpathos group. Yet in only two cases have subspecies corresponding to these separations been differentiated. The western Cretan *Campanula saxatilis* L. is represented on Kithira and Andikithira by subsp. *cytherea* Rech.f. & Phitos (Phitos, 1965); and the eastern Cretan *Silene ammophila* Boiss. & Heldr. is represented on Karpathos by subsp. *carpathae* Chowdhuri (Chowdhuri, 1957). The remaining 21 species are apparently uniform throughout their range.

Twenty-one species are endemic to the area which I have called (Greuter, 1971b) "Cardaegean", and cross the old separation line between Crete and the Cyclades. In not a single one of these cases has it been possible to attribute the Cretan and Cycladean populations to different subspecies. Thus, *Anthemis ammanthus* subsp. *ammanthus* is found both on the south-eastern Cyclades and in north-easternmost Crete (and on the Karpathos group); while in adjacent areas of Crete, in spite of the obvious lack of isolating barriers, there is a different taxon, subsp. *paleacea* Greuter (Greuter, 1968).

If we try to draw some conclusions from the observed patterns of geographical differentiation, we can state the following:

(a) Several, but still comparatively few taxa are endemic to the old isolated areas (Crete or one of its Pliocene subdivisions). One can suppose that these taxa have originated in their present area through differentiation under the conditions of isolation. This cannot, however, be proved; the probability that taxa with restricted areas already existed in the Miocene, and have been since preserved without appreciable evolutionary change, cannot be discarded.

(b) A great number of uniform, old taxa extends beyond the isolated area of Crete. In all these cases, the isolation, in spite of its constancy and duration, has not produced enough differentiation to warrant the description of distinct taxa. Many of these plants do not look, however, like monomorphic, invariable types which have "lost" their faculty to evolve. They show a distinctive geographical pattern of variation which seems to be partly independent of the known present and past isolation barriers. They look, in fact, as if they had differentiated prior to the splitting up of their territory and had since been conserved, virtually fossilized, in their isolated, restricted habitats. This puzzling and rather unorthodox possible explanation will be discussed later on.

MOSAIC VARIATION IN SMALL-POPULATION SYSTEMS

Small-population systems are a rather common and very characteristic feature in the Aegean area. They are not only met with in the numerous groups of small islands of the Cyclades. They are also found on greater land-masses like Crete, if the species grow in sufficiently specialized, isolated and long-established habitats; this means, in particular, the vertical calcareous cliffs scattered throughout the area, and the plants in question are called chasmophytes.

The seemingly random polymorphism of these plants has been a challenge to many botanists; their adequate systematic understanding and treatment sets very serious problems, which have been solved, in a few cases, with the help of biosystematic methods based on abundant material. There are two pilot studies in this field, both centred on the Cyclades but including Crete; these are the monograph of *Erysimum* sect. *Cheiranthus* by Snogerup (1967a, b) and the revision of Aegean *Nigella* by Strid (1970). In both cases, an underlying geographical variation could be traced, which was more or less concealed by a striking random polymorphism. The geographical variation, which can be treated in an orthodox way by describing species and subspecies, is certainly older than the splitting up into small populations; from the distribution maps given, it appears that it is even older, at least in some cases, than the isolation of the areas by the sea. This confirms fully what we have stated above.

The individual small populations are, as a rule, quite invariable and each presents a peculiar combination of characters. Theoretically it is possible to give them recognition as separate taxa, since they represent homogeneous monophyletic units. This point has recently been thoroughly discussed (Runemark, 1971); it was realized that where the populations were numerous, such distinctions were of little practical value, and that they were problematical if some of the populations, especially the large continental ones, showed a high degree of intrinsic variability.

One of the most interesting problems of small-population systems is the role which they may play in evolution. It has been said that differentiation processes are more rapid in such systems than in medium and large-sized populations. If this were true, it would help to explain why much of the geographic differentiation observed in nature is based, apparently, on non-adaptive characters; the random processes of genetic drift can be invoked as the causes of the evolution of small populations.

Unfortunately, I cannot believe in the effectiveness of such a system, which is contradicted both by the observed facts and the mathematical probability. It seems to me that "population polymorphism" is just the phenotypic expression, in the homozygous state, of the normal genetic variability of a large parental taxon split up into a mosaic of micro-populations. Actual evolution of the latter is apparently lacking altogether; not even the faint, very old original geographic differentiation has yet been obscured.

Mathematically, we can take the relatively easy example of an adaptively neutral mutation arising anew in a finite, stable population and being finally fixed. The probability for such an event is independent of the size of the population (in small populations, the higher probability of fixation is counterbalanced by the smaller probability of new mutations). It depends only on the mutation rate, to which it is proportional. If the latter is appreciably high, then we expect to find the corresponding allele already in the big parental population; if it is extremely low, then the probability of fixation, even in a small-population system, is practically nil.

The conclusion from this discussion is that small-population systems, far from presenting opportunities for rapid, effective evolution, produce a high degree of stability over long periods of time; they are among the most instructive examples of non-evolving complexes.

POSSIBLE INTERPRETATIONS

It should be realized that, while areas and variation patterns are objective more or less well-known facts, their interpretations are, at best, good guesses. If we are trying to discuss possible interpretations and the extent to which they can explain our observations, we must always be aware of the high degree of uncertainty prevailing in all these questions, and of the tremendous gaps in our knowledge of many background phenomena, especially for past eras. We may eventually come to the conclusion that some of the possible schemes fit best with what we know about the facts; and we may then suggest that these schemes be accepted as working hypotheses and tested against other sets of facts. Whatever we may conclude must not be taken as an apodictic statement of belief, but as a suggested basis for further research.

One of the possible theories, to which I have already alluded, is that geographic differentiation takes place previous to the isolation; subsequently, the isolated populations do not evolve any more (except for the cases of ecological differentiation). The isolation will tend to preserve the established geographical patterns, preventing gene-exchange with nearby vicarious taxa, and will protect many plants from being eliminated by expanding, more competitive forms. The richer geographical differentiation and endemism which one observes in areas with pronounced isolating barriers would thus be due to conservation rather than to creative evolution.

The acceptance of such a theory involves two main difficulties which will be discussed below. The first is the need to postulate a prolonged evolutionary standstill (in the case of Crete it would have to last about ten million years), in spite of obvious and considerable changes in climatic and other environmental conditions. The second is the difficulty of explaining how the observed differentiation could have taken place; that is, what apparently exceptional conditions must be postulated, in the pre-isolation phase, to allow for an evolutionary change.

Before we go into these matters we must ask, however, if there are no other possible schemes of explanation. One suggestion would be that the period of isolation might have been very much shorter than we are now thinking. It cannot be denied that our present knowledge of Cretan palaeogeography is still very fragmentary and sometimes subject to doubt. The results of different approaches, as synthesized in a recent multidisciplinary symposium (Strid, 1971), agree however as to the early date of the isolation of Crete.

A second possibility would be to make human action responsible for the present distribution patterns, especially for those areas crossing old lines of isolation. Most of the members of the Cretan flora not endemic to the island would thus have been imported there by man, or else be former endemics with artificially widened areas. The human influence on the composition of the flora has admittedly often been underrated; I have estimated that about one third of the wild flora of Crete has been introduced by man (Greuter, 1971a). It is obvious that no sensible evolutionary theory can be built on the present distribution of weedy annuals! The detailed data underlying the present discussion have been drawn, however, from carefully chosen species whose relict nature, in Crete and elsewhere in their area, is beyond any doubt.

Natural long-distance dispersal is a much more serious hypothesis. The fossil plant records are extremely scanty, and the great majority of the apparently relict species with which we are dealing grow in dry habitats not suitable for fossilization. It is impossible, therefore, to prove the presence on Crete, at any time in the past, of a given species. Furthermore, we have already admitted the possibility of long-distance dispersal for a part of the Cretan mountain flora.

Runemark (1969) has dealt with the problems of the long-range migration of plants. He comes to the conclusion that the establishment of a newcomer is an extremely unlikely event in areas with a closed, reasonably complete and balanced vegetation. We have presumed that the Cretan mountain flora was greatly impoverished at some time in the past, so that immigration over long distances (and from the surrounding lowlands as well) became possible. We have not the slightest evidence, however, for a parallel impoverishment of the lowland flora. The present relict communities are rich in species and well-balanced, and it is hard to believe that part or all of them could have settled there by chance long-distance migration. A further clue is the almost total absence of phenomena comparable to "adaptive radiation", which we should expect as a corollary of an unbalanced impoverished flora. To adopt a very prudent attitude, if long-distance migration ever occurred to an appreciable extent in the Cretan lowland relict communities, it must have been very long ago and cannot, at any rate, solve the problem of the non-evolution of many species.

A fourth theory, that of parallel evolution in Crete and elsewhere, under similarly changing environmental conditions, need not be seriously discussed. These conditions and changes were certainly far from being uniform and parallel within the big area concerned; and many of our relict taxa are clearly rather ill-adapted to their present environment.

Since we have thus failed to find a satisfactory alternative to the postulate of non-evolution under the conditions of isolation, we shall now try to discuss this last possibility.

NON-EVOLUTION OF "STATIC POPULATIONS"

Let us define a "static population" as a population which is prevented from invading sizeable new areas and from colonizing new habitats, but which may fluctuate considerably as to its size and move rather freely within a circumscribed territory. Then we have, I think, a portrait of the essential features of the non-evolving Cretan relict taxa.

Animals of such "static populations" are known to evolve. The deer, elephant and hippopotamus of the Cretan Pleistocene all formed dwarf endemic island taxa; nevertheless, with further changes of the climate, they disappeared, and were later replaced by the newly immigrating species of the modern mammal fauna. Is it reasonable, then, to think that plants did not adapt themselves to climatic changes, and yet were able to survive when the adapting animals were eliminated?

I think that there are at least two basic differences, in this context, between higher plants and animals. The first is the enormous ecological tolerance of the former, which contrasts strikingly with the well-defined, often very special habitat conditions under which they are actually growing. For example, *Dianthus creticus* Tausch and many other Cretan plants which appear to be highly specialized to the habitats of vertical calcareous cliffs, do very well in British gardens if protected from competing species. Ecologists have long been aware of the fact that competition is by far the most effective factor limiting the distribution of plants.

In a closed biosystem like that of an island, climatic changes will have a rather limited influence on the plant cover. They will cause vegetation belts and plant communities to shift about, and some of these will eventually be eliminated. But for the remaining plants, the basic ecological factor, i.e. the surrounding competing species, will remain practically the same and present no stimulus to fresh adaptation.

Now there is no doubt that the strongest competitors, for plants, must be other plants of the same species; they have exactly the same exigencies and are always certain to grow in the immediate surroundings. Since plants are apparently unable to develop social mechanisms—and here I see the second main difference from animals—they will be permanently exposed to a very high selection pressure by conspecific individuals. It is generally admitted that permanent strong selection pressure prevents genetic rearrangements linked with evolutionary change. This makes it rather plausible, indeed, that plants in static populations are subject to an evolutionary standstill.

EVOLUTIONARY PROCESSES IN "DYNAMIC POPULATIONS"

We must now look for the conditions under which evolution can actually take place. Let us consider in this respect the contrary of our "static population", which we shall call a "dynamic population". This is one which is either colonizing new habitats or invading sizeable new areas.

In the first case, i.e., in new habitats, we expect that ecological differentiation will take place, a strongly adaptive process which has already received much attention and need not be discussed any more in this context. In the second case, the newly invaded territories, we shall try to show how geographical differentiation may occur.

An invading species may behave in very different ways, and the process of invasion may be extremely complex. I will limit my considerations to a single model which may well match the actual processes which took place in the Greek area, at some time in the Miocene. My example will deal with "steppe" plants, i.e. plants of dry, open, unforested areas, since a majority of the Cretan relict species belong to this type.

Let us imagine that a period of drought sets in and that the hitherto prevailing forest climax is replaced by a steppe climax. Huge areas of land will be freed rather suddenly, mainly by fire action, from their plant cover and will await new occupants. The occupants come indeed, starting from the former boundaries of the forest area or from local treeless enclaves, and covering long distances at a single step. This, in my opinion, is a situation extremely favourable to differentiation, and for the following reasons:

(a) The establishment of numerous new, independent colonies, and then in their turn daughter colonies, produces a multiple "founder effect". Gene frequency distribution will be different from one colony to another, and many new gene combinations can be tried out. Eventually some colonies will meet and give rise to a still greater variety of descendants.

(b) Selective pressure both within and between species will be greatly reduced. The colonies, at least in the initial period, can expand freely, and all individuals will have approximately equal chances of multiplication.

(c) Selection will operate along totally new lines. Many features which may be absolutely useless later on may have a great adaptive value in the colonizing stage. This offers an explanation for many apparently non-adaptive characters of present-day relict species, as for example devices for anemochory in local island endemics.

(d) A final stabilization phase, which will lead to the establishment of balanced plant communities, will imply severe selection with elimination of many populations and unsuitable gene combinations. This will represent the truly adaptive phase of the whole differentiation process.

This picture is still very sketchy and oversimplified in many respects. Several features not discussed here might have a great deal of influence on the presumed differentiation processes. The size of a colonizable area and the speed with which it is freed for colonization must be important factors, especially if compared with the individual size and duration of the colonizing plants. It may be significant, in this context, that the most salient examples of geographical differentiation within the Greek flora are found among perennial herbs and small shrubs, with a mean generation time ranging from a few years to a few decades.

Another important factor is hybridization, which is likely to occur if discrete yet interfertile taxa invade a new area simultaneously from different starting points. An alternation of colonizing and recessive phases would presumably afford ideal conditions for repeated hybridization and segregation, and might lead to almost explosive speciation. Such could be the essential cause of the very rich differentiation of many genera in areas like Anatolia and the Iranian highlands; a shifting balance between steppe and forest communities may have led, through the climatic fluctuations of the past, to a frequent alternation of steppe and forest climax situations. Similarly, it is probable that in the Cyclades, where the separating sea is often rather shallow, repeated fusion and isolation of islands and island groups has occurred, mainly as a result of the eustatic sea level fluctuations of the Pleistocene. In fact, the investigations of both Snogerup (1967) and Strid (1970) showed a higher degree of differentiation in the central Aegean area than in the comparatively stable south Aegean area.

I tend to believe that the present Greek flora can be largely explained by the succession of a few short dynamic phases involving differentiation, and of long static phases in which isolation (which may not be of the insular type) predominated. The extension of analogous considerations to other areas and plant groups is certainly tempting. It should be stressed once again, however, that geographical differentiation models such as those proposed here cannot be applied universally! In particular, it is evident that in many cases (not only on oceanic islands) ecological differentiation is by far the most important process. Still, even here one might suspect that relatively short dynamic phases have alternated, in many cases, with longer phases of conservation. This is corroborated, to a certain extent, by fossil evidence.

SUMMARY

Ecological and geographical differentiation, including mosaic variation in small-population systems, is discussed with reference to the phanerogamic flora of Crete. Apparently a great number of species have undergone no evolutionary change since the mid-Tortonian, when the island was isolated from the continents. As a suitable hypothesis to explain this situation, it is suggested that "static" populations are subject to an evolutionary standstill, and that the observed (mainly geographical) differentiation is due to relatively short "dynamic" colonization phases.

REFERENCES

Borrill, M. (1961). *J. Linn. Soc. (Bot.)* **58**, 87–93.

Bothmer, R. von (1970). *Bot. Not.* **123**, 519–551.

Chowdhuri, P. K. (1957). *Notes R. Bot. Gdn Edinburgh* **22**, 221–278.

Davis, P. H. (1953). *Notes R. Bot. Gdn. Edinb.* **21**, 101–142.

Favarger, C. and Contandriopoulos, J. (1961). *Ber. schweiz. bot. Ges.* **71**, 384–408.

Greuter, W. (1965). *Candollea* **20**, 167–218.

Greuter, W. (1967). *Candollea* **22**, 233–253.

Greuter, W. (1968). *Candollea* **23**, 143–150.

Greuter, W. (1971a). *Boissiera* **19**, 329–337.

Greuter, W. (1971b). *Op. bot. Soc. bot. Lund.* **30**, 49–64.

Greuter, W. and Rechinger, K. H. (1967). *Boissiera* **13**, 1–206.

Meulenkamp, J. E. (1971). *Op. bot. Soc. bot. Lund.* **30**, 5–12.

Phitos, D. (1965). *Öst. Bot. Z.* **112**, 449–498.

Rechinger, K. H. (1943a). *Akad. Wiss. Wien Math.-Naturwiss. Kl. Denkschr.* **105/1**, 1–924.

Rechinger, K. H. (1943b). *Akad. Wiss. Wien Math.-Naturwiss. Kl. Denkschr.* **105/2/1**, 1–184.

Rechinger, K. H. (1947). *Öst. Bot. Z.* **94**, 152–234.

Rechinger, K. H. (1949). *Phyton (Horn)* **1**, 194–228.

Rechinger, K. H. (1965). *Öst. Bot. Z.* **112**, 186–187.

Runemark, H. (1969). *Bot. Not.* **122**, 90–129.

Runemark, H. (1971). *Boissiera* **19**, 169–179.

Snogerup, S. (1967a). *Op. bot. Soc. bot. Lund.* **13**, 1–70.

Snogerup, S. (1967b). *Op. bot. Soc. bot. Lund.* **14**, 1–86.

Strid, A. (1965). *Bot. Not.* **118**, 104–122.

Strid, A. (1970). *Op. bot. Soc. bot. Lund.* **28**, 1–169.

Strid, A. (1971). "Evolution in the Aegean". *Op. bot. Soc. bot. Lund.* **30**, 1–83.

Wagenitz, G. (1970). *Willdenowia* **6**, 115–138.

11 | The Origin of Endemics in the California Flora

HARLAN LEWIS

University of California, Los Angeles, California, U.S.A

INTRODUCTION

The flora of California is very diverse with respect to vegetation types, composition of plant communities, and number of taxa of vascular plants (*ca* 7000; Munz and Keck, 1959). This diversity is associated with extreme differences in topography and climate, differences that frequently occur within very short distances. Ecological gradients are often very steep and the ecotones between contrasting vegetation types are in many instances very narrow.

Many species are endemic to California, a feature of the flora that has been recognized since the earliest studies. By conservative estimate about one-third of the indigenous species are endemic (Noldke and Howell, 1960) and many of these are very local in distribution; in some cases they are known only from one or two colonies. Some of the endemic species are known from fossil records to have had a much wider distribution in the Tertiary and are demonstrably relicts or palaeoendemics (e.g. *Sequoiadendron giganteum* (Lindl.) Buchh.) whereas many other species with no close relatives in the area can be assumed with confidence to be palaeoendemics even though no fossil record exists. A few species are known from cytogenetic and other evidence to be neoendemics of relatively recent origin and a number of other species with a small area of distribution are almost certainly neoendemics to judge from their ecological and morphological relationship to similar species in the same area. The number of neoendemics has made California an outstanding place for studies of evolution and speciation. In some instances one can see evolution and, with a little imagination, speciation taking place before his eyes.

In a recent comprehensive survey and review of endemism in California by Stebbins and Major (1965), the present distribution of apparent palaeoendemics, other relict species, and neoendemics were plotted in order to identify areas of high concentration. By comparing the climatic, topographic and edaphic attributes of areas differing in frequency and kinds of endemism they were able to identify those conditions that seem to be conducive to the persistence of relict species and those most favorable to speciation. Their conclusions are as follows: "In regions with ample moisture, as well as in regions with a marked deficiency of moisture, floras are likely to be relatively stable, and most of the endemic species are ancient or at least not recent. This is due to the fact that most of the climatic shifts which occur during geological periods of time and which affect moisture availability will keep the climate within the range of tolerance of the resident species, and will allow them to persist in the same area with little or no change in their genetic makeup. On the other hand, in regions on the borderline between zones of adequate moisture and of deficient moisture, even small climatic shifts will change local conditions beyond the limits of tolerance of the resident species, so that they must either migrate or evolve new ranges of tolerance. . . . In many instances both climatic shifts and species migrations will bring together related species which previously were isolated from each other, and so will promote hybridization between differently adapted types. Hybrid swarms derived from such crossings are particularly favorable gene pools from which new adaptive gene complexes may be constructed. In this way, climatic and edaphic diversity, occurring on ecotones or border regions between different biotic provinces are factors which most actively promote the evolution and differentiation of species of higher plants."

My own studies and knowledge of the flora and area lead me to agree in general with Stebbins and Major. The purpose of this paper, therefore, is not to take exception to their major conclusions but rather to support and extend them.

THE ORIGIN OF ENDEMICS FOLLOWING HYBRIDIZATION

1. *Speciation without Change in Chromosome Number*

Many California endemics are products of hybridization between morphologically and ecologically well-differentiated species, to judge from very good circumstantial evidence and the experimental demonstration that stable intermediate lines can be extracted from fertile, semi-fertile and even nearly sterile hybrid combinations.

Hybrid swarms associated with open or disturbed habitats created by human activities have become well known, especially through the attention given to them by Edgar Anderson. Few botanists seem to realize, however, how frequently in California one can find local hybrid swarms that are not associated with human disturbance. A few years ago, for example, I photographed in one afternoon three such hybrid swarms within less than a mile of one another on the east side of the Sierra Nevada (Lewis, 1969) in an area where several distinct vegetation types come into contact with one another (i.e. Subalpine Coniferous Forest, Alpine Fell-fields, Sagebrush Scrub, Subalpine Meadows, stream banks and lake shores). The genera involved were *Aquilegia*, *Castilleja*, and *Pentstemon*.

The significance of such hybridization in producing neoendemics at the diploid level in California has been admirably shown in *Pentstemon* by Straw (1955) who presented convincing evidence for the origin of *P. spectabilis* Thurb. from hybridization between *P. centranthifolius* Benth. and *P. grinnellii* Eastw. followed by the origin of *P. clevelandii* A. Gray from hybridization between *P. spectabilis* and one of its parents, *P. centranthifolius*. A parallel situation is found among the Californian species of *Delphinium* (Lewis and Epling, 1946, 1954, 1959) in which a number of endemic diploid taxa are undoubtedly of relatively recent hybrid origin to judge from morphology, ecology, geographic distribution and segregation of hybrid progenies. Depending largely on the sharpness of the morphological discontinuity between the hybrid derivatives and their putative parents, some of the derivatives have been designated subspecies (e.g. *D. parryi* A. Gray subsp. *seditiosum* (Jepson) Ewing from *D. parryi* × *D. hesperium* A. Gray; *D. parishii* A. Gray subsp. *subglobosum* (Wiggins) Lewis & Epling and *D. parishii* subsp. *purpureum* Lewis & Epling from *D. parishii* × *D. parryi*) whereas others have been designated species (e.g. *D. gypsophilum* Ewan from *D. hesperium* × *D. recurvatum* Greene, *D. umbraculorum* Lewis & Epling from *D. parryi* × *D. patens* Benth. and *D. inopinum* (Jepson) Lewis & Epling from *D. hansenii* Greene × *D. parishii*). As in *Pentstemon*, at least one diploid species of hybrid origin, *D. gypsophilum*, has in turn hybridized with one of its putative parents, *D. recurvatum*, to produce a large homogeneous population of morphologically and ecologically intermediate individuals (Lewis, unpub.). This derivative population is confined to a small area in the basin surrounding Soda Lake in eastern San Luis Obispo County. Although it has not been described, it unquestionably warrants taxonomic recognition as an endemic of very recent origin.

Most of these examples and others from California are included in a recent review by Stebbins (1969) in which he effectively points out the deleterious as well as the stimulating effects of hybridization in relation to the formation of neoendemics without a change in chromosome number. For this reason the few illustrations I have given are intended only to impress upon one the dynamic role that hybridization has played and is currently playing in California in the formation of endemics without change in chromosome number. As Stebbins (1969) and others have pointed out, however, hybridization is merely a precondition for reticulate evolution and whether or not new stable races or species are produced depends on the availability of a habitat or ecological niche, different from that of either parent, to which the derivative is adapted and where it will become stabilized and maintained by selection pressures different from those operating on the parental populations. Although a significant number of Californian species have undoubtedly had a hybrid origin, the number must be extremely small compared to the frequency with which local hybrid swarms have been formed.

A special kind of presumed hybrid derivative is represented in California by *Gayophytum heterozygum* Lewis & Szweyk., a complex heterozygote which like some of its relatives in the genus *Oenothera* regularly forms a ring of 14 chromosomes. On the basis of morphology, Lewis and Szweykowski (1964) suggested that *G. heterozygum* may have had its origin from hybridization between *G. eriospermum* Coville, which is endemic to the Sierra Nevada, and *G. oligospermum* Lewis & Szweyk., an endemic to the mountains of southern California. Since the distribution of the presumed hybrid derivative is more widespread than its putative parents in California and extends into Oregon and Washington, we can think of the parents as being comparable to diploid species that are more restricted in distribution than their polyploid derivatives, species which Favarger and Contandriopoulos (1961) called patroendemics. One of the putative parents, *G. eriospermum*, which is the only out-crossing diploid species in the genus, may be thought of as patroendemic not only on the basis of its relation to *G. heterozygum* but also to the widespread polyploid complex *G. diffusum* Torr. & Gray. The distribution of this patroendemic coincides very closely with that of the big tree, *Sequoiadendron giganteum*, and is often found in and near groves of that famous relict species. Of interest is that Stebbins and Major (1965), who were unaware of any endemic species of *Gayophytum* at the time, noted that the highest concentration of patroendemics in the Sierra Nevada is in an area that is almost coincident with the range of distribution of *Sequoiadendron*.

Whether or not *Gayophytum oligospermum*, the other species which Lewis and Szweykowski (1964) suggested as a parent, is also a relatively old endemic or a recent species is debatable. The relationship between *G. heterozygum* and *G. oligospermum* is very close. From pollinating *G. heterozygum* with *G. oligospermum*, Thien (1969) obtained two classes of hybrids. One of these formed 7 pairs of chromosomes and set a full complement of seeds but had only 50% good pollen (*G. heterozygum* has 50% seed set and 50% good pollen). This means that the two species have a genome in common and that the oligospermum-like genome in *G. heterozygum* carries the pollen lethal. The other hybrid, as expected, formed a ring of 14 chromosomes. These results can be interpreted to mean that *G. oligospermum* is one of the parents of *G. heterozygum* as earlier postulated or alternatively that *G. oligospermum* arose from *G. heterozygum* through the loss of the lethal in the genome that came to characterize *G. oligospermum*. If the latter proves to be true, *G. oligospermum* is an endemic of relatively recent origin.

2. *Polyploidy*

It will come as no surprise that many endemics in California are polyploid. Some are unquestionably very old relictual species that fit Favarger's (1961) definition of palaeopolyploids; they have no close diploid relatives. At the other extreme are polyploid species that are almost certainly of recent origin. These polyploids are very local in distribution relative to their demonstrable or putative parents, species which still occur in close proximity to one another. In the terminology of Favarger and Contandriopoulos (1961) these polyploids are apoendemics. Stebbins and Major (1965) list 77 from the Californian flora, but this list is certainly not exhaustive and will undoubtedly be extended with further biosystematic investigation. For example, the recent work of Raven (1969) on *Camissonia* would add several species to this list. I doubt, however, that additions will change the correlation noted by Stebbins and Major (1965) between areas of highest frequency of apoendemics and areas of abrupt gradients between different vegetation types, the same areas which according to my observations are those in which one is most likely to find local hybrid swarms other than those associated with human disturbance.

Polyploidy in *Clarkia*, a genus in which most species are endemic to California, has been of particular interest to me because of the role it has played in the phylogeny of the genus by combining some of the most diverse sections (Lewis and Lewis, 1955). When the genus was monographed in 1955, ten of the 33 species recognized were polyploid, five of which were endemic to California, and of these, three are apoendemics, although none of these [*C. delicata* (Abr.) Nels. & Macbride, *C. similis* Lewis & Ernst, and *C. prostrata* Lewis & Lewis] are necessarily very recent.*

Biosystematic studies since the monograph was published have resulted in the recognition of nine additional species (Vasek, 1964a, b, 1968; Lewis and Raven, 1958; Mosquin, 1962; Parnell, 1968, 1970; Davis, 1970; Small, 1971a, b), several of which appear to have been very recent in origin, but every one of which is diploid. By contrast, no new polyploid has come to light since Wallace Ernst discovered, about 1950, that an apoendemic subsequently described as *C. similis* (Lewis and Ernst, 1953) was distinct from the diploid *C. epilobioides* (Nutt.) Nels. & Macbride. To judge from the few apoendemics in *Clarkia*, polyploidy does not seem to have been a major source of endemics in recent times.

The two tetraploid apoendemics of *Clarkia* (*C. similis* and *C. delicata*) listed by Stebbins and Major (1965) are noted by them to be exceptional, together with those in *Microseris*, in that the parental species are sympatric over much of their range of distribution and occur in the same plant communities; in other genera the putative parents are generally adapted to different climatic regions and are essentially allopatric. To understand the situation in *Clarkia* one must consider the breeding habit of the parental species concerned in relation to barriers to hybridization. In *Clarkia* several diploid species are often found growing together on the same hillside. For example, at one of our experimental sites in the foothill woodland on the San Joaquin Experimental Range, five diploid species are found growing in mixed colonies or within ten meters of one another within an area of 100 sq m. All of these species (*C. williamsonii* (Dur. & Hilg.) Lewis & Lewis, *C. speciosa*, *C. unguiculata*, *C. dudleyana* (Abr.) Macbride and *C. modesta* Jepson) are normally outcrossed and have broadly overlapping flowering periods. No hybrids among these species have been found on this hillside although we have studied the site for a decade, and

* *Clarkia prostrata* (6*x*) a narrow endemic of the sea bluffs of San Luis Obispo County derived from hybridization between *C. davyi* (Jeps.) Lewis & Lewis (4*x*) and *C. speciosa* Lewis & Lewis (2*x*) was omitted from a list of apoendemics by Stebbins and Major (1965, Table 4) perhaps because it was once thought to have a more extensive range (Lewis and Lewis, 1955) than subsequently proved to be the case (Lewis and Raven, 1960). In this same list, *C. deflexa* (Jeps.) Lewis and Lewis is erroneously given as one of the parents of *C. delicata*; the correct species is *C. unguiculata* Lindl. (Lewis and Lewis, 1955).

we have been unable to produce hybrids between any of them in the garden. However, unpublished data show that when some of these same species from widely separated areas are grown together, spontaneous hybrids are formed (*C. unguiculata* from the Sierra Nevada × *C. dudleyana* from southern California) or are readily produced in the garden (*C. williamsonii* from the Sierra Nevada × *C. speciosa* from the Coast Ranges). This suggests that allopolyploidy from hybridization between outcrossing species of *Clarkia* must occur soon after two species come into contact, for if they remain in contact for a long period, polyploidy will be precluded by the evolution of strong barriers to hybridization through the process that Grant (1966) has called the "Wallace Effect." The exception to this rule is found when one of the sympatric species giving rise to the polyploid is normally self-pollinating. In the cases cited by Stebbins and Major (1965) one parent of both *C. similis* and *C. delicata* is *C. epilobioides* which is habitually self-pollinating and frequently cleistogamous. This autogamous species forms hybrids when artificially crossed with individuals from sympatric populations of *C. unguiculata* (the other parent of *C. delicata*) although natural hybrids have never been observed and would not be expected to occur often, not only because *C. epilobioides* is self-pollinating but because it normally flowers several weeks earlier than *C. unguiculata* where they occur together. Similarly, hybrids between *C. modesta* and *C. epilobioides* are readily produced in the garden but have not been observed in the several mixed populations that have been carefully examined for possible hybrids. Although these two species have broadly overlapping flowering periods and have probably been sympatric over a wide area in the south Coast Ranges for a long time, they are still able to cross readily. One may assume that there has been no selection for barriers to crossing because the frequency of hybridization is extremely low, due primarily to the autogamy of *C. epilobioides* and some degree of autogamy in many populations of *C. modesta*.

Natural hybridization between diploid species of *Clarkia* is rare but one case deserves to be watched for the possibility of the formation of a new polyploid species. *Clarkia cylindrica* (Jeps.) Lewis & Lewis (section *Peripetasma*), an outcrossing species, and *C. exilis* Lewis & Vasek (section *Phaeostoma*), which is facultatively autogamous but is about 43–45% outcrossing (Vasek, 1964c, 1967), have distributions that overlap in Deer Creek in Tulare County. At a few sites in this area vigorous but very sterile hybrids are quite frequent, which suggests that these two species may have come into contact relatively recently. Since hybrids between members of the two sections to which these species belong gave rise to the tetraploid *C. delicata*, the possibility of the formation of a new polyploid species in nature while it is being observed is an intriguing prospect.

THE ORIGIN OF ENDEMICS WITHOUT HYBRIDIZATION

Many endemics in California have had no apparent history of hybridization. Those endemics that are geographically separated from their nearest relatives have had no opportunity for recent hybridization and are probably in most, if not all cases, relictual species or races. In contrast, several genera, predominantly annuals, have endemic diploid species that are morphologically very similar to neighboring relatively widespread diploid species which occupy a similar habitat. Hybrids between these pairs of very closely related species have low fertility which is usually attributable to extensive chromosomal repatterning and in some cases to aneuploid differences in basic number. A notable example is *Clarkia*. In those cases that have been studied most carefully in this genus, hybridization as a factor in speciation is at most a very remote possibility and in most instances can be ruled out with confidence. Almost certainly, these local endemics arose relatively recently through chromosome reorganization in extremely small ecologically marginal populations of the more widespread species. This mode of speciation and the arguments for it have been discussed elsewhere (Lewis and Raven, 1958; Lewis, 1962, 1966) and have been very succinctly presented by Grant (1971) under the heading of "Quantum Speciation" where he adds examples from *Gilia*, which he interprets to have had this origin.

How many endemics in California have arisen recently through quantum speciation is unknown because most of them probably have not yet been discovered. The derivative species are so similar morphologically to the parental species that they are usually indistinguishable from minor morphological variants of the parental species in the absence of chromosome counts that show an aneuploid difference or hybrids of low fertility in which meiosis can be studied. The examples of this mode of speciation in *Clarkia* have only gradually been uncovered over a period of 25 years. The first example to suggest this origin to me was *C. lingulata* Lewis & Lewis which I gathered thinking that its entire petals probably represented a single gene difference from the notched petals of *C. biloba* (Dur.) Nels. & Macbride that was growing adjacent to it. Had these two species differed by a less conspicuous trait, I doubt that I or anyone else would have found it. Of the eleven species of *Clarkia* that most clearly have arisen from rapid chromosome reorganization, only four were known when the genus was monographed in 1955. In that year, Peter Raven located living material of *C. franciscana* Lewis & Raven which had been mentioned in the monograph as an aberrant *C. rubicunda* (Lindl.) Lewis & Lewis, the species from which it is almost certainly derived (Lewis and Raven, 1958). Only through the extensive collecting and hybridization programs of Vasek (1958, 1964a, 1968) in section *Phaeostoma* and of Mosquin (1962) and Small (1971a, b) in section *Myxocarpa* have six additional endemic species been discovered, one of which has not yet been formally described. Considering the effort required to discover them, particularly in genera which unlike *Clarkia* are difficult to work with experimentally, a great many neoendemic species will long remain unknown. This is unfortunate, for each new one that has been discovered has added to our understanding of conditions conducive to quantum speciation, conditions which in California are primarily associated with exceptional drought during the growing season in ecologically marginal sites.

CONCLUDING REMARKS

Many aspects of the origin of endemism in California have not been considered in this paper, perhaps most notably those associated with disjunct patterns of parent soil material (e.g. serpentine outcrops) and little attention has been given to the origin of endemic relict species. The latter omission undoubtedly reflects my interest in the origin of neoendemics by processes which can be duplicated or at least simulated experimentally. We have seen, however, that a biosystematic study of *Gayophytum* has added support to the conclusion of Stebbins and Major (1965) that the highest concentration of relatively old species (i.e. patroendemics) in the Sierra Nevada occurs in an area that is nearly coincident with the area of distribution of the well known relict species *Sequoiadendron giganteum*. Unquestionably some habitats in California have persisted relatively unchanged for very long periods of time and serve as repositories of endemic relict species.

Speciation in California has been progressing at a relatively rapid rate in many genera and this has given rise to a large number of neoendemics. In most instances interspecific hybridization has apparently been involved either in the production of allopolyploids or in the formation of new stable populations with the same number of chromosomes as the parental species. Where no change in chromosome number is involved, it is frequently arbitrary as to how many of the products should be recognized in a study of endemism and at what taxonomic level, because at one extreme the resulting population may be scarcely distinguishable from one of the parents and in some areas may intergrade morphologically and ecologically with it. At the other extreme the products are very distinct from both parents (e.g. *Pentstemon spectabilis*). Furthermore, taxa of hybrid origin may hybridize with one of the parents or with another species to give rise to further derivatives, derivatives which may tend to obscure the morphological and ecological gap between the initial hybridizing species.

Endemics that arise without hybridization may be the products of gradual speciation but of particular interest in California are those species that have arisen from chromsome reorganization in presumably very small surviving or founding populations at the ecological margin of established species. The examples that have been studied are among annuals although I believe that under some circumstances species of herbaceous perennials and woody plants have had a similar origin (Lewis, 1966). How prevalent the phenomenon is and how many narrowly endemic species in California have had such an origin in recent time is not easily established, because in most cases hybrids must be produced and studied in order to distinguish derivative species separated by strong chromosomal barriers from population to population variation within species.

Although the flora of California is relatively well known compared to that of many parts of the world, there is still much to be learned about the actual frequency of endemism and the factors responsible for it.

REFERENCES

DAVIS, W. S. (1970). *Brittonia* **22**, 270–284.

FAVARGER, C. (1961). *Ber. Geobot. Inst. Eidg. Techn. Hochschule* **32**, 119–146.

FAVARGER, C. and CONTANDRIOPOULOS, J. (1961). *Bul. Soc. Bot. Suisse* **71**, 384–408.

GRANT, V. (1966). *Am. Natur.* **100**, 99–118.

GRANT, V. (1971). "Plant Speciation". Columbia University Press, New York and London.

LEWIS, H. (1962). *Evolution* **16**, 257–271.

LEWIS, H. (1966). *Science, N.Y.* **152**, 167–172.

LEWIS, H. (1969). *Taxon* **18**, 21–25.

LEWIS, H. and EPLING, C. (1946). *Am. J. Bot.* **33**, 22s–23s (abstr.).

LEWIS, H. and EPLING, C. (1954). *Brittonia* **8**, 1–22.

Lewis, H. and Epling, C. (1959). *Evolution* **13**, 511–525.
Lewis, H. and Ernst, W. R. (1953). *Madroño* **12**, 89–92.
Lewis, H. and Lewis, M. (1955). *Univ. Calif. Publs Bot.* **20**, 241–392.
Lewis, H. and Raven, P. H. (1958). *Evolution* **12**, 319–336.
Lewis, H. and Raven, P. H. (1960). *Leafl. West. Bot.* **9**, 94–95.
Lewis, H. and Szweykowski, J. (1964). *Brittonia* **16**, 343–391.
Mosquin, T. (1962). *Leafl. West. Bot.* **9**, 215.
Munz, P. A. and Keck, D. D. (1959). "A California Flora". University of California Press, Berkeley and Los Angeles.
Noldke, A. M. and Howell, J. T. (1960). *Leafl. West. Bot.* **9**, 124–127.
Parnell, D. R. (1968). *Brittonia* **20**, 387–394.
Parnell, D. R. (1970). *Madroño* **20**, 321–323.
Raven, P. H. (1969). *Contr. U.S. natn Herb.* **37**, 161–396.
Small, E. (1971a). *Evolution* **25**, 330–346.
Small, E. (1971b). *Can. J. Bot.* **49**, 1211–1217.
Stebbins, G. L. (1969). *Taxon* **18**, 26–35.
Stebbins, G. L. and Major, J. (1965). *Ecol. Monogr.* **35**, 1–35.
Straw, R. M. (1955). *Evolution* **9**, 441–444.
Thien, L. B. (1969). *Evolution* **23**, 456–465.
Vasek, F. C. (1958). *Am. J. Bot.* **45**, 150–162.
Vasek, F. C. (1964a). *Evolution* **18**, 26–42.
Vasek, F. C. (1964b). *Madroño* **17**, 219–221.
Vasek, F. C. (1964c). *Evolution* **18**, 213–218.
Vasek, F. C. (1967). *Evolution* **21**, 241–248.
Vasek, F. C. (1968). *Am. Natur.* **102**, 25–40.

12 | Endemism in the Montane Floras of Europe

C. FAVARGER

Institut de Botanique
Université de Neuchâtel, Neuchâtel, Switzerland

INTRODUCTION

The question of endemism in the mountain ranges of Europe is an extremely complicated one, depending as it does on a large number of variable factors, some ecological, some historical. For this reason, we shall limit ourselves in this paper to a few specific problems.

STATISTICAL DATA ON ENDEMISM

It would be extremely interesting to compare the percentages of endemic species in each of the different mountain ranges, but though this appears at first sight to be easy, it is in fact extremely difficult, because authors do not use the same criteria. Some use only Linnean species, others take account of varieties or microspecies, and this at once causes trouble, as the exact distribution of microspecies or chromosome races is rarely known. Some phytogeographers, e.g. Pawłowski (1970), lump together endemics of the montane and lower levels; others, e.g. Anzalone and Bazzichelli (1959, 1960), express the number of endemics as a percentage of the total flora, including the species of the plains.

Ideally, a single author would calculate the percentage of endemics in the various mountain floras; but no botanist could be equally well acquainted with the floras of the Alps, the Sierra Nevada, the Carpathians and the Greek mountains. The compilation of Floras is likewise difficult. Mountain chains, unfortunately, often form political boundaries and the state of knowledge on the two sides of the boundary is not always equal. Synonymy is also a complication. Flora Europaea will eventually solve some of these problems, though the indications it gives of altitude are often extremely brief. In the meantime, we have made an attempt at a statistical treatment (Table I, Fig. 1). Following the example of Turrill (1929) for the Balkan Peninsula, we have summed up the endemic taxa, including species, subspecies and occasionally varieties, and expressed these as a percentage of the total number of montane taxa; the latter include those species which grow exclusively or mainly in the alpine or subalpine levels of the Central European mountains, or in the high mountain mediterranean level (in the sense of Emberger (1930)) of the mediterranean mountains.

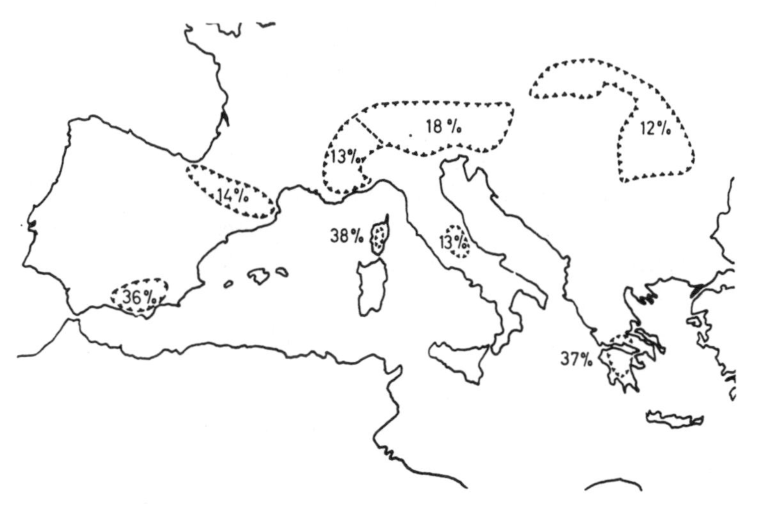

FIG. 1. Endemic taxa in montane floras of Europe as a percentage of total number of montane taxa (details in Table I).

Our sources have been:

(1) Recent lists and floras (Apennines, Alps, E. Pyrenees).
(2) Phytosociological data, where they cover the whole of a range [Quézel (1953) for the Sierra Nevada; Braun-Blanquet (1948) for the E. Pyrenees; Quézel (1964) for the mountains of S. Greece].
(3) The classical work of Turrill (1929) for the Balkan mountains; and the very stimulating paper of Contandriopoulos (1962) for Corsica.
(4) The recent publication of Pawłowski (1970) (Alps, Carpathians). For the Alps we have made much use of Pawłowski's excellent study, but have included only the high mountain species, and for the subendemics, only those which do not significantly extend beyond the alpic range.
(5) Our own personal experience (Alps, Apennines) and that of our collaborator, Ph. Küpfer (Pyrenees, Sierra Nevada).

TABLE I.

Mountain chain	Number of mountain taxa	Number of endemic mountain taxa	% of endemics	% of taxa shared with the Alps*	Source of data
Baetic–nevadan range	349	125	36%	30%	Quézel, 1953, 1957; Rivas-Martinez, 1969; Soutadé, Baudière, 1970; Küpfer (unpublished)
High mountains of Corsica	142	54	38%	44%	Contandriopoulos, 1962, 1971
Balkan mountains	1193†	404	33·9%	?	Turrill, 1929
	529†	203	38·5%	?	
Mountains of S. Greece	475	174	37%	10%	Quézel, 1964
W. Alps	805	105	13%		Pawłowski, 1970
E. Alps	866	154	18%		
Whole Alpine range	1049	331	31%		
C. Apennines	288	36	13%	49%	Furrer and Furnari, 1960
Pyrenees	720	103	14%	67%‡	Gaussen and Leredde, 1949
				60%	Gaussen, 1953–1971
Carpathians	*ca* 900	*ca* 110	*ca* 12%	?	Pawłowski, 1970

* Taxa found in the Alps at the subalpine and alpine levels.

† Plants growing in the zones 5b and 5c of Turrill (probably corresponding to the high mountain level). Unfortunately, Turrill did not give the percentage of taxa which are common to levels 5b and 5c, so that the figures cannot be added.

‡ Upper figure is for the W. Alps, the lower for the E. Alps.

We are very conscious of the provisional nature of our figures, and we know that they will be modified as our knowledge of taxonomy and phytogeography increases. Certain defects are apparent at once; thus the percentage of endemics in the Abruzzi is probably somewhat low, although close to those given by Furrer (1961) which were 11%, 13% and 18%.* Again the percentage for the Baetic range is probably somewhat high. Here, to the species noted by Quézel (1953) or observed by Küpfer in the Sierra Nevada, we have added recent records by Rivas Martinez (1969) and Küpfer in the high mountain level of the Sierras of S.E. Spain, but these lists are certainly not complete, and some non-endemic mountain species are certainly missing.

We do not know for certain the total number of species at the subalpine and alpine levels of the Carpathians, nor the percentage of species which these mountains have in common with the Alps. We can thus only make estimates here, based on the work of Pawłowski (1970).

INTERPRETATION OF THE STATISTICAL DATA

In spite of these difficulties, some of the data of Table I are very interesting. First, the southern mountains (except the Abruzzi) *are characterized by a high percentage of endemics* (30–40%).

Secondly, the mountains further north (Pyrenees, E. and W. Alps, Carpathians) as well as the C. Apennines, have a much lower percentage of endemic taxa, ranging from 12 to 18%.

Thirdly, the Alps have to be considered as a special case. It has been known for a long time (and it has been confirmed by the work of Schmidt (1944) and Pawłowski (1970)), that there are two centres of endemism in the Alps, one in the S.W. Alps, the other in the S.E. Alps and the N.E. calcareous Alps (Merxmüller, 1952). Now, as shown in Table II, most of the endemic Alpic taxa are confined either to the western or to the eastern Alps;† but *most of the non-endemic taxa occur throughout the whole area of the Alps.* If we add together the numbers of E. and W. Alpic endemics and the pan-alpic endemics, we obtain the high figure of 31% of endemics; but this we feel is artificially high, because the non-endemic

* The percentage varies according to whether one considers the species common to the Gran Sasso and the Terminillo, those common to the Gran Sasso and the national park of the Abruzzi, or the species of the Gran Sasso only.

† In dividing the Alps, we follow Merxmüller (1952), who proposes 2 provinces, the Gallic (W. Alps) and the Helvetic-noric (C. & E. Alps). The boundary is approximately the line from Lac Léman to Lac Majeur.

TABLE II.

	Pan-alpic	Limited to the W. Alps	Limited to the E. Alps
Endemic mountain plants of the Alps (331 taxa)	22%	32%	46%
		78% (W. + E.)	
Non-endemic mountain plants of the Alps (718 taxa)	76%	11%	13%
		24% (W. + E.)	

mountain flora of the two regions of the Alps is broadly the same. Accordingly, one could perhaps leave out of account the pan-alpic endemics, or at least some of them, because they do not fit the definition of Good (1947).* This would give us a percentage of about 25% for the Alps as a whole.

We prefer to think of the E. and W. Alps as two distinct centres of endemism. The comparative independence of the two centres is supported by the fact that there are a dozen species common to the Pyrenees and the E. Alps which are absent from the W. Alps (e.g. *Primula integrifolia* L., *Willemetia stipitata* (Jacq.) Cass.). This comparative independence is due, in our opinion, to the effects of the quaternary glaciations, during which alpine taxa were concentrated in two main refugia to the S.W. and S.E. of the Alps. These refugia became the main sources of Alpic endemics, as is well shown by the map published by Pawłowski (1970, p. 189). This example shows how important it is, where endemics are concerned, to choose correctly the area of study. Fixing the boundaries of the area is equally important; for the limit of the Alpic region is by no means clear, especially towards the mountains of the Balkan peninsula. In our statistics, we have included as Alpic endemics those taxa which reach the Jura or the N. Apennines, but not those which extend to the Tuscan Apennines, the Vosges, or Croatia.

There are several possible explanations of the data of Table I. The first postulates that there exists, in general, a north–south gradient of endemism, as suggested by Contandriopoulos (1962) for island floras, the

* If this is done, it would be necessary, for consistency, to omit the Greek endemics from the figures for the S. Greek mountains (cf. Quézel, 1964).

cause being the more favourable climatic conditions in the Mediterranean region. According to Emberger (1946), the flora of the Moroccan Grand Atlas has 75% of endemics; it would thus seem that there is also a north–south gradient in the mountain floras of W. Europe and N. Africa. On the other hand, Pawłowski (1970) has considered and rejected this explanation; one of the reasons against it is the fact that the summer drought of the Mediterranean, which is effective even at high altitudes, is not favourable to plant growth; besides, the low figure for endemics in the C. Apennines is an exception to the rule.

Secondly, it can be supposed that the mountains of S. Europe, having been only slightly affected by glaciation, have provided refugia for species which have become extinct elsewhere. This is the explanation favoured by Pawłowski, and it would probably be acceptable if all the endemic taxa were relics (passive endemism); but there is also an active endemism (Favarger and Contandriopoulos, 1961). According to Turrill (1929) most of the endemics of the high Balkan mountains are related to taxa at lower altitudinal levels, and have thus been formed by active endemism. However only cytotaxonomic studies can prove this hypothesis, because it is possible that some species are patroendemics (Favarger, 1967). Now the part played by the glacial periods in active endemism seems to be important. On the one hand their effect in producing isolation has favoured the differentiation of alpic montane taxa in their refugia in the S.W., the S.E. or the N.E. Alps. On the other hand, they have indirectly stimulated the production of apoendemics, through their effect on the migration and the mixture of floras.

Thirdly, we can suppose that the mountains of C. Europe have a relatively low proportion of endemics because their floras *have a high proportion of mountain species in common*. This high coefficient of similarity may be due to four factors:

(1) The existence of a common stock of mountain species which originated in Tertiary times, and which have shown little divergence since the various mountain ranges are completely separate.
(2) Invasion at the end of the Tertiary (Diels, 1910) by Asiatic mountain species, which have not had time to differentiate.
(3) Invasion by species of Arctic origin during the glacial periods; again, these species have had no time to differentiate into endemics, or at least not beyond the stage of microspecies.
(4) The exchange of floras between mountain ranges, which could have occurred during the glacial periods.

For the Carpathians, the E. and W. Alps and the Pyrenees, all four factors must have played a part, and this explains their high coefficients of similarity; thus the Pyrenees have 67% of their flora in common with the W. Alps. For the C. Apennines,* we consider that only factors (1) and (4) were operative; the orientation of the mountain ranges will have facilitated the migration of alpic taxa in the Quaternary period. On the other hand, very few alpine taxa have been able to reach the Sierra Nevada or the Greek mountains, where the proportion of species shared with the Alps is 30% and 10% respectively.

Corsica is a special case. The percentage of mountain endemics is high (38%), yet at the same time, there is a high percentage of taxa which are shared with the Alps (44%). These percentages are not entirely significant, as they are based on only 142 species. In absolute figures, the number of species common to Corsica and the Alps is small (62 taxa), and it is much higher for the C. Apennines (142 taxa) and even for the Sierra Nevada (105 taxa).

Land connections between Corsica and other mountain areas were broken in the Miocene, hence its high proportion of mountain endemics. According to Contandriopoulos (1971), the species which the mountains of Corsica share with the Alps (and which constitute the arctic-alpine and boreal-montane elements) are in a "precarious equilibrium", which would not have favoured their evolutionary differentiation. For Corsica, the high percentage of species shared with the Alps is to be explained by factor (1); in any case, as we have already mentioned, the absolute number of such species is small.

CONNECTIONS BETWEEN ALPINE AND PYRENEAN FLORAS

It is worth looking in a little more detail at factor (4), with special reference to the connections between the Alpic and Pyrenean floras. Three categories of data have a bearing on the hypothesis that there have been exchanges during the Quaternary period between the Alps and the Pyrenees.

(1) The existence of isolated stations for Pyrenean taxa in the W. Alps, and Alpic taxa in the Pyrenees (especially in the east). Supporting data (the lists are not exhaustive) are given in Table III.

* The mountain flora of Gran Sasso has about 7% of Balkan taxa. These perhaps reached the Abruzzi at the end of the Tertiary through connections between the Gargano promontory and Dalmatia, and there has not been enough time for endemics to be differentiated.

TABLE III.

Species widely distributed in the Pyrenees, but isolated or rare in the W. Alps	Species widely distributed in the Alps, but local or rare in the Pyrenees
Potentilla nivalis Lapeyr.	*Arenaria biflora* L.
Hypericum nummularium L.	*Pedicularis rostrata* L.
Vicia pyrenaica Pourret	*Polygala chamaebuxus* L.
Adonis pyrenaica DC.	*Dianthus neglectus* Loisel.
	Listera cordata (L.) R. Br.
	Minuartia liniflora Schinz. & Thell.
	Ranunculus oreophilus Bieb.
	Sedum anacampseros L.
	Hugueninia tanacetifolia (L.) Rchb.

(2) The presence of isolated relict colonies of Alpine or Pyrenean species in the Massif Central of France (Braun-Blanquet, 1923). We can add *Colchicum alpinum* Lam., recently discovered by R. Nozeran (unpublished) in l'Aubrac, to the list of alpine species which occur in the Massif Central. Other examples are *Minuartia liniflora*, *Ligusticum mutellina* (L.) Crantz, *Carduus personata* (L.) Jacq., *Cirsium erisithales* (Jacq.) Scop., *Hieracium aurantiacum* L. etc. Pyrenean species, which also occur as isolated colonies in the same area are *Lilium pyrenaicum* Gouan, *Fritillaria pyrenaica* L., *Campanula speciosa* Pourret, *Silene ciliata* Pourret, *Sagina pyrenaica* Rouy, etc. (Braun-Blanquet, 1923, p. 215 et seq.).
(3) Cytotaxonomical investigations. The cytotaxonomical work of our collaborator Ph. Küpfer (Küpfer, 1968, 1969; Küpfer and Favarger, 1967) has shown that certain taxa which are diploid in the alpine level of the Pyrenees are always polyploid in the Alps; this is true for *Paradisia liliastrum* (L.) Bertol. and alpine varieties of *Koeleria vallesiana* (All.) Bertol. Conversely, *Draba subnivalis* Br.-Bl. and *Minuartia sedoides* (L.) Hiern are diploid in the Alps, but always polyploid in the Pyrenees (*Draba subnivalis* is the Pyrenean pseudo-vicariant of *D. tomentosa* Clairv.). It is difficult to explain these facts without postulating migration between the Alps and the Pyrenees during the glacial epoch.

In yet other cases the diploid taxon occurs in the E. Pyrenees and the S. Alps, while the corresponding polyploid is found in the C. Pyrenees and the N. or N-W. Alps (the Jura included), e.g. *Bupleurum ranunculoides* L.,

Anthyllis montana L. Again, in *Veronica bellidioides* L., *Agrostis rupestris* All. (Björkman, 1960 and Ph. Küpfer, unpublished) and *Ranunculus parnassiaefolius* L.,* the E. Pyrenean populations are diploid, and those of the C. Pyrenees and Alps polyploid. For these 5 taxa it is possible to think of a double migration, starting from a refugium which included the S. Alps and E. Pyrenees (or the E. Pyrenees alone) and leading both to the C. Pyrenees and the C. or N. Alps.

Finally, in some cases, migration accompanied by polyploidization has worked in one direction only, either towards the C. Pyrenees (as in *Avena montana* Vill. (Gervais, 1966) and *Anthemis montana* L.) or towards the C. and E. Alps (as in *Senecio doronicum* L. and *Ranunculus pyrenaeus* L.).

All these examples illustrate the importance of migration, during the glacial epoch, both between the Alps and the Pyrenees and within each of the mountain ranges. It is clear, too, that while such migrations may have tended to reduce the proportion of endemics in each range and increased the number of species in common, they have also produced apoendemic taxa, though these are only very slightly differentiated morphologically (they have not been taken into account in the statistics of Table I).

It is not possible at present to derive much more information about endemism from statistical methods, though Pawłowski (1970) has some interesting data on supraspecific endemic taxa, and on the altitudinal distribution of endemics in the Alps and Carpathians.

CYTOTAXONOMIC METHODS AND ENDEMISM

With Dr Contandriopoulos (Favarger and Contandriopoulos, 1961), we have tried to apply cytotaxonomic methods to the study of endemism. These are particularly important in giving us information about the relative age and mode of formation of endemic taxa. Our knowledge is still too fragmentary to permit us to classify the endemics of the montane flora of Europe as palaeo-, schizo-, patro- and apoendemics; and we must limit ourselves here to mentioning some interesting data on the subject of active endemism.

Schizoendemism is the main process in Alpine endemism. The vicariant taxa of the W. and E. Alps are found in other parts of the Alps, in the Pyrenees, and less often in Corsica, the Carpathians and the Balkan mountains; and these taxa themselves are often endemic. In fact there are

* This example is complicated, as the diploid is also found in the Sierra de Cabrera (Küpfer, 1971b).

examples of endemovicariants in all the mountains of Europe. Cases of ancient schizoendemism, i.e. when the mountain ranges concerned are far apart (Table IV), are particularly interesting. Sometimes, the disjunctions may be even larger, as in *Rhodothamnus* (Pawłowski, 1970) or *Cerastium pyrenaicum* Gay, which is related to species of the Caucasus (Favarger and Küpfer, 1968).

These data indicate that the Tertiary floras of the mountain ranges of Europe had a certain amount in common; and this is perhaps a little difficult to accept. Nevertheless, there are still several mountain taxa with a large European area which are, comparatively speaking, not much differentiated (e.g. *Draba* sect. *Aizopsis pro parte*), *Vitaliana primuliflora* Bertol.); and there are some species with a wide, very disjunct distribution which show little or no variation (e.g. *Viola parvula* Tineo, Favarger and Küpfer, 1969).

Two recent pieces of work, one on *Iberis* species of the group *spathulata* (Küpfer, 1971b) and the other on *Viola* section *Melanium* (Küpfer, 1971a), indicate one of the ways in which schizoendemism in the mountain floras of W. Europe may have arisen in the Tertiary. In *Iberis*, Küpfer has shown that 3 endemic races from the Iberian peninsula, with $n = 7$, are quite independent of the endemic races of the W. Alps, which have $n = 9$. Here there has been presumably a primary differentiation, with dysploidy between the Alpine and Iberian stocks, followed by a second differentiation within each region.

In the mountain Violas of the section Melanium, Küpfer (1971) has shown that *V. nummulariifolia* Vill., a species of the Alpes Maritimes, has $n = 7$. This discovery has illuminated the whole question of the origin of those species with $2n = 34$ which occur in the Iberian peninsula on the one hand and in Italy on the other; they are probably amphidiploids which have arisen from ancestors with $n = 7$ and $n = 10$. This example shows that the Alpes Maritimes have been important as a centre of differentiation for species of this section, even before the glacial period. In other complex taxa, such as *Vitaliana primuliflora*, schizoendemism has probably originated from isolation of Nevadan, Pyrenean, Alpine and Italian portions of an original montane syngameon.

As regards apoendemism, we only have enough cytological data to discuss it for the Alps, and even here with a certain amount of reserve; but one fact is striking. *Apoendemic taxa are two to three times as numerous in the E. Alps as in the W. Alps.* This in part comprises what we have called (Favarger, 1962, 1967) pseudo-vicariance of the W.–E. type, but with

TABLE IV.

Sierra Nevada	Pyrenees (and Cantabrian Mts)	Western Alps	Eastern Alps	Balkan mountains or Eastern Carpathians
Artemisia granatensis Boiss.		*Artemisia glacialis* L.		
	Ramonda myconi (L.) Schultz			*Ramonda nathaliae* Panč. & Petrov.
	Veronica nummulariaefolia Gouan.			*Veronica satureioides* Vis.
	Aquilegia pyrenaica L.		*Aquilegia einseleana* F. W. Schultz	
	Cirsium rufescens Ramond		*Cirsium carniolicum* Scop.	
		Potentilla valderia L.		*Potentilla haynaldiana* Janka

the difference that it applies also to non-endemic taxa and to chromosome races. All this indicates more active abrupt speciation in the E. than in the W. Alps. (In the Pyrenees, we have so far recorded 7 cases of apoendemism, of which 2 are dubious.)

It would seem that gradual speciation has also been more frequent in the E. Alps. In fact, the generic coefficient of Jaccard (1929) is significantly lower in the endemic flora of the E. Alps than in that of the W. Alps (Table V).

TABLE V.

	W. Alps	E. Alps
Generic coefficient of Jaccard	58%	42%
No. of apoendemic taxa recorded	9 (4 dubious)	22 (3 dubious)

Several alpine apoendemics are of hybrid origin, e.g. *Draba ladina* Br.-Bl., *D. dolomitica* Buttler (Buttler, 1967, 1969), and it is tempting to attribute their success to the ecological conditions of high mountains, which favour hybridization and the establishment of hybrids (Stebbins, 1954). Nevertheless, it is difficult to understand the lower proportion of apoendemics in the W. Alps. Perhaps further work will show that the number is greater than it now appears.

In the meantime, we have postulated pathways of migration from west to east in interglacial or postglacial times, and this involves the hypothesis of a more extensive or prolonged glaciation in the E. Alps (Favarger, 1971). As for the more active gradual-speciation, this could have been caused by a wider variety of ecological conditions in the E. Alps, or possibly by fewer connections between the different ranges of the E. Alps at the time when the Alps were being formed (a hypothesis based on the work of Scharfetter, 1929).

To conclude, we can state, without gainsaying the importance of ecological factors, that historical factors have played an important part in mountain endemism in Europe.

ACKNOWLEDGEMENTS

We should like to offer our sincere thanks to M. Ph. Küpfer for his help in producing the statistical data.

REFERENCES

ANZALONE, B. and BAZZICHELLI, G. (1959–60). *Annali Bot.* **26**, 1–182.

BJÖRKMAN, S. O. (1960). *Symb. bot. upsal.* **17**, 1–112.

BRAUN-BLANQUET, J. (1923). "L'origine et le développement des flores dans le Massif Central de France". Lhomme, Paris; Beer, Zürich.

BRAUN-BLANQUET, J. (1948). "La végétation alpine des Pyrénées orientales". Cons. sup. de invest. cient. Barcelona.

BUTTLER, K. P. (1967). *Mitt. bot. St. Samm. München* **6**, 275–362.

BUTTLER, K. P. (1969). *Mitt. bot. St. Samm. München* **8**, 539–566.

CONTANDRIOPOULOS, J. (1962). "Récherches sur la flore endémique de la Corse et sur ses origines". Thèse, Montpellier.

CONTANDRIOPOULOS, J. (1967). *C.r. 91ème Congr. soc. sav. Rennes* **3**, 271–280.

CONTANDRIOPOULOS, J. (1971). *Ann. litt. Univ. Besançon.* Les Belles Lettres, Paris, pp. 205–222.

DIELS, L. (1910). *Beibl. Bot. Jb.* **102**, 7–46.

EMBERGER, L. (1930). *Rev. gén. bot.* **42**, 641–662; 705–721.

EMBERGER, L. (1946). *Vol. Jub. Soc. Sci. nat. Maroc.* 1–11.

FAVARGER, C. (1962). *Rev. Cytol. Biol. vég.* **25**, 397–410.

FAVARGER, C. (1967). *Biol. Rev.* **42**, 163–206.

FAVARGER, C. (1971). *Ann. Litt. Univ. Besançon*, 139–164.

FAVARGER, C. and CONTANDRIOPOULOS, J. (1961). *Bull. Soc. bot. suisse* **71**, 383–408.

FAVARGER, C. and KÜPFER, PH. (1968). *Coll. bot.* **7**, 325–352.

FAVARGER, C. and KÜPFER, PH. (1969). *Bol. Soc. Brot.* **43**, 315–331.

FURRER, E. (1961). *Ber. Geobot. Forsch. Inst. Rübel* **32**, 70–83.

FURRER, E. and FURNARI, F. (1960). *Boll. Ist. Bot. Univ. Catania* Sér. (2) **2**, 143–202.

GAUSSEN, H. (1953–71). "Le Monde des Plantes" No. 293–369.

GAUSSEN, H. (1964). *C.r. somm. Séanc. Soc. biogéogr.* **356**, 13–19.

GAUSSEN, H. and LEREDDE, C. (1949). *Bull. Soc. bot. Fr.* **96**, (Session extraord.), 57–83.

GOOD, R. (1947). "The Geography of Flowering Plants". Longmans, London.

JACCARD, P. (1929). *Bull. Soc. bot. Fr.* **76**, 47–66.

KÜPFER, PH. (1968). *Bull. Soc. neuchât. Sci. nat.* **91**, 87–104.

KÜPFER, PH. (1969). *Bull. Soc. neuchât. Sci. nat.* **92**, 31–48.

KÜPFER, PH. (1971a). *C.r. hebd. Séanc. Acad. Sci., Paris* **272**, 1085–1088.

KÜPFER, PH. (1971b). *Ann. Litt. Univ. Besancon*, 167–185.

KÜPFER, PH. and FAVARGER, C. (1967). *C.r. hebd. Séanc. Acad. Sci., Paris* **264**, 2463–2465.

MERXMÜLLER, H. (1952). "Untersuchungen zur Sippengliederung und Arealbildung in den Alpen". Thèse, München.

PAWŁOWSKI, B. (1970). *Vegetatio* **21**, 181–243.

QUÉZEL, P. (1953). *Mem. Soc. Brot.* **9**, 5–77.
QUÉZEL, P. (1957). "Peuplement végétal des Hautes montagnes de l'Afrique du Nord". Lechevalier, Paris.
QUÉZEL, P. (1964). *Vegetatio* **12**, 289–385.
RIVAZ-MARTINEZ, S. (1969). *5ème Symp. Flora Europaea, Sevilla*, pp. 53–80.
SCHARFETTER, R. (1929). *Bot. Jb.* **62**, 524–544.
SCHMIDT, A. (1944). "Zur Kenntnis der Endemismen der Alpen". Thèse, München.
SOUTADÉ, G. and BAUDIÈRE, A. (1970). *Annales Géogr.* 709–736.
STEBBINS, G. L. (1954). *In* "Étude botanique de l'étage alpin" (P. Chouard, ed.), pp. 135–140. R. P. Colas, impr., Bayeux.
TURRILL, W. B. (1929). "The Plant Life of the Balkan Peninsula". Clarendon Press, Oxford.

13 | The Role of Hybridization in the Evolution of the Hawaiian Flora

GEORGE W. GILLETT*

University of California, Riverside, California, U.S.A.

INTRODUCTION

It is now generally accepted that the seed plant flora of the Hawaiian Islands is a disharmonic assemblage of genera selected for their capacity to accomplish long-distance dispersal. These have gained ingress from diverse source regions, the affinities being Indo-Pacific (40%), Austral (16·5%), American (18%), and pantropic (12·5%), with the remaining portion either obscure or boreal in origin (Fosberg, 1948). As a strongly isolated archipelago, the Hawaiian Islands would be expected to have received a small flora through the agencies of chance dispersal, and a small flora it is, comprised of only 216 (Fosberg, 1948) to 238 (Balgooy, 1960) genera, all angiosperms. By comparison, there are some 451 genera in the flora of Fiji (Smith, 1951), an archipelago of nearly the same area, but much closer to the New Guinea source region.

The diversity of the Hawaiian flora is conspicuously limited at the generic level, but this is largely compensated by the existence of an unusually broad range of complex morphological diversity at the species level. Nearly all genera, both large and small, are characterized by remarkable specific and trans-specific variability.

Many Hawaiian genera have been revised or monographed over the past 60 years. This work has been accomplished by the use of traditional descriptive methods, not infrequently with a typological emphasis that gave scant recognition to variability, if not ignoring it altogether. Some of the work was accomplished without field studies, with predictably mediocre results, but other contributions have been enhanced with unusually energetic and perceptive field work (Rock, 1919; Fosberg, 1937).

* Supported by Grant GB-8608, National Science Foundation.

However, most of the revisionary and monographic work has received scant praise, including the verdict (Fosberg, 1948) that some of it is so poor as to necessitate redoing. Competent students find that many identifications are very difficult to arrive at in terms of definitive morphological criteria. Many determinations are made by the circuitous and unreliable methods of extrapolating quantitative "key" characters, or by relating the locality of the specimen in question to a cited locality in the particular reference.

Systematic problems in these groups invariably center on extensive variation that is inadequately delineated in descriptions, and on a complex mixing of "key" characters, suggesting natural hybridization, that compromises the best efforts to devise a definitive diagnostic "key".

The taxonomic methods utilized in producing these references have been generally satisfactory as applied to many continental floras. Therefore, their failure to generate satisfactory results as applied to the Hawaiian flora suggests that there is a relative difference, in terms of evolutionary processes, between the flora of the Hawaiian Islands and continental floras.

An attempt will be made here to account for at least a portion of this disparity through the proposition that natural hybridization has been a significant evolutionary process in the development of the Hawaiian flora, using selected examples. Each of these examples involves a polymorphic genus that is highly refractive to the traditional, descriptive approach.

SURVEY OF SELECTED GENERA

1. Bidens

Hawaiian *Bidens* are hexaploids, with a diploid meiosis and chromosome complements of 36 pairs. The experimental work of Gillett and Lim (1970) has emphasized that the adaptive radiation of the genus in Hawaii has occurred with a minimum of internal genetic barriers. The total of 23 experimental interspecific hybrids reported by them has been increased by 7 additional experimental hybrids in 1971. One of the latter hybrids, between *B. skottsbergii* Sherff and *B. menziesii* A. Gray var. *filiformis* Sherff, combines two contrasting populations on the island of Hawaii. Over 25 vigorous, fertile, F_1 hybrids of this cross have been grown in the greenhouse. This material, combined with the evidence from a biosystematic study (Mensch and Gillett, 1972), involving *B. menziesii* var. *filiformis* and two populations intermediate between it and *B. skottsbergii*, provided the experimental basis for the ensuing discussion of this complex.

Because of the extensive variation and taxonomic confusion in these four populations, they will be designated by the letter symbols "A", "B", "C" and "D". Their relationship to formal taxonomic descriptions are as follows:

"A" *Bidens menziesii* var. *filiformis*
1800 m, saddle road 31 miles east of Kailua-Kona and 27 miles east of population "C" (map, Fig. 1). Leaves bipinnate, with filiform divisions. Achenes awnless, or very nearly so.

"B" *Bidens menziesii* var. *filiformis*
Bidens micrantha Gaud.
Bidens campylotheca Schz. Bip.
Bidens hawaiiensis A. Gray
Bidens ctenophylla Sherff
600–1000 m, Puu Waawaa Ranch, 14 miles west of population "A" and 13 miles east of population "C". Bipinnate to simple leaves, awnless to strongly awned achenes.

"C" *Bidens hawaiiensis*
Bidens ctenophylla
Bidens skottsbergii
150 m, Huehue Ranch, four miles N.E. of Kailua-Kona. Trifoliate to simple leaves, shortly awned to strongly awned achenes.

"D" *Bidens skottsbergii*
12 m, two miles east of Black Sand Beach, south coast of Hawaii (Fig. 1). Simple leaves, strongly awned achenes.

The difficulty of using formal taxonomic references for the above populations is obvious. Of the four populations, the morphological and ecological extremes are reflected by "A" and "D".

In Hawaiian *Bidens*, the "ray" florets of the capitula bear no sex organs and function only for display. The "disc" florets may be either hermaphroditic or female. In some species, only hermaphroditic disc florets are known. However, in other species, including the *B. menziesii–B. skottsbergii* complex, plants may be either hermaphroditic or female, thus portraying the gynodioecious expression. Self-compatibility is a general rule for hermaphroditic individuals, but the paucity of fruits on unattended hermaphroditic greenhouse plants suggests that a measure of outcrossing probably operates in nature where abundant fruit is produced.

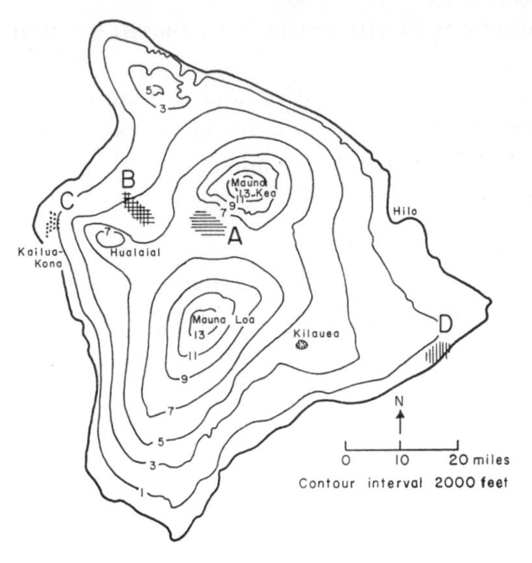

FIG. 1. Map showing the locations of populations studied in the *Bidens menziesii* var. *filiformis* × *B. skottsbergii* hybrid complex. "A", *B. menziesii* var. *filiformis*; "B", and "C", intermediate populations; "D", *B. skottsbergii*.

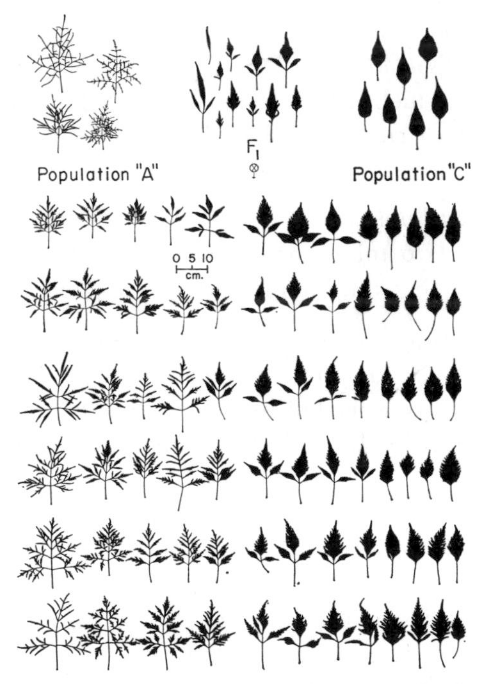

FIG. 2. Leaf form in population "A", upper left; population "C", upper right; experimental F_1 hybrid, upper center; and F_2 hybrids, below.

In gynodioecious groups, outcrossing would operate at a much higher level, and it undoubtedly has played an important role in generating the high level of diversity in this complex.

Experimental crosses involving populations "A" and "C" have produced F_1 and F_2 hybrids (Mensch and Gillett, 1972). The segregation of leaf form in the F_2 hybrids (Fig. 2) was closely approximated in three progenies grown from wild parents selected from population "B". This is convincing evidence of the origin of population "B" through natural hybridization. However, it is augmented by studies of the inheritance of the apical awns of the fruits and the variation and segregation in that character.

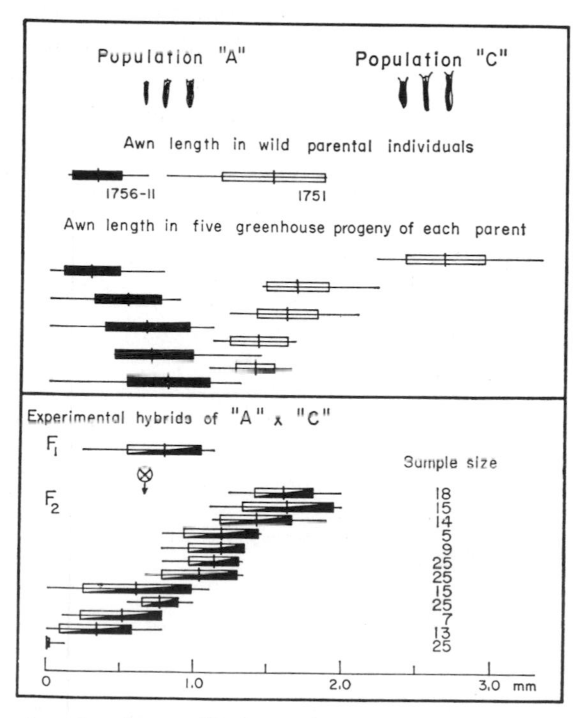

FIG. 3. Above: Awn length in wild plants of population "A" (1756–11), population "C" (1751), and their respective progeny (samples of 25 achenes per plant). Below: Awn length of the F_1 (25 achene sample) and F_2 (sample size as indicated) hybrids. Awn length for the sample of an individual plant is given as range (horizontal line), mean (vertical line), and two standard deviations (bar). All graphs are plotted on the scale at the bottom of the illustration.

Each mature fruit of *Bidens* bears two, or more, apical, retrorsely-barbed awns, or these may be lacking. In the total complex, involving populations "A", "B", "C" and "D", a complete spectrum of variation occurs from the awnless to the strongly awned expression. Population "A" has fruits that are awnless, or very nearly so. The other three populations have gradations extending between the awnless and the strongly awned condition. The segregation of this expression in the F_2 population produced by crossing "A" with "C" is shown in Fig. 3. Segregation in two progenies grown from individuals of population "B" is shown in Fig. 4. The comparison of these two illustrations shows a remarkable similarity and leaves little doubt that population "B" is derived through natural hybridization. Population "D" has the strongest, highly uniform, expression of simple, unlobed leaves and prominent achene awns. Therefore, there is the possibility that this population, rather than the less extreme, variable population "C", is the ultimate source of genes for simple leaves and awned achenes. Such a hypothesis is supported by the vigorous F_1 hybrids produced from the experimental cross between "A"

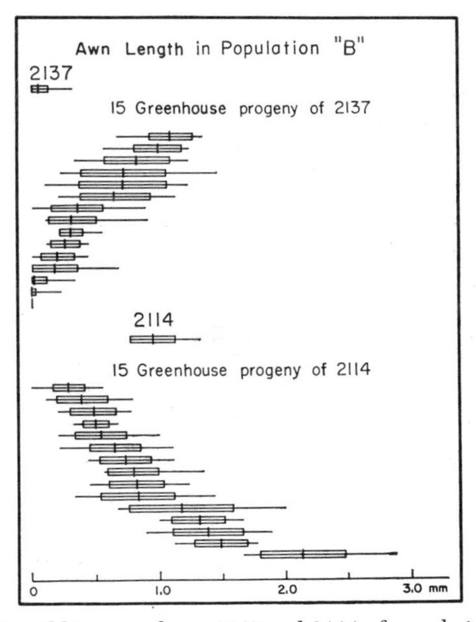

FIG. 4. Awn length in wild intermediates 2137 and 2114 of population "B", and in their respective progeny (sample size: 25 achenes per plant). Awn lengths for each sample of an individual plant given as range (horizontal line), mean (vertical line), and two standard deviations (bar).

and "D", enumerated earlier. The ease of this cross indicates the absence of genetic barriers between the extreme forms of the complex and provides an explanation for the intermediate status and variability of populations "B" and "C". There is a strong indication that natural hybridization has occurred on a broad scale between "A" and "D" and that there has been gene flow between these extremes. This hypothesis is advanced by the graphic portrayal of variability in the complex (Fig. 5). Figure 5

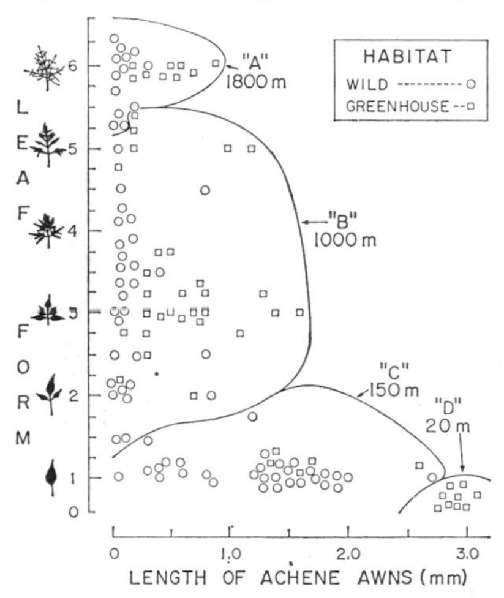

FIG. 5. Scatter diagram portraying clinal variation in leaf form (vertical scale) and length of achene awns (horizontal scale) expressed by populations "A", "B", "C", and "D" of the *Bidens menziesii* var. *filiformis* × *B. skottsbergii* hybrid complex.

shows a clinal variation for leaf form and for length of achene awns. These clines extend from 12 meters to 1800 meters, from very recent lava formations, at the lowest elevation, to much older ones at the highest elevation. The illustration also explains the taxonomic confusion in this hybrid complex and the difficult problem it presents to the systematist who seeks to delineate discrete taxa.

There are similar instances of natural hybridization involving *Bidens menziesii* on the island of Maui, and several additional ones for the genus elsewhere in Hawaii (Gillett and Lim, 1970). The widespread occurrence of hybridization in Hawaiian *Bidens* suggests that this process has a role in the evolutionary strategy of the genus.

2. Scaevola

This genus of the Goodeniaceae provides the next example of an extremely polymorphic group, highly refractive to traditional descriptive methods.

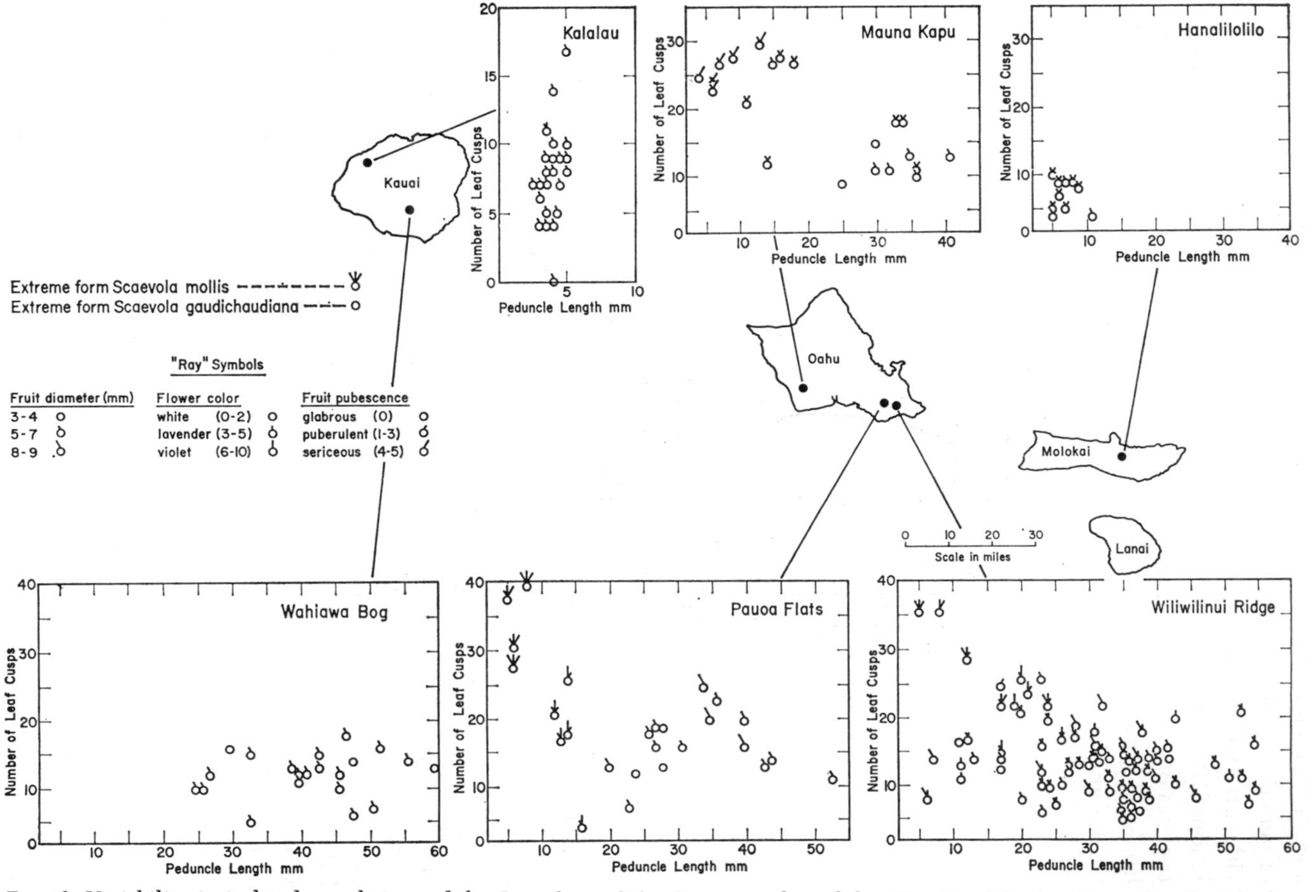

FIG. 6. Variability in isolated populations of the *Scaevola gaudichaudiana* complex of the Hawaiian Islands. (Reproduced with the permission of the editor of *Evolution*.)

The five upland species of Hawaiian *Scaevola*, including the *S. gaudichaudiana* complex, are characterized by a diploid chromosome complement of 8 pairs. The flowers are protandrous. Prior to anthesis the pollen is transferred from the anthers to a hairy stylar sheath projecting just above the unexpanded stigmatic lobes. From this site it is picked up by insects en route to the nectar at the base of the corolla tube, so that a good measure of outcrossing is promoted. Dispersal in these species is accomplished by a fleshy, purple drupe, eaten by birds.

A study recently carried out on the *Scaevola gaudichaudiana* complex (Gillett, 1966) centered on an extremely variable population, accorded the status of natural hybrids, on the island of Oahu. Other highly variable populations were studied on Oahu, Molokai and Kauai (Fig. 6), each with a hybrid background. This complex reflects the operative dynamics of outcrossing, hybridization and subsequent dispersal of heterozygous products. In this example, hybridization and back-crossing have been so extensive as to blur the expressions of the parental lines, the montane, purple-flowered, pubescent *Scaevola mollis* H. & A. and the foothill, white-flowered, glabrous *Scaevola gaudichaudiana* Chamisso. However, in the process this evolutionary stock has developed many populations that have successfully exploited an array of different, isolated habitats on three islands.

A total of 19 formal taxonomic descriptions have been accorded this hybrid complex, including 6 species names, 7 varietal names, and 6 forms.

3. Pipturus

Hawaiian *Pipturus* (Urticaceae) undoubtedly have arisen from a single invasive stock that has subsequently developed into an extremely variable series of populations that have been accorded no less than 13 specific names (Skottsberg, 1934). A total of 23 populations, occurring on five islands, are characterized by a chromosome number of 14 pairs (Nicharat and Gillett, 1970). *Pipturus* is a monoecious to dioecious, wind-pollinated genus, so that a high level of outcrossing prevails. The recombination potential is further enhanced by a highly dispersible, small achene surrounded by a white, fleshy receptacle, the whole simulating a drupaceous fruit. The achenes probably are transported by birds (Skottsberg, 1934).

The current systematic treatment of this genus hinges almost entirely on the texture, pubescence, shape and color of the leaves. However, a thorough study of these characters, including the investigation of leaf texture through anatomical criteria (Nicharat and Gillett, 1970), brought forth the conclusion that they are of little to no value in advancing definitive species limits. Three natural hybrids are admitted in the literature (Skottsberg, 1934; Nicharat and Gillett, 1970). In one area on the island of Hawaii four "species" and nearly all possible intergrades occur in an extraordinary expression of variability. Gehring (1967) made a biometric study of *Pipturus* on Kauai and Oahu and found evidence of introgressive hybridization in populations from the uplands of Kauai and from the Waianae and Koolau ranges of Oahu. This was confirmed by the segregation of inter-specific characters in 150 greenhouse progeny of a single tree growing in the southern Koolau range. The very extensive morphological diversity in this genus confirms the predicted high level of recombination. Here we are dealing with an evolutionary stock that has maintained a high level of outcrossing and hybridization and has at the same time accomplished remarkable ecological divergences in which habitats have been exploited from 100 to 1900 meters on 6 islands.

In dealing with this complex, Hillebrand (1888) recognized only one species, while Rock (1913) recognized two species. However, the later descriptive work of Skottsberg (1934) presented a confusing assemblage of 13 species. This great disparity in the works of highly competent taxonomists would suggest that the traditional descriptive method is inadequate to the problem at hand.

4. Gouldia

In his monograph of this Hawaiian endemic of the family Rubiaceae, Fosberg (1937) dealt with a vast array of variation. It is now known to be a high polyploid (Skottsberg, 1953), with a chromosome number of $2n = 72–105$. The plants are gynodioecious, but the hermaphroditic individuals are self-fertile (Fosberg, 1937), so that there is a high level of outcrossing combined with a measure of inbreeding. The fleshy fruits are bird dispersed, this promoting a high level of recombination.

Gouldia is comprised of only three species, yet it is one of the most frequently encountered groups in the Hawaiian flora. Fosberg's extensive citation of natural hybrids makes it quite clear that hybridization has played a very important role in the development of the extreme polymorphism in the genus. The genus has established populations over a broad range of rain forest habitats on the 6 major islands, and is encountered with a frequency that leaves little doubt as to its highly successful evolutionary projection and ecological divergence. There can be little doubt that hybridization and a continued high level of recombination have played significant roles in this success story.

Fosberg's monograph (1937) presents an example of the application of traditional descriptive methods and formal Latin trinomials and quadrinomials to a broad array of natural hybrids. In dealing with only one species, *Gouldia terminalis* (H. & A.) Hillebrand, he recognized 34 varieties, 50 forms, and cited a total of 58 putative natural hybrids of a wide range of combinations, some of them interspecific.

These four cases of hybridization each involve sizeable populations of hybrid background. They represent somewhat of a random sample of angiosperm families, so that a broad expression of this process is suggested for the flora as a whole. Several additional cases of hybridization can be cited from the literature, and this has been done (Carlquist, 1966b). However, the description of a specimen as a hybrid does not in itself constitute evidence that hybridization is operative as a significant evolutionary dynamic.

DISPERSAL

Each of the four genera cited above is distinguished not only by a remarkable propensity for evolutionary change, but also by an equally remarkable capacity for inter-island dispersal. *Bidens* has established populations on 7 islands, and the other three genera occur on the 6 major islands. There is little evidence of depression in dispersal potential for these genera.

The evidence against loss of dispersibility in these four genera is paralleled in Hawaiian Drosophilidae (H. L. Carson, personal communication) in which a very high level of speciation is correlated with the retention of wings and an impressive capacity for flight. While the evolutionary dynamics of Drosophilidae involve abundant speciation through the frequent development of genetic barriers (as opposed to the infrequent genetic barriers in angiosperms), dispersal in Drosophilidae and angiosperms probably serves a common purpose—the expansion of territory through the exploitation of additional habitats.

The clinal diminution of achene awns in *Bidens* as populations ascend above the coastal habitat would suggest a progressive loss of dispersibility (Carlquist, 1966a). However, the upland *B. menziesii*, with awnless to nearly awnless achenes, has an inter-island distribution that does not support this concept. *Bidens menziesii* is, in fact, one of the most widely dispersed species, with populations on Hawaii, Maui and Molokai. In addition to these, there are putative hybrid populations on Oahu and

Kauai on which *B. menziesii* has left its mark (Gillett and Lim, 1970). Therefore, the earlier distribution of this species may have included Oahu and Kauai. Populations of *B. menziesii* cover extensive territory—on Hawaii extending from 1000 to 1800 meters over an area of many square miles. There is no evidence of precinctiveness for this species.

One possible explanation for the broad distribution of *Bidens menziesii* is that the genes for the awnless condition are carried in achenes produced by hybrids. Since the awned condition is dominant, the awnless recessive (Gillett and Lim, 1970), such achenes would bear awns and possibly could be widely dispersed, the awnless expression segregating out in subsequent generations. While this might explain the dispersal of some populations, it is excessively speculative to assign a hybrid origin to the many populations of *B. menziesii* on Molokai, Maui and Hawaii.

The evidence forces us to concede that the dispersal of Hawaiian *Bidens* is poorly correlated with the presence or absence of the retrorsely-barbed terminal awns of the achenes. Absence of achene awns is not correlated with geographical restriction. The evolutionary role of the loss of achene awns therefore remains obscure, and is not explained by the theory of loss of dispersal (Carlquist, 1966a).

SELECTIVE VALUE OF HYBRIDIZATION

Rattenbury (1962), and Carlquist (1966b) each present a discussion of the selective value of hybridization. Comments of the former are directed to the extensive array of natural hybrids in the flora of New Zealand, while those of the latter are more broadly assigned to floras of oceanic islands in general.

The ingress of a species to an oceanic island or archipelago begins with the chance operation of long-distance dispersal. A successful dispersal event is a potential basis for an isolated founder population with limited genetic resources, comprised of a small, random sample of the total gene pool of the species. The founder population, unless it is a recent product of hybridization, has a gene pool that is precariously restricted in its capacity for conveying diversity to this small assemblage on which the selective forces of the new habitat are immediately operative.

In successful establishment after long-distance dispersal there is an undeniable selective value for self-compatibility (Baker, 1955). Therefore, in self-compatible plants, of which there are many in the Hawaiian flora, the initial stages of the small founder population will be marked by generations of inbreeding and a further depletion of the limited genetic resources, until such time as outcrossing provisions are established.

The destiny of a founder population hinges on several factors, including the facility with which it maximizes its genetic resources, the availability of "open" habitats, and the accessibility of available habitats by the adaptive potential. Therefore, survival centers largely on the capacity of the population to accomplish an adaptive shift that will project it within the extremes of the new habitat. This first step requires a fairly rapid change in the breeding system. In a self-compatible outcrosser such as *Pipturus*, with a founder population of wind-pollinated, monoecious individuals, there is an immediate response to selective pressure for outcrossing, and continued selection has developed the dioecious to near-dioecious habit common in current populations of this genus.

In hermaphroditic, entomophilous genera such as *Scaevola*, outcrossing hinges upon the selection of protandry. In many other hermaphroditic lines, more drastic changes have been selected, these projected to diclinous flowers. The gynodioecious habit is a common end product in the latter group. The high incidence of dicliny in the Hawaiian flora is discussed by Carlquist (1966b), and the very high selective pressure required for the development of unisexual flowers is stressed by Grant (1958) and Prout (personal communication).

The adaptive change in the founder population is thus seen to have been relatively drastic in genera in which dicliny has evolved and less so in others, such as *Pipturus*, where outcrossing provisions were present from the beginning. In either instance there can be little doubt that there is a relatively intense selective force operative for the retention and increase of genetic diversity.

These dynamics are functional in the exploitation of new habitats after the successful establishment of the founder population. As this expansion of territory develops, the failure of selection to produce genetic barriers preserves compatibility between populations (Rattenbury, 1962) so that the gene pool is of a given invasive stock related to all populations. Experimental evidence for this is meagre in terms of number of genera, although in one of these (*Bidens*) it is particularly convincing. A successful cross has been accomplished between two species of *Santalum*, *S. album* (Lesser Sunda Islands) and *S. freycinetianum* (Hawaii) by Charles Poole, of the University of Hawaii. Derral Herbst, of the Pacific Tropical Botanical Garden, has successfully crossed two species of *Schiedea* (Caryophyll.), *S. globosa* (Oahu) and *S. verticellata* (Nihoa). Both of the latter crosses produced vigorous F_1 hybrids, and several *Bidens* crosses have produced vigorous F_2 hybrids (Gillett and Lim, 1970).

The above dynamics are operative throughout the career of a given evolutionary line that accomplishes ecological divergence on an island within an oceanic archipelago. The available habitats often occur over climatic and edaphic gradients of an impressive scale, so that there is continued selective pressure for the diversity provided by outcrossing and hybridization. An invasive stock with the capacity for successful ingress to an archipelago as isolated as the Hawaiian Islands is almost certain to be successful in colonizing additional islands in the group. The latter process is infinitely simpler and carries more favorable odds. While ingress to Hawaii might occur no more frequently than once every 20,000 years (Fosberg, 1948), successful ingress from one island to another might conceivably take place within the time span of a few thousand years. The failure in the development of genetic barriers facilitates the influx of genes through hybridization, these conveying flexibility for the exploitation of the spectrum of habitat gradients in the new insular situation.

The selective pressure for diversity would be intensified by secular climatic changes in the environment, and in Hawaii these have not been of small magnitude. Some have been imposed by Pleistocene cycles, although these have been less intense than those of New Zealand, related to cyclic hybridization by Rattenbury (1962). Broad, profound secular changes were imposed as the topography was changed by volcanic eruptions. Thus the island of Hawaii, through successive lava flows, has been built from near sea-level to over 13,000 feet in approximately one million years. Among other changes, this has created marked temperature and rainfall gradients, including areas of very high precipitation, "rain shadows" where precipitation is less than 10 inches per year, and a broad range of territory and elevations with intermediate rainfall.

Hybridization in the Hawaiian flora may thus be viewed not as an infrequent phenomenon related to a few genera of a peculiar nature, but as an evolutionary dynamic of fundamental importance and with a broad relationship to the flora, probably occurring in many genera. The lack of recognition accorded natural hybridization derives from the infrequent application of experimental methods. The question of the occurrence of natural hybridization, or its absence, often demands an objective, experimental approach and cannot be answered by subjective assertions. The application of traditional descriptive methods to such problems tend to conceal and intensify, rather than to clarify them.

REFERENCES

BAKER, H. G. (1955). *Evolution* **9**, 347–348.

BALGOOY, M. M. J. VAN (1960). *Blumea* **10**, 385–430.

CARLQUIST, S. (1966a). *Evolution* **20**, 30–48.

CARLQUIST, S. (1966b). *Evolution* **20**, 433–455.

FOSBERG, F. R. (1937). "The Genus *Gouldia* (Rubiaceae)." Bishop Museum Bulletin 147.

FOSBERG, F. R. (1948). *In* "Insects of Hawaii" (E. C. Zimmerman, ed.), Vol. 1, pp. 107–119. University of Hawaii Press, Honolulu, Hawaii.

GEHRING, PHILIP E. (1967). A study of variation in Hawaiian species of the genus *Pipturus* of Kauai and Oahu. Unpublished M. S. dissertation, Univ. of Hawaii, Honolulu.

GILLETT, G. W. (1966). *Evolution* **20**, 506–516.

GILLETT, G. W. and LIM, E. K. S. (1970). "An experimental study of the genus *Bidens* (Asteraceae) in the Hawaiian Islands". *Univ. Calif. Publs Bot.* **56**, 1–63.

GRANT, V. (1958). *Cold Spr. Harb. Symp. quant. Biol.* **23**, 337–363.

MENSCH, J. A. and GILLETT, G. W. (1972). *Brittonia* **24**, 57–70.

NICHARAT, S. and GILLETT, G. W. (1970). *Brittonia* **22**, 191–206.

ROCK, J. F. (1913). "The Indigenous Trees of the Hawaiian Islands". Published by patronage, Honolulu.

ROCK, J. F. (1919). "A monographic study of the Hawaiian species of the tribe Lobelioideae, family Campanulaceae". *Mem. B. P. Bishop Mus.* **7** (2).

SKOTTSBERG, C. (1934). "*Astelia* and *Pipturus* of Hawaii". Bishop Museum Bulletin 117.

SKOTTSBERG, C. (1953). *Ark. Bot.* **3**, 63–70.

SMITH, A. C. (1951). *Sci. Month.* **73**, 3–15.

14 | Phytogeography of the West African Mountains

J. K. MORTON

University of Waterloo, Waterloo, Canada

INTRODUCTION

Mountains, like oceanic islands, have long been a focal point of interest to biogeographers. They provide a window into nature's laboratory, through which we can see the processes and products of evolution. Here evolutionary forces are concentrated and focused onto a smaller number of biota in simpler communities. Environmental extremes, such as temperature, wind, radiation, rainfall and drought operate on small isolated populations of plants and animals, often with an abundance of open habitats, where competition is at a minimum; for those same extremes of environment reduce the inflow of competitors from lower altitudes. Under such conditions evolution is accelerated, and the origin and survival of new biotypes is encouraged. Here too the effects of genetic drift are sometimes apparent. Chance survival of characters may occur in small populations when these characters are not subject to strong selective pressures. Hence mountains provide an excellent open-air laboratory in which to study evolution. This is, no doubt, one of the reasons why they have attracted the attention of so many biologists. However, a further reason, I am sure, is the beauty and grandeur of many mountains, and the peculiar fascination which they hold for so many of us.

The African mountains are remarkable because many of them lie on or near the equator, so that tropical and arctic conditions occur in close association, with tropical forest and savanna at the foot and arctic ice-fields and tundra at the top of the higher peaks.

LOCATION OF HIGH GROUND

Figure 1 shows the location of the higher land in Africa and of the alpine peaks. Land over 10,000 ft is restricted to a few more or less isolated

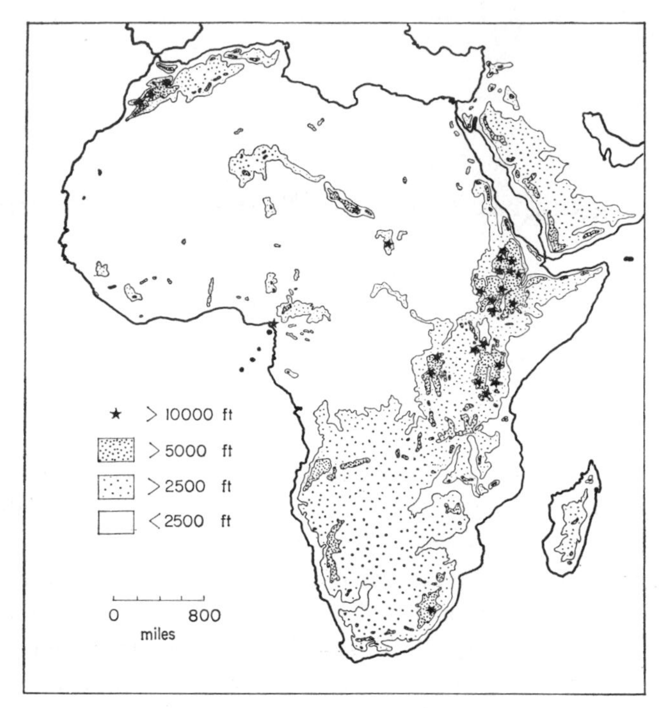

FIG. 1. Areas of high land in Africa.

peaks, mainly in the East African and Ethiopian regions. These peaks are the home of what Hedberg (1959) appropriately terms the afro-alpine flora. Mountains of lower elevation are much more frequent and many peaks, plateaux and massifs occur between 5000 and 10,000 ft. These are the home of montane floras and faunas, and present much of great interest to the biologist. At lower elevations the effects of altitude are also apparent in the floristic composition, if not in the physiognomy of the vegetation.

In West Africa (Fig. 2) the areas of higher land are fewer, lower and more distantly spaced, but each holds a characteristic association of montane species. Lower hills also occur which, in a few cases, hold remarkable assemblages of plants normally found at higher altitude. Such areas are of major importance to the biogeographer, for in times of changing climate they could form stepping stones for the migration of plants and animals, or refugia for the remnants of former floras and faunas. They emphasize that altitude is not the only factor determining the occurrence of montane species. Other factors, both environmental and historical, can partially compensate for it.

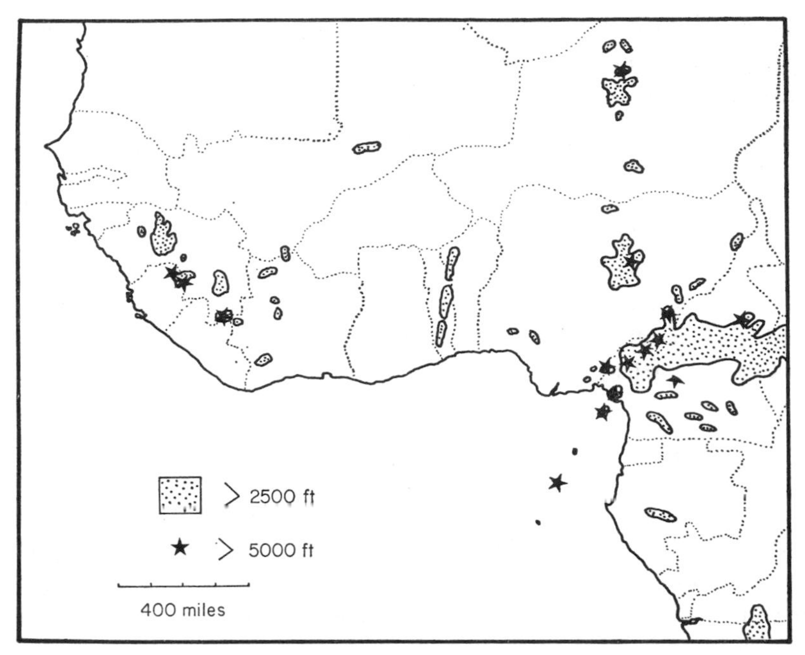

FIG. 2. High land in West Africa.

STATISTICS OF THE WEST AFRICAN FLORA

West Africa has a flora of some 7500 species. It has been the subject of intensive study over the last 20 years with the preparation of the revised edition of the Flora of West Tropical Africa. As a result our knowledge of the taxonomy, nomenclature and distribution of the vascular plants of this region is probably more complete than that of any other major region in tropical Africa—perhaps even in the tropical world. The following analysis deals only with the dicotyledons (the revision of the monocotyledons is not yet published) and covers both mainland West Africa and the offshore islands of Principé, S. Tomé and Annobon (data from Hepper and Keay, 1954, 1963; Exell, 1944, 1956). Out of 5091 species of dicotyledons 718 (14%) are montane. I have defined montane species as those predominantly or exclusively associated with land over 2500 ft. Land of this elevation is markedly disjunct in West Africa and can be grouped into 4 major systems.

(1) The Cameroons system—i.e. Cameroons Mt., Fernando Po (only 40 miles offshore), the Adamaoua and Mandara ranges extending inland

to near Lake Chad; the Bauchi plateau, and isolated peaks in southern Nigeria—Oban and the Idanre Hills.

(2) The Togoland system—the Togoland Hills and Atakora Mts.

(3) The Guinean system including the Nimba and Loma Mts, Tingi Hills and the Fouta Djalon, etc.

(4) The oceanic islands of Principé and S. Tomé (Annobon is too low to have a montane flora).

Within the montane flora of West Africa 47% of the species are confined to West Africa—i.e. are West African endemics; 53% are species which also occur outside West Africa, mainly on the mountains of East and Central Africa, i.e. are widespread species.

ENDEMISM

Most of the endemics are confined to only one of the four mountain systems which I have described; only 15% occur on two or more systems. The Cameroons system is by far the richest in endemics with 48%. The Guinean system has 20%. The islands have 15% and the Togoland system only 2%.

The high incidence of endemism (47%) in the mountains of West Africa is remarkable and is an indication both of the active evolution which has occurred, and is still occurring, in these mountains, and also of the considerable degree of isolation to which this flora has been subject. Many of these endemics are species of apparently recent origin with close relatives either in the montane vegetation or in the lowland forest and savanna.

As shown in Table I, several genera and families have been particularly successful in exploiting the opportunities afforded by these mountains and islands, thereby filling the ecological niches with new species.

Equally remarkable is the dearth of endemics in most of the larger mainly tropical families which make up a major part of the lowland forest floras. The numbers of endemic species in these families are as follows: Annonaceae 3; Menispermaceae 3; Woody Papilionaceae 3; Caesalpiniaceae 0; Mimosaceae 0; Moraceae 1; Meliaceae 3; Connaraceae 1.

A comparison of endemics in the mainland mountains with those of the islands (Table II) provides an explanation of these data.

The significance of these figures lies, not in the dearth of montane savanna endemics on the islands (this is to be expected, for there is no montane savanna on them), but in the high percentage of such endemics

TABLE I. Endemic montane species in W. Africa.

	No. of montane species in West Africa	No. of these which are endemics	Habitat, etc.
Dissotis	17	15	forest edge, rocks, grassland
Impatiens	17	13	mainly forest
Vernonia	38	14	forest edge and grass
Lobelia	12	6	mainly grassland
Plectranthus	12	7	forest and grassland
Solenostemon	8	7	grassland
Begonia	7	7	6 from islands
Compositae	138	47	—
Melastomataceae	33	30	—
Acanthaceae	24	18	—
Herbaceous Papilionaceae	67	20	—
Rubiaceae	57	29	14 from islands

TABLE II. Habitat and location of endemic montane species.

	Forest	Savanna (inc. woodland and rocky scrub)
Mainland	71 (26%)	201 (74%)
Islands	48 (95%)	2 (5%)

on the mainland montane savannas. There is no lack of montane forest on the mainland, and in fact its area probably exceeds that covered by montane savanna; but only one third as many endemics have evolved in it as in the savanna. The reasons for these differences become clear when we consider the past history of these mountains and islands. As Exell (1944, 1956) observed, the islands of S. Tomé, Principé and Annobon have been isolated from the mainland since their volcanic origin during the Tertiary. Their present flora must have arisen from chance arrivals across the ocean from mainland Africa. S. Tomé (the largest of the islands, with the major area of montane habitat) lies 250 miles from Fernando Po and from the Niger Delta, and 150 miles from the Gabon coast. With such distances, and with the large expanse of coastline formed by the arc of the mainland from which migrants could originate, the ocean does not form a complete barrier to plant dispersal—rather it is an effective means of ensuring that some dispersal occurs. Anyone who has lived on the coast of Africa near the mouths of some of the major rivers, is familiar with the considerable quantities of plant material which are brought down into the sea. This ranges from whole trees and branches, complete with epiphytes, to bulbs, rootstocks, fruits and seeds. These are distributed along the coast, often carried considerable distances by tides and currents. That most of this living material perishes in the salt water is undoubtedly true, but it is equally obvious from the plants which begin to grow on the shoreline and in small lagoons etc, that many survive. Hence I am of the opinion that distances across the sea of 150 or 250 miles are insufficient to form a complete barrier to plant migration. It is by such ocean transport, of the normal forest flora of the coast of the Gulf of Guinea, that these islands were colonized. Their flora is almost wholly derived from that of the lowland forests. Most of the endemics are closely related to, and presumably evolved from, these immigrants from the mainland. The high proportion of forest endemics in the mountains of these islands is merely a reflection of the natural vacuum that existed on these islands after their volcanic origin and of the forest habitat which dominates them. Chance arrivals would find a wide range of niches in which to grow, with rich volcanic ash and abundant rain. Under these favorable conditions, with a lack of competition and with small isolated populations, rapid evolution would be likely to occur. Thus these islands became populated by species from the lowland forests of the mainland and by endemics which evolved from them and were able to colonize the montane habitats. The montane flora of these islands consists of 84 species (dicotyledons only) of which 50 are endemic. The other 34 occur on the mainland, having either migrated to or from the islands. All except 3 have obvious possible means of dispersal over the distances involved. Fourteen have seeds or fruits eaten by birds or bats; 10 have sticky or small seeds which could become attached to birds; 3 have efficient wind-borne fruits (with pappus); and 4 have large floating fruits which may have been water-borne at a time when vegetation belts were lower than they are today.

In contrast to the islands, the montane forests of the mainland have not shown the same degree of speciation because they have always presented a closed system, which has usually been continuous with the lowland forests. Any changes in climate which occurred would merely elevate or depress the upper limit of the forest. Hence no ecological vacuum existed in which accelerated speciation would be favoured, and as a result few montane forest endemics have evolved.

The montane grasslands, on the other hand, present a situation comparable to that on the islands. They were colonized by species from the lowland savanna and by migrants from other mountain systems. Changes in climate, both temperature and rainfall, caused advances of the forest, and most of the areas of montane grassland were probably greatly reduced or eliminated, except on the high alpine peaks of the Cameroons. However, cliffs and rock outcrops formed refugia in which dense forest could never grow, regardless of climate. Here many montane grassland species survived, though doubtless many others became extinct. Such refugia within the limits of the present range of forest can be seen today in many parts of West Africa, e.g. the Nkawkaw Scarp in Ghana, the Idanre Hills of Nigeria, the hills of Bafodia and the Freetown Peninsula in Sierra Leone. Further oscillations in climate caused recession of the forest, leaving considerable areas on the mountain tops suited for grassland species, but with relatively few such species available to colonize them. A partial ecological vacuum was created, similar to that on the islands after their volcanic origin. In this vacuum rapid speciation occurred. New types arose from species which moved into the montane savanna habitat from the refugia, the surrounding forest and from further afield, and adapted to it.

This I believe to be the explanation for the high endemism in the montane grassland floras of the mainland and the montane forest flora of the islands. Both have been areas of evolutionary opportunity. Speciation was stimulated by the creation of ecological vacua which could not be filled from surrounding sources. In the case of the mainland mountains, climatic changes were the cause, through a shift in the limits of the forest; in the case of the islands it was their volcanic origin, *de novo* out of the sea.

The process of speciation on the mountains of the mainland can be likened to an evolutionary pump. Climatic changes drive the pump. The piston is the forest which advances and retreats on the mountain sides. The montane grasslands on the summits are thus alternatively compressed and allowed to expand. New types of plants evolve to fill the expanded grassland habitats. The next stroke of the pump will not destroy all these new types. In many cases it will merely compress them and bring them into close proximity in the refugia. This will stimulate the production of more variation through hybridization which will provide the raw material for another spurt of evolution during the next expansion of the grassland habitats when the piston again descends.

WIDESPREAD SPECIES AND DISJUNCTION

The remaining 53% of the montane flora of West Africa consists of widespread species, i.e. those which also occur in areas outside West Africa, and in almost all cases on the mountains of East and Central Africa. They represent but a small proportion of the total montane flora of tropical Africa. Hooker (1864) in the first account of the flora of Cameroons Mountain pointed out that it was but an impoverished version of the flora of the other high mountains of Africa. It is often assumed, no doubt correctly, that most of these plants migrated into West Africa, though I am sure some also migrated in the other direction to enrich the floras of East and Central Africa. Analysis of these widespread species shows that 36% only get as far as the Bauchi Plateau and interior part of the Cameroons system (southwards to the Bamenda Highlands); 29% penetrate to Cameroons Mountain and Fernando Po; 9% have reached the islands of the Gulf of Guinea; 2% occur on the Togoland system, but not further west; and 20% extend to the Guinean system.

A further 4% have reached the Guinean system but are absent from the intervening mountains of West Africa, i.e. they either made a long distance hop from East or Central Africa to the Guinean system, or they migrated across the intervening area and then died out (or collectors have yet to find them!).

ADAPTATIONS TO THE AFRICAN MONTANE ENVIRONMENT

What characteristics have enabled plants to migrate into montane habitats in Africa and to speciate there? Information is fragmentary, but several aspects warrant comment, particularly those related to daylength, reproductive biology and polyploidy.

1. Daylength

Many of the afro-alpine species (which occur on the high peaks of the East African mountains and the Cameroons) belong to genera, and in some cases species, mainly found in the northern hemisphere, where they flower under the long day conditions of the northern summers. Of 54 species of afro-alpine plants which I have successfully grown to flowering and which belong to these high latitude genera, all appear to be daylength neutral—none are short-day species. This is in marked contrast to many of the species belonging to lowland tropical genera, which are short-day plants. It would appear that species which migrated from the Northern regions into afro-alpine habitats have not undergone any basic change in the physiology of flowering; they have not become short-day plants. Thus it may be presumed that they were pre-adapted to the daylength conditions of the African mountains.

2. *Reproductive Biology*

Most of the African montane species which I have grown in cultivation have proved to be self-compatible and in many cases self-pollinating. This applies to all the species of the genera (Table III) which I have been able to study.

TABLE III. Self-compatible genera of the montane flora.

Family	Genus*
Gramineae	*Poa*, *Agrostis*, *Aira*, *Deschampsia*, *Pennisetum*
Cyperaceae	*Carex*, *Cyperus*
Juncaceae	*Luzula*
Caryophyllaceae	*Uebelinia*, *Stellaria*, *Cerastium*, *Sagina*, *Silene*
Labiatae	*Satureja*, *Salvia*, and most species of *Solenostemon* and *Plectranthus*
Compositae	*Crassocephalum*, *Dichrocephala*, *Laggara*, *Aspilia*, *Crepis*, *Sonchus*, many species of *Helichrysum* and *Vernonia*
Rubiaceae	*Galium*, *Anthospermum*
Umbelliferae	*Sanicula*, *Caucalis*, *Cryptotaenia*, *Pimpinella*
Geraniaceae	*Geranium*
Rosaceae	*Rubus*, *Alchemilla*
Papilionaceae	*Trifolium*, *Indigofera*, *Desmodium*, and at least some species of *Crotalaria* and *Vigna*
Urticaceae	*Parietaria*

* *Note:* It is possible that a few of these species may be apomicts.

In marked contrast is the large-flowered genus *Dissotis*, which has speciated so freely in the West African mountains and produced many endemics. These are mostly insect pollinated and in many cases self-incompatible. Few members of the montane flora are obligate inbreeders and most have retained an unspecialized reproductive biology—a valuable adaptation for species migrating into new areas.

3. *Polyploidy*

Polyploidy has also played a role in the evolution of these montane floras. Several years ago (Morton, 1966) I reported that the incidence of polyploidy in the flowering plants of Cameroons Mountain was 49%, almost the same as that in the East African mountains (45%). This is in marked contrast to the lower percentage (26%) found in the flora of West Africa as a whole. The association of polyploidy with montane habitats is well seen in a number of species which occur in both the lowlands and the mountains (Table IV).

TABLE IV. Intraspecific polyploidy and the montane habitat.

Name	Somatic chromosome number
Platostoma africanum P. Beauv.	14, 28, 42
Crassocephalum crepidioides (Benth.) S. Moore	20, 40
Commelina benghalensis L.	22, 44, 66
Commelina diffusa Burm. f.	30, 60
Aneilema beniniense (P. Beauv.) Kunth	52, 78
Aneilema umbrosum (Vahl) Kunth	20, 40, *ca* 100
Crinum ornatum (Ait.) Bury	22, 44, 88
Eupatorium africanum Oliv. and Hiern.	22, 44
Hyptis lanceolata Poir.	32, 64

TABLE V. Interspecific polyploidy and the montane habitat.

Species	Somatic chromosome no.	Species	Somatic chromosome no.
Rubus fellatae A. Char., *R. pinnatus* Willd.	28	*R. exsuccus* Steud.	42
Vernonia undulata Oliv. & Hiern.	18	*V. blumioides* Hook. f.	36
Achyrospermum oblongifolium Bak.	28	*A. africanum* Hook. f.	84
Microglossa volubilis DC.	18	*M. densiflora* Hook. f.	36
Cardamine africana L., *C. hirsuta* L.	16	*C. trichocarpa* Hochst.	32
Geranium simense Hochst. ex. A. Rich.	28	*G. ocellatum* Cambess.	56

Many of the polyploids of Table IV are probably incipient new species and several which are morphologically distinct have been given infraspecific recognition. Cases involving distinct but related species are also frequent (Table V).

Hence it is clear that polyploidy has been a significant factor in the evolution of the montane floras.

THE SIGNIFICANCE OF DISJUNCTION

Fifty-three per cent of the montane species of West Africa also occur outside this region and as a result have a markedly disjunct distribution.

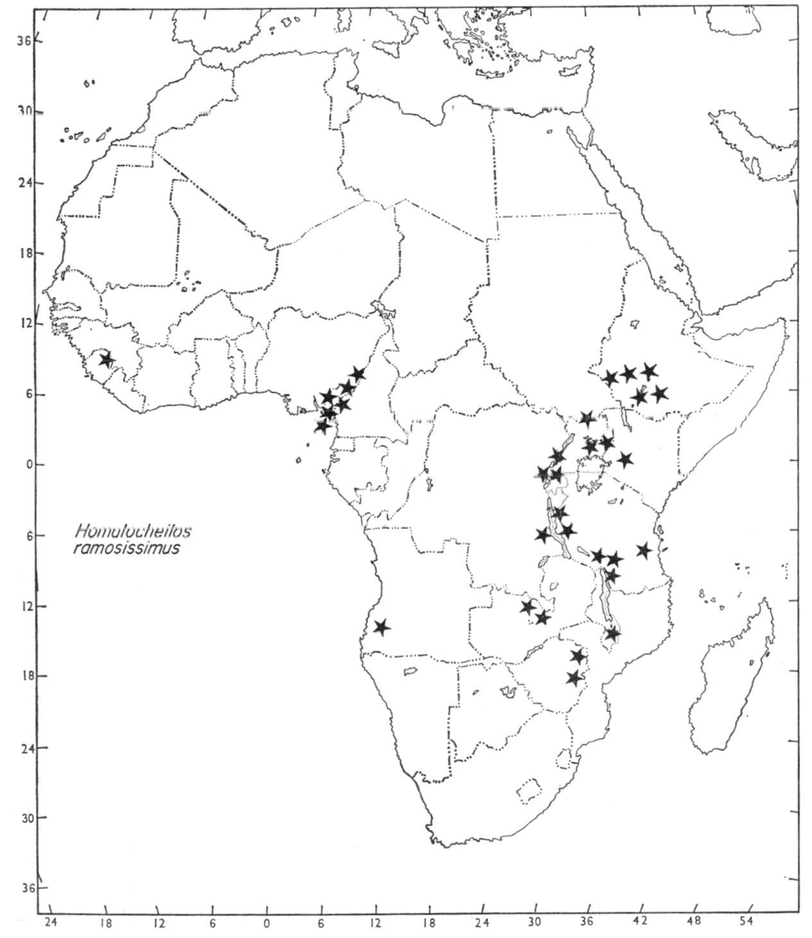

FIG. 3. Distribution of *Homolocheilos ramosissimus*, showing a typical pattern of disjunction.

Homalocheilos ramosissimus (Hook.f.) J. K. Morton (Fig. 3) is a typical example. Many of their stations are separated by considerable distances. I have previously suggested (Morton, 1961) that these disjunctions could only be explained on the basis of past climatic changes in Africa, which would have extended the areas occupied by montane species, though not necessarily making them continuous. Data on present day altitudinal ranges of many of the montane species suggests that a lowering of vegetation belts of up to 3000 ft (i.e. equivalent to a drop in temperature in the region of 4° to 6°C) must have occurred in relatively recent times. (Most of the montane species have diverged but little in their disjunct populations and hence have probably not been isolated for very long.)

Why postulate climatic change? Could not long-distance dispersal account for these disjunctions? That some long-distance dispersal has occurred I do not doubt. The small seeds of such plants as *Drosera* and *Radiola* could become attached to the feet or feathers of birds, which might carry them appreciable distances. Similarly the spores of ferns and the minute seeds of orchids would appear to be well adapted to long distance dispersal in the upper atmosphere. However, many heavy-seeded plants have no conceivable means of traversing great distances across the land, where flotation on ocean currents cannot be involved—e.g. *Crossandra massaica* Mildbr. and *Ochna ovata* F. Hoffm., both of which are confined in West Africa to the Shai Hills of Ghana but occur in east or central Africa—a disjunction of over 2000 miles. To discount wholly climatic change we must be able to show that such plants have a means of dispersal across the distances involved. In many cases no such means exist. An analysis of dispersal adaptations in the flora of one of the mountain groups in West Africa, the Loma Mountains and Tingi Hills (Table VI), gives some indication of how insufficient are theories of long-distance dispersal in accounting for present-day distributions patterns. These mountains have 183 montane species of flowering plants (including monocotyledons).

It appears that less than half of this flora has any apparent possible means of traversing appreciable distances, and even amongst these it is doubtful whether many could travel from one side of the continent to the other—distances of 1000 to 2000 miles. However, given climate change, these distances would be greatly reduced. Upland areas lying between the main mountain systems could support montane communities and become stepping stones for the spread of the montane flora. In West Africa, lowering of vegetation belts by about 2000 ft would create a situation in which montane communities were rarely if ever more than 100 miles apart,

TABLE VI. Dispersal of seeds in 183 montane species of the Loma Mountains and Tingi Hills

Mode of seed dispersal	Percentage of flora
Small seeds, possibly suited for dispersal by wind or on the feet and feathers of birds	22
Succulent fruits; or seeds eaten by birds or bats	12
With effective means of wind dispersal (e.g. pappus)	6
Total adapted to some form of long-distance dispersal	40
With sticky seeds or fruits, suited for dispersal by mammals	9
Inefficient means of wind dispersal suited for local transport only	6
Explosive mechanisms	6
Adapted for dispersal by ants	2
Total adapted for local dispersal	23
With large seeds or fruits with no apparent means of dispersal	37

well within the range of dispersal by roving animals, birds and wind; the constipated duck or elephant then becomes a real factor in plant dispersal! A lowering of vegetation belts by 2000 to 3000 ft in the area separating the Cameroons system and the mountains of East and Central Africa would have a similar effect and a series of short hops would be all that was necessary to get the present montane flora across the continent. It should be noted that even today, few of the afro-alpine species of the Cameroons are confined to the summit peaks. Most can be found as low as 6000 ft and in many cases much lower. The afro-alpine element does not extend west of the Cameroons system.

In West Africa many upland areas exist between the main mountain systems, e.g. Idanre and Ado in Nigeria, the Shai, Atewa and Banda Hills and the Gambaga Scarp in Ghana, the Gbengbe and Kuru Hills in Sierra Leone. Most of these already hold a small and presumably relict montane flora, indicating that they have functioned as such stepping stones for migration under former climates which were more favourable to montane species. A corridor of high land connects the mountains of Central and

East Africa with the Cameroons system, and out of this, smaller peaks rise which would have formed the links in a chain of montane communities across the continent.

That climatic changes of the size envisaged have occurred has been demonstrated in recent years by many workers in a variety of fields. The evidence has been ably collated by Coetzee and Zinderen Bakker (1970) and Zinderen Bakker (1969), and reviewed by Hedberg (1969). It comes from numerous studies on fossil pollen in many parts of Africa; from fossil snow lines and glacial drift on the high mountains of East Africa; from fossil evidence of frost activity at low altitudes in North and southern Africa; from lake deposits in the Sahara; and from studies on other groups of organisms including birds (Moreau, 1966), butterflies (Carcasson, 1964) and the tropical forest (Aubréville, 1949). As observed by Livingstone (1967) and Hedberg (1969), some of the data from pollen analysis and glacial phenomena may be interpreted in terms of either a depression in temperature or a change in rainfall and humidity. However, much of the other evidence clearly indicates a lowering of temperature. The evidence from these sources indicates that 2 major cool periods existed, together with a third more moderate period. These extended between 50,000–43,000 years ago, 30,000–12,000 years ago and 5500–4700 years ago. The depression in temperature involved in the first two has been variously estimated at 4° and 9°C in tropical Africa, greater in North Africa. The more recent cool spell was apparently less severe, but it too would have an effect on plant distribution. The last major change in Africa, that between 30,000 and 12,000 years ago, was contemporaneous with the Würm glacial advance of Europe (which is equated with the Wisconsin of North America) and has been termed the Mount Kenya Hypothermal (Zinderen Bakker, 1969).

Though both Exell (1944, 1956) and Hedberg (1969) tend to discount climatic change as a factor in the distribution of the African flora, they worked on very specialized floras, Exell on oceanic islands and Hedberg on the alpine flora of the high mountains. In both cases long-distance dispersal must have played a major role in producing the present day floras, though the distances involved range only from about 100 to 300 miles. In the case of the East African mountains it could have been less under different climatic conditions. It would be wrong to extrapolate from the peculiar and very specialized problems of these island and high alpine floras, to the whole of the African upland floras. Biogeographers are frequently guilty of such extrapolations; hence we have those who would explain all distributions in terms of, on the one hand long distance dispersal, or on the other of climatic change and land bridges. If there is any consistency in Nature it is in her diversity—diversity both in form, function and process—and this is no less true of distribution methods.

CONCLUSIONS

It is my belief, on the basis of the evidence presented here, that the present-day distribution of the montane floras of Africa is the result of:

(1) Climatic change which has created major advances and recessions of the montane (and lowland) vegetation.

(2) Dispersal, most of it over relatively short distances at periods of maximum extent of the montane vegetation, when most of the distances involved would have been in the region of 50 to 100 miles—well within the capacity for dispersal by wind, birds and wandering animals. Longer distance dispersal has also doubtless occurred, particularly of species with minute seeds which might be carried by wind and migrating birds. In the case of the oceanic islands dispersal by the sea has been the major factor.

(3) Rapid speciation in (a) the montane habitats resulting from a type of "evolutionary pump" operated by climatic change; (b) the forested islands where an ecological vacuum was created by their volcanic origin in the Gulf of Guinea.

REFERENCES

AUBRÉVILLE, A. (1949). Climats, forêts et désertification de l'Afrique tropicale. Société d'Editions Géographiques Maritimes et Coloniales, Paris.

CARCASSON, R. H. (1964). *Wild Life* **2**, 122–157.

COETZEE, J. A. and ZINDEREN BAKKER, E. M. VON (1969). *S. Afr. J. Sci.* (March, 1970), 78–84.

EXELL, A. W. (1944). Catalogue of the Vascular Plants of S. Tomé. British Museum, London. Supplement, 1956.

HEDBERG, O. (1959). *Symb. bot. upsal.* **15**, 1.

HEDBERG, O. (1969). *Biol. J. Linn. Soc.* **1**, 135–148.

HEPPER, F. N. and KEAY, R. W. J. (1954, 1963). *In* "Flora of West Tropical Africa" (J. Hutchinson and J. M. Dalziel, eds), Edition 2, London.

HOOKER, J. D. (1864). *J. Linn. Soc.* **7**, 171–240.

LIVINGSTONE, D. A. (1967). *Ecol. Monogr.* **37**, 25–52.

MOREAU, R. E. (1966). "The Bird Faunas of Africa and Its Islands". Academic Press, London and New York.

MORTON, J. K. (1961). *In* Comptes rendus de la IVe réunion plenière de l'AETFAT, 391–409, Lisbon.

MORTON, J. K. (1966). *In* "Chromosomes today" (C. D. Darlington and K. R. Lewis, eds), Vol. 1, pp. 73–76. Oliver and Boyd, Edinburgh.
ZINDEREN BAKKER, E. M. VON (1970). *Acta bot. neerl.* 18(**1**), 230–239.

Section IV

GEOGRAPHICAL EVOLUTION IN GENERA AND FAMILIES OF SPECIAL INTEREST

15 | The Distribution and Variation of some Gesneriads on Caribbean Islands

BRIAN MORLEY

National Botanic Gardens, Glasnevin, Dublin, Eire

INTRODUCTION

Of the 54 genera and about 1000 described species of tropical American gesneriad, 12 genera and 125 species occur on Caribbean islands (Wiehler, 1970a). Some Caribbean gesneriad genera are found almost exclusively in this area such as *Gesneria* or *Rhytidophyllum* while others are equally or better represented in Central and South America, such as *Alloplectus* or *Columnea*.

Gesneriads are well known for their endemic and restricted geographical occurrence in both mainland and island habitats, and for this reason, and for others concerning their pollen vectors, the distribution and variation of the 17 Caribbean *Columnea* taxa and the 5 *Alloplectus* taxa repay study. The groups from the Greater Antilles (i.e. Cuba, Jamaica, Hispaniola and Puerto Rico) and from the Lesser Antilles (excluding Trinidad and Tobago), differ phenotypically in ways which may help towards an understanding of the evolution of these gesneriads.

The Greater and Lesser Antilles have different geological histories and occupy different loci along "plant migration" corridors which are thought to have provided communication between the Caribbean and continental American floras of the Oligocene. To quote Briggs (personal communication), "The forces that created the Greater Antilles as a more or less continuous chain in Eocene or possibly early Oligocene time were tectonic in nature, signifying an end to all but the most limited volcanism, after 100 million years or more of extensive volcanism along the north edge of the Caribbean basin. The Lesser Antilles combine volcanic and tectonic origins, for they mark a line of underthrusting of Atlantic oceanic crust as well as a line of Eocene (or earlier) to Holocene volcanism. Compressive buckling owing to underthrusting and immense quantities of volcanic detritus built the Lesser Antilles island arc of today." Palaeobotanical and geological evidence lends support to the existence of two distinct migration routes, one connecting the Greater Antilles to Mexico and Central America, and the other connecting the Lesser Antilles to northern South America. To oversimplify, the floras of the two Antillean areas reflect appropriate affinity with either Central or South America (Graham and Jarzen, 1969; Howard, 1970). Whether tectonic forces alone account for the development of the necessary migration routes, or whether some form of Caribbean continental drift can have been involved, as suggested by some workers, it is not yet possible to say (Dietz and Holden, 1970). The past history of the Antilles fits in well with the "new global tectonics" of sea-floor spreading and continental drift.

To illustrate the results of speciation which has occurred in the Greater Antilles, a brief description of the often erect, sub-shrubby Jamaican Columneas is given, as this island possesses the highest number of endemic Columneas anywhere; 100% are endemic. To illustrate speciation in the Lesser Antilles the sole representatives chosen are, *C. scandens* L. and *Alloplectus cristatus* (L.) Mart. Mention will also be made of *C. tocoensis* Britt. from Trinidad, *C. tulae* Urb. from Puerto Rico and Haiti, and *C. tomentulosa* Mort. from Panama, Costa Rica and Nicaragua.

An assumption is made throughout that significant phenotypic, biochemical, anatomical and karyological differences are a reflection of past speciation and "evidence of a greater informational content in the organism's DNA, and therefore of evolutionary progress" as Stebbins (1971) has put it. A taxonomic species concept is implied whenever the term "species" is used. The herbarium specimens studied are listed in a taxonomic revision of the West Indian Columneas (Morley, in preparation).

SPECIATION OF *Columnea* IN THE GREATER ANTILLES

1. *Observations*

There are nearly 1000 vegetated islands in the West Indies and of these Cuba has roughly the same area as Honduras, and Jamaica is only a little smaller than Panama. Howard (1970) states that Cuba has some 50% endemic species and approximately 41 endemic genera, while Jamaica has 20% endemic species and 3 endemic genera. When the *Columnea* floras of Cuba and Jamaica are compared the position is reversed with Jamaica having 11 endemic species and Cuba only 2. Hispaniola, so far as present exploration shows, has no endemic *Columnea* but shares one of its two species with Puerto Rico, a fact which supports the concept of an "ancestral Puerto Rico landmass" which included present day Hispaniola (Graham and Jarzen, 1969).

The distribution maps in Fig. 1 show the restricted ranges of most Jamaican Columneas; and in relatively well botanized Jamaica this kind

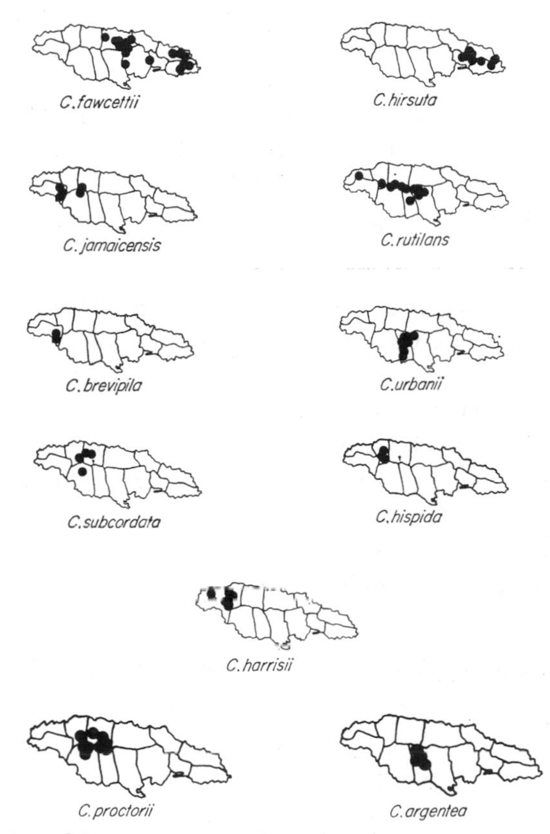

FIG. 1. Distribution of Jamaican species of *Columnea*.

of distribution is real and not due to restricted collecting. It demonstrates the endemism which so often characterizes *Columnea* species and numerous other gesneriads.

Two species such as *C. argentea* Griseb. and *C. hispida* Sw. have dissimilar phenotypes and differ in many taxonomically reliable characters for which no satisfactory adaptive significance has been found (Table I). These species have 65% phenetic association for a numerical classification

TABLE I. Phenotypic differences between *C. argentea* and *C. hispida* (All measurements in cm)

Character	*C. argentea*		*C. hispida*	
	No. in sample	Mean and s.d.	No. in sample	Mean and s.d.
Leaf pair ratio	11	1·5 ± 0·4	6	3·3 ± 1·5
Leaf shape	narrowly oblong-elliptic		broadly oblong-elliptic	
Leaf length	25	12·1 ± 1·9	9	10·3 ± 2·6
Leaf width	14	2·8 ± 0·3	6	4·8 ± 0·9
Adax. leaf-hairs	sericeous; 713/sq cm		hispid; *ca* 60/sq cm	
Petiole length	14	1·0 ± 0·2	6	1·8 ± 0·5
Corolla length	5	4·9 ± 0·3	1	3·4
Corolla tube length / Galea length	5	0·9	1	1·4

of 51 assorted characters as described by Stearn (1969). Genetic drift is a plausible but in some ways suspiciously accommodating explanation for the origin of the phenotypes of Columneas in western Jamaica shown in Fig. 2, based on data in Morley (1968).

Other Jamaican species are more similar to one another such as *C. argentea* and *C. brevipila* Urb. with 85·3% phenetic association, and *C. hirsuta* Sw. and *C. fawcettii* (Urb.) Mort. with 83·2% phenetic association. *C. hirsuta* and *C. fawcettii* are important taxa because they exhibit distributions and characters for which some adaptive significance can be argued (Table II).

TABLE II. Phenotypic differences between *C. hirsuta* and *C. fawcettii* (All measurements in cm)

Character	*C. hirsuta*		*C. fawcettii*	
	No. in sample	Mean and s.d.	No. in sample	Mean and s.d.
Leaf pair ratio	43	1·8 ± 0·5	68	1·9 ± 0·6
Leaf shape	obovate-elliptic		elliptic	
Leaf length	81	8·3 ± 1·6	113	8·4 ± 1·9
Leaf width	37	2·9 ± 0·5	45	3·1 ± 0·4
Adax. leaf hairs	pilose; 254/sq cm		strigose; 114/sq cm	
Petiole length	37	1·4 ± 0·5	45	1·1 ± 0·3
Corolla length	9	4·7 ± 0·2	18	4·6 ± 0·3
Corolla tube length / Galea length	9	0·7	18	0·7

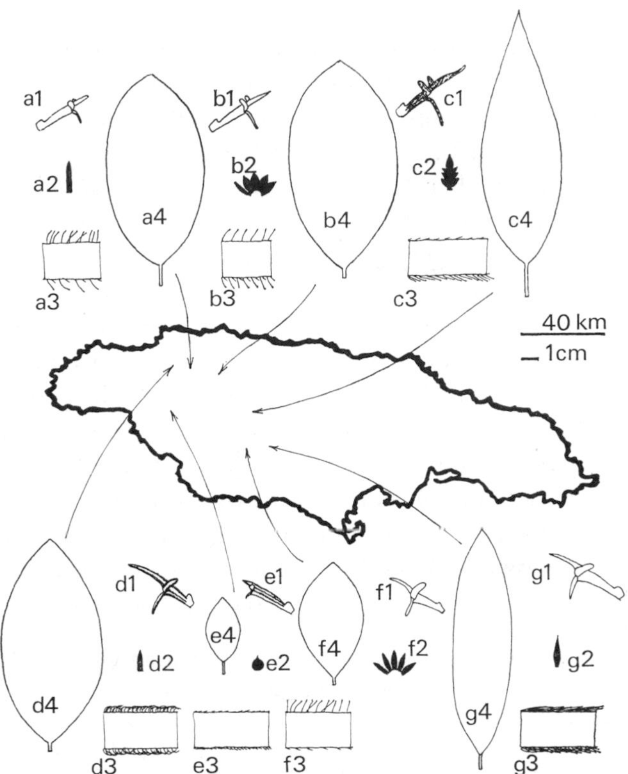

FIG. 2. The phenotypes of some Columneas from western Jamaica. a. *Columnea hispida* Sw.; 1 corolla, 2 sepal, 3 section through leaf to show vestiture, 4 leaf shape. b. *C. subcordata* Mort.; 1 corolla, 2 showing fusion of four sepals, 3 leaf section showing vestiture, 4 leaf shape. c. *C. rutilans* Sw.; 1 corolla, 2 sepal, 3 leaf section showing vestiture, 4 leaf shape. d. *C. harrisii* (Urb.) Britton ex Mort.; 1 corolla, 2 sepal, 3 leaf section showing vestiture, 4 leaf shape. e. *C. jamaicensis* Urb.; 1 corolla, 2 sepal, 3 leaf section showing vestiture, 4 leaf shape. f. *C. urbanii* Stearn; 1 corolla, 2 sepals almost free, 3 leaf section showing vestiture, 4 leaf shape. g. *C. argentea* Griseb.; 1 corolla, 2 sepal, 3 leaf section showing vestiture, 4 leaf shape. All dimensions and vestiture density are to scale except leaf thickness. Data from Morley (1968) where black and white on corollas represents red and yellow respectively.

(a) *Eastern Jamaica*. Eastern Jamaica is dominated by the climatic influence of the east–west trend of the Blue Mountains and the north–south axis of the John Crow Mountains in the extreme east of the island. *C. hirsuta* and *C. fawcettii* are the only Columneas which grow there, and there is no climatic gradient of comparable magnitude to be found in western Jamaica, which is a dissected limestone plateau. *C. fawcettii* is found between altitudes of about 80 and 950 m while *C. hirsuta* occurs at about 300 to 1700 m and only grows at lower altitude in the wet habitats of the northern John Crow Mountains. Both taxa are clearly related on the basis of phenotypic, leaf flavone, and karyotypic characters (Morley *In* Stearn, 1969) and an apparently derived concentric distribution of *C. fawcettii* around that of *C. hirsuta* but at lower altitudes (Fig. 1).

It is debatable whether *C. fawcettii* should be regarded as a "newly fledged" taxonomic species or a still diverging intraspecific taxon of *C. hirsuta*. In the juvenile condition the two taxa cannot be easily distinguished, but when mature the flowers of *fawcettii* are more brilliant brick red, and leaf vestiture and thickness differ from those of *hirsuta*. Short, adpressed 3-celled hairs occur on the adaxial leaf surface of *fawcettii*, but long, erect 6- to 10-celled hairs, as well as short adpressed hairs, occur adaxially on leaves of *hirsuta*. The layers of leaf hypodermis in Jamaican Columneas appear to serve for water storage (Morley, 1968) in agreement with the observations of Wiehler (1970b) who worked with chiefly Central American species. The layers of hypodermis vary in number from one to three as shown in Table III. The leaves of the phenotypically dissimilar *C. argentea* and *C. hispida* are seen to be more histologically

TABLE III. Histological measurements of Jamaican *Columnea* leaves (All measurements in mm)

Species	Leaf thickness	Thickness of hypod.+epid.	No. of hypod. layers	Palisade thickness
C. hirsuta	0·25	0·13	1	0·04
C. rutilans	0·36	0·13	2	0·04
C. argentea	0·49	0·13	2	0·06
C. urbanii	0·43	0·14	1 (2)	0·05
C. brevipila	0·43	0·15	2	0·04
C. subcordata	0·46	0·18	2	0·03
C. harrisii	0·37	0·20	1 (2)	0·04
C. hispida	0·45	0·22	1 (2)	0·04
C. proctorii	0·47	0·23	2	0·03
C. jamaicensis	0·45	0·24	2 (3)	0·05
C. fawcettii	0·64	0·40	2 (3)	0·05

similar than those of *C. fawcettii* and *C. hirsuta*, so great is the difference in leaf tissues between the last two species. As *C. fawcettii* grows at lower altitudes and often in drier places than *C. hirsuta*, it is logical to suggest that adaptation to drier habitats is shown in the thicker leaves of *C. fawcettii*. Rainfall data for *C. fawcettii* habitats are not available, but neighbouring stations receive 59·6 inches (Worthy Park, alt 335 m) and 51·6 inches (Ewarton, alt 259 m); precipitation in a *C. hirsuta* habitat is 104·6 inches (Hardwar Gap, alt 1220 m). *C. hirsuta*, with the thinnest leaves of all Jamaican species and with the least developed hypodermis, is the only Jamaican *Columnea* to occur in evermoist habitats above about 1100 m.

(b) *Western Jamaica*. No comparable adaptive significance has been recognized for characters of species in western Jamaica; and the absence of conspicuous ecological adaptations could relate to the lack of any pronounced environmental gradient in the west of the island. However, local climatic diversity is seen in meteorological data. Critical ecological studies have not been made on *Columnea* habitats and it is too early to dismiss speciation involving ecological selection.

The restricted distribution of some Columneas may be partly produced by physical dissection of the Jamaican landscape dating from Pliocene block faulting, tilting, and subsequent erosion. For example, the distribution of *C. brevipila* and *C. urbanii* Stearn, and intermediate taxa between the two may indicate a past relationship between the two species; they are now 30 miles apart and are separated by the Pliocene or early Pleistocene block fault valley of the Black River (Fig. 3). Near Newport in Manchester Parish there are true breeding plants of *C. urbanii* which resemble *C. brevipila* in many ways, but have calyx characters which are typically *C. urbanii* (Table IV). The proximity of the Newport specimens to the ranges of *C. urbanii* and *C. brevipila*, and similarities between the

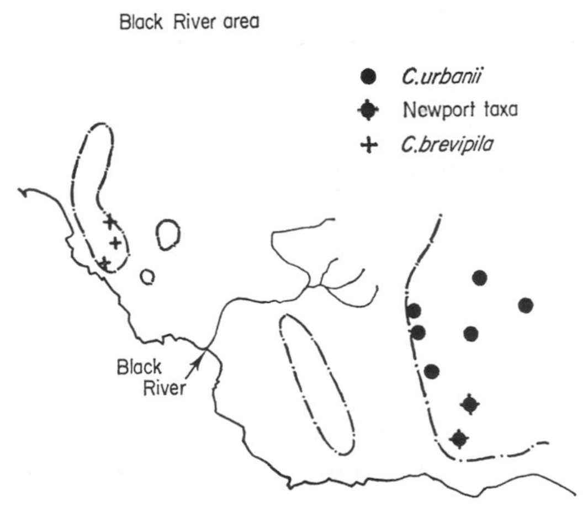

FIG. 3. Distribution of *C. brevipila* Urb., *C. urbanii* Stearn and Newport taxa.

TABLE IV. Comparison between *C. urbanii*, *C. brevipila* and Newport taxa

Character	*C. brevipila*	Newport taxa	*C. urbanii*
Leaf-hairs	2- to 3-celled, adpressed	3- to 4-celled, adpressed	5- to 7-celled, erect
Petioles	long (mean length 1·1 cm, $n = 16$)	short	short (mean length 0·4 cm, $n = 32$)
Hairs of calyx-lobes	adpressed	erect	erect
Leaves	large (mean length 9·9 cm, $n = 33$)	small	small (mean length 6·9 cm, $n = 65$)

karyotypes and leaf flavones of the two species (Morley *In* Stearn, 1969), suggests possible past conspecificity.

2. *Discussion*

Both western and eastern Columneas in Jamaica occur terrestrially and as epiphytes. When the habitat is epiphytic, as is more usual, niche availability is observed to maintain density of breeding populations at a low level, so that dense and extensive stands do not occur. When the habitat is terrestrial, there are usually more aggressive ground plants which also maintain *Columnea* populations at a low level, often along the tops of walls or in the centre of bushes. Quantitative measurements are required on the community structure of *Columnea* populations.

The majority of western Jamaican species do not cohabit with the exception of *C. proctorii* Stearn and *C. rutilans* Sw.; *C. hispida* and *C. rutilans*; and *C. urbanii*, *C. rutilans* and *C. argentea*. *C. rutilans* may be expected to overlap in range with a number of other species in western Jamaica because it has a wide distribution. Hummingbirds pollinate at least three Jamaican Columneas, and possibly all except *C. jamaicensis*, and their feeding territories, which require study, could theoretically confine gene-flow to Columneas growing within such a territory if rigidly guarded. The combination of some degree of pollen immobilization between jealously guarded hummingbird territories, low breeding population density, and topographical isolation of groups of breeding populations, could be the model which would account for the appearance of apparently non-adaptive characters in western species by a genetic drift process. Such a breeding system might also provide suitable conditions for the accumulation and fixation of structural chromosome rearrangements, which have accumulated in all Jamaican species, especially the western, and also in Columneas from elsewhere (Morley, 1972). Mayr (1963) has noted that a possible origin of truly non-adaptive characters could be from pleiotropic gene-effects which could result from structural chromosome changes. Those species which cohabit and have synchronous flowering form hybrids readily (Morley, 1971), so that hybridization may have played some part in the speciation of western species (e.g. the creation of *C. proctorii*).

All Greater Antillean islands have at least one endemic species of hummingbird (Table V), and Jamaica, with two of its three hummingbirds endemic, has completely endemic Columneas and Alloplecti. The high endemism of *Columnea* pollen-vectors in the Greater Antilles could partly explain the occurrence of reproductive isolation between the Columneas on different islands, reproductive isolation which must have been a prerequisite for subsequent intense speciation, the results of which are now to be seen in Jamaica.

Speciation of Jamaican Columneas may be summarized as involving (1) some degree of adaptation to ecological factors in at least *C. hirsuta* and *C. fawcettii*; (2) the appearance of apparently non-adaptive characters in other species by a combination of topographical isolation, structural chromosome rearrangements, and possible restriction of gene flow by the size and structure of the breeding population and the behaviour of the pollen vector in relation to the breeding population.

SPECIATION OF *Columnea* IN THE LESSER ANTILLES

1. *Observations*

This Antillean arc has been called the principal migration route between the north and south American biotas during most of the Tertiary. This is because the Panama–Costa Rica region was submerged from at least Palaeocene until late Miocene, and Panama was separated from Colombia until Miocene and possibly into Pliocene by the sea-covered Atrato trough in western Colombia (Woodring, 1966). The present-day richness of the Panamanian *Columnea* flora (43 described taxa) must therefore be due to Pliocene or post-Pliocene "migrations" from South America, or a consequence of great evolutionary activity after completion of the isthmus. Present research on Pliocene deposits in the Atrato trough should help

TABLE V. Distribution of West Indian Hummingbirds (after Bond, 1960)

Area	Species 1	2	3	4	5	6	7	8	9	10	11	12	13	14	15	16	17
N. America				x											x		
Bahamas				x										o			
Mexico						x									x		
C. America	x					x									x		
Cuba				x											x	o	
Hispaniola			o					x							x		x
Puerto Rico		o						x	o		x	x					
Jamaica							o						o				x
Lesser Antilles	x				x					(x)	x	x					
Trinidad/Tobago	x																
S. America	x					x											

x = resident or casual inhabitant; o = endemic; (x) = endemic to Lesser Antilles.

1 = *Glaucis hirsuta*; 2 = *Chlorostilbon maugaeus*; 3 = *Chl. swainsonii*; 4 = *Chl. ricordii*; 5 = *Cyanophaia bicolor*; 6 = *Anthracothorax prevostii*; 7 = *An. mango*; 8 = *An. dominicus*; 9 = *An. viridis*; 10 = *Eulampis jugularis*; 11 = *Sericotes holosericeus*; 12 = *Orthorhynchus cristatus*; 13 = *Trochilus polytmus*; 14 = *Calliphlox evelynae*; 15 = *Archilochus colubris*; 16 = *Mellisuga helenae*; 17 = *Mel. minima*.

TABLE VI. Comparison of selected attributes in the *C. scandens* complex and related taxa

Species	*scandens*				*tocoensis*	*fendleri*	*tulae*	*tomentulosa*
Variety	*scandens*	*vincentina*	*aripoensis*	*australis*				
Mean leaf length	3·8	2·8	2·1	2·4	3·9	4·7	4·0	2·4
	(3·1)	(2·9)	(2·6)		(3·1)		(3·3)	((to 2·5))
Mean leaf width	1·8	1·3	0·8	0·7	1·0	2·1	1·4	1·0
	(2·3)	(1·4)	(1·0)		(0·9)		(1·3)	((0·9–1·2))
Mean sepal length	1·1	1·2	1·0	0·8	0·8	1·4	1·0	0·9
	(1·1–1·3)	(0·9–1·5)	(1·0–1·0)		(0·7–1·0)		(1·0–1·3)	((0·7–1·0))
Corolla colour	Red	Red	Red or red and yellow	Red	Yellow	Red and yellow	Red, pink or yellow	Red and yellow

( ) = dimensions given by Morton (1944); (( )) = dimensions given by Morton (1971); unbracketed dimensions obtained from specimens listed by Morley (in preparation).
All measurements in centimetres.

to determine when the connection was made between Panama and Colombia. As it is almost certain that Columneas occurred in South America in pre-Pliocene times, Panama cannot be regarded as a centre of origin for *Columnea* in the sense of Vavilov, despite its impressive gesneriad statistics. Colombia is a more likely centre of origin of the genus, as this area is physiographically older and rich in Columneas.

Where Greater Antillean species have distributions which are often less than 10 miles across, *C. scandens* occurs at intervals over more than 1000 miles and on into Venezuela. *C. scandens* is phenotypically similar to *C. tocoensis* from Trinidad and *C. fendleri* Sprague from Venezuela; this illustrates the tendency for Lesser Antillean plants to have floristic affinity with South rather than Central America. At the same time, *C. scandens* also has some phenotypic similarity with *C. tulae* from the Greater Antilles and *C. tomentulosa* from Central America. The taxonomic difficulties of species with extensive distributions are well known; and *C. scandens* presents problems both in terms of the taxonomic status to be attributed to its variation, and the relationship of the *scandens* complex to other Caribbean taxa such as *C. tulae*, *C. tocoensis*, and to the west, *C. tomentulosa* (Table VI).

The variation of characters in *C. scandens* is illustrated in Fig. 4 and Table VI. Morton (1944, 1953) described four varietal taxa, based chiefly on calyx characters, to accommodate the variation found in *C. scandens*: var. *scandens*, var. *aripoensis* (Britt.) Mort., var. *vincentina* Mort., and var. *australis* Mort. Leeuwenberg (1958) rejected Morton's taxa on the basis that their distinguishing characters intergrade. Leeuwenberg also regarded *C. fendleri* as a synonym of *C. scandens* although mitotic karyotypes of cultivated *fendleri* have since been shown to differ from those of cultivated *scandens* (Morley, 1972). The taxonomic treatment of geographical variation in *C. scandens* illustrates how systematic information can be lost in the process of taxonomic reassessment. While variation of *C. scandens* might be discounted taxonomically, it must not be overlooked in evolutionary studies, as it shows the presence of partial geographic isolation between island populations which have different modal values for leaf shape, vestiture and floral attributes.

2. *Discussion*

Unlike Jamaica, with its numerous endemic species of *Columnea*, *C. scandens* is the only species throughout the Lesser Antilles. The paucity of Columneas in the Lesser Antilles is a subject for speculation at present

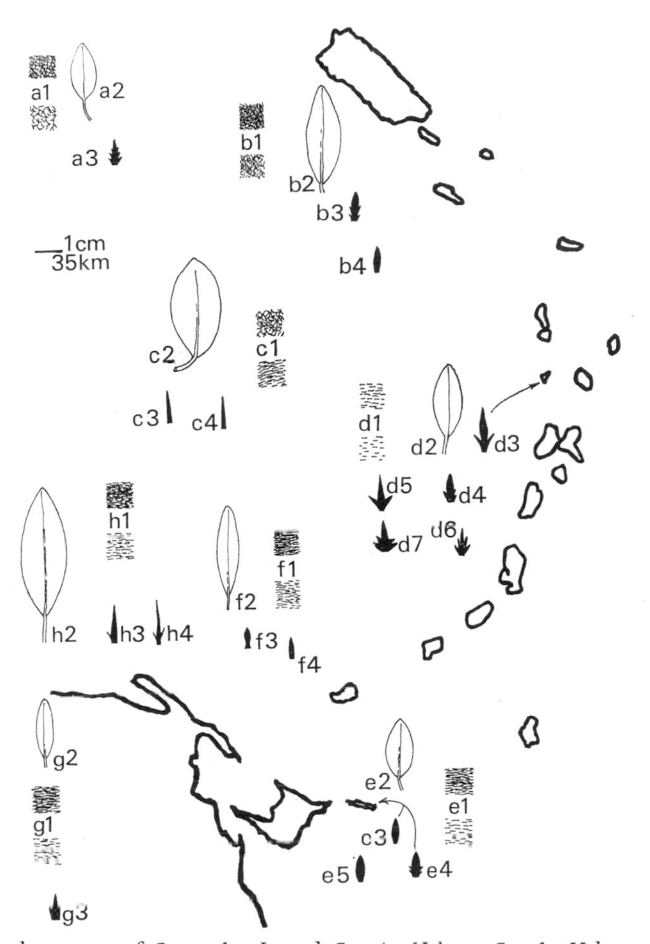

FIG. 4. The phenotypes of *C. scandens* L. and *C. tulae* Urb. a. *C. tulae* Urb. var. *tomentosa* (Oerst.) Morley; 1 abaxial vestiture (upper diagram) adaxial vestiture (lower diagram), 2 leaf shape, 3 sepal (Wedel 38 Panama). b. *C. tulae* Urb. var. *tulae*; 1 vestiture, 2 leaf shape, 3 sepal (Ekman 15299 Dominican Republic), 4 sepal (Britton & Earle 6089 Puerto Rico). c. *C. scandens* L. var. *scandens*; 1 vestiture, 2 leaf shape, 3 sepal (Zusi 323 Dominica), 4 sepal (Imray 194 Dominica). d. *C. scandens* L. var. *scandens* (syn. var. *vincentina*); 1 vestiture, 2 leaf shape, 3 sepal (Proctor 19067 Montserrat), 4 sepal (Proctor 17897 St. Lucia), 5 sepal (Herb. Anderson St. Vincent), 6 sepal (Howard 11149 St. Vincent), 7 sepal (Beard 1241 Grenada). e. *C. scandens* L. var. *scandens* (syn. var. *aripoensis*); 1 vestiture, 2 leaf shape, 3 sepal (Sandwith 1837 Tobago), 4 sepal (Baker 15298 Tobago), 5 sepal (Broadway 9856 Trinidad). f. *C. scandens* L. var. *scandens* (syn. *C. tocoensis* Britt.); 1 vestiture, 2 leaf shape, 3 sepal (Broadway 6176 Trinidad), 4 sepal (Baker & Simmonds 15004 Trinidad). g. *C. scandens* L. var. *scandens* (syn. var. *australis*); 1 vestiture, 2 leaf shape, 3 sepal (Steyermark 62229 Venezuela). h. *C. scandens* L. var. *fendleri* (Sprague) Morley; 1 vestiture, 2 leaf shape, 3 sepal (Dunsterville s.n. Venezuela), 4 sepal (Aristeguieta 3855 Venezuela). All dimensions and vestiture density to scale.

but may be partly due to the poorer *Columnea* flora of north-eastern South America, where "introductions" of the ancestors of *C. scandens* appear to have originated. Movements of hummingbird pollen vectors, or seed dispersal agents, from island to island in the Lesser Antilles would also decrease reproductive isolation between gene-pools on neighbour islands, and thus slow down speciation. Support for this possibility comes from the fact that no individual Lesser Antillean island has an endemic hummingbird (Table V). This is in contrast to several islands in the Greater Antilles. The Lesser Antilles possess two species of hummingbird confined to the island system, another which extends its range into South America and two more which extend into the Greater Antilles. In consequence, birds such as *Glaucis hirsuta* which occur in C. America, the Lesser Antilles, Trinidad and South America may provide opportunity for gene flow between *C. scandens* populations of these areas, not necessarily frequently, but sufficiently to impede speciation.

The origins of *C. scandens* and phenotypically similar taxa in the Caribbean area is at present obscured by our ignorance of how "plant migrations" occurred and which routes were taken. The presence of *C. scandens* along the Antilles must imply some sort of aerial migration of the tiny seeds, either by birds or wind, in the 9–10 million years since the Antilles first appeared, for as Briggs says, ". . . the Lesser Antilles never constituted one continuous landmass." Some *C. scandens* populations may have inhabited certain islands longer than others for, to quote Tomblin (personal communication) ". . . some of the eruptions, e.g. one in St. Vincent dated by radiocarbon at about 14,000 years before present, have been so large that they have covered entire islands with a thick layer of ash which must have destroyed all animal and plant life." While there appear to be no missing species on St. Vincent, so that restocking of the island must have occurred easily enough from nearby islands, such periodic devastation would tend to impede speciation. By contrast, the origin of Greater Antillean species must be further back in the past, probably in pre-Miocene times, according to geological information and the apparent absence of aerial migration between Greater Antillean islands.

C. scandens var. *aripoensis*, found in Trinidad with *C. tocoensis*, is in many ways intermediate between *scandens* and *tocoensis*, and if the varietal distinctions of Morton are to be dismissed, then *tocoensis* could be regarded as synonymous with *scandens*. *C. fendleri* differs from the *C. scandens* complex by having larger leaves, notably subulate sepals, larger corollas and a distribution confined to the mainland of South America. For these reasons *C. fendleri* should be taxonomically regarded as a variety of *C. scandens*, and var. *scandens* should have the synonyms var. *vincentina*, var. *aripoensis*, var. *australis*, and *C. tocoensis*.

The origin of *C. tulae* from *C. scandens* or its ancestors is possible, on a superficial comparison of phenotypes, only if aerial migration occurred from the Lesser to the Greater Antilles, for to quote Briggs once more, ". . . in the sense of more or less present geographic relations of crustal blocks and on the basis of present plentiful data, no part of the Lesser Antilles ever has been joined by land to the Greater Antilles." The distribution of *C. tulae* on both Puerto Rico and Hispaniola could date from a time when, "the crustal blocks that now contain Puerto Rico and Hispaniola . . . supported one continuous large landmass in the late Eocene or perhaps early Oligocene." The distribution could equally date from a time of aerial spread when the two islands had become separated (by middle Miocene at the latest). The other possibility exists that *C. tulae* evolved from ancestors which migrated from Central America rather than the Lesser Antilles, and such a theory is upheld by the presence of the remarkably similar *C. tomentulosa* on the isthmus. So phenotypically alike are *C. tulae* and *C. tomentulosa* that it is taxonomically desirable to make them conspecific.

In summary, the distribution of *C. scandens* indicates that its ancestors were South American plants which "migrated" along the Lesser Antilles by aerial means from Pliocene times onward. The variation of *scandens* has an island–mainland component as illustrated respectively by var. *scandens* and var. *fendleri* and an island–island component as illustrated by the taxa var. *vincentina* and var. *aripoensis*. *C. tulae* has island and mainland populations in a similar way, the mainland populations having been known as *C. tomentulosa*; but in *C. tulae* the ancestry is traceable to Central America. Pollen vector and/or seed dispersal agents may have maintained some level of continuing gene flow between island and mainland populations, which slowed down speciation in *scandens*.

Alloplectus IN THE CARIBBEAN

1. *Observations*

Alloplectus cristatus and other Caribbean Alloplecti with berry fruits are perhaps better placed in the genus *Columnea*, but for the present current nomenclature will be used. The variation of *A. cristatus* resembles that of

C. scandens; it occurs throughout the Lesser Antilles into South America as a scrambling or climbing plant with weak stems. Morton (1944) expressed this variation by the intraspecific taxa var. *brevicalyx* Mort. from the Lesser Antilles north of Martinique, var. *cristatus* from Martinique and St. Lucia, and var. *crenatus* Mort. from Grenada and St. Vincent. Leeuwenberg (1958) accepted these varieties and added to them var. *epirotes* from Guyana, a variety which is close to var. *brevicalyx*. The phenotypic relationship between *A. cristatus* and endemic Greater Antillean Alloplecti, viz. *A. domingensis* Urb. from Hispaniola, *A. ambiguus* Urb. from Puerto Rico, *A. pubescens* (Griseb.) Fawcett from eastern Jamaica, and *A. grisebachianus* (Kuntze) Urb. from western Jamaica, is much closer than that of *C. scandens* with the Greater Antillean Columneas. However, Jamaica has the largest number of endemic Alloplecti in the Caribbean. Figure 5 illustrates the variation of critical characters of *A. cristatus* and other Caribbean Alloplecti.

2. *Discussion*

A. cristatus has a distribution pattern and variation which indicates the existence of partial geographic isolation between island populations. How complete this geographic isolation is, remains to be studied but on inspection it appears to be greater than that found in *C. scandens*, as the sepal morphology of *A. cristatus* is diverse and pronounced. Like *C. scandens*, *A. cristatus* probably arose from South American ancestors.

CONCLUSIONS

Greater Antillean Columneas are mostly endemic, and speciation has been associated with complete geographic and topographic isolation so far as phenotypes and their distributions are concerned. The products of speciation often have local distribution and show large phenotypic differences in Jamaica, where structural chromosome change has also been associated with speciation. Speciation in Jamaica has been intensive, with the creation of multiple endemics. Certain Jamaican Columneas appear to have evolved in response to ecological selection, while others have apparently non-adaptive characters. Topographic isolation of Columneas in Jamaica may have been profoundly influenced by the faulting and rapid erosion which followed Pliocene times.

Lesser Antillean Columneas and Alloplecti exhibit non-endemic distribution and are more variable. It is suggested that partial geographic

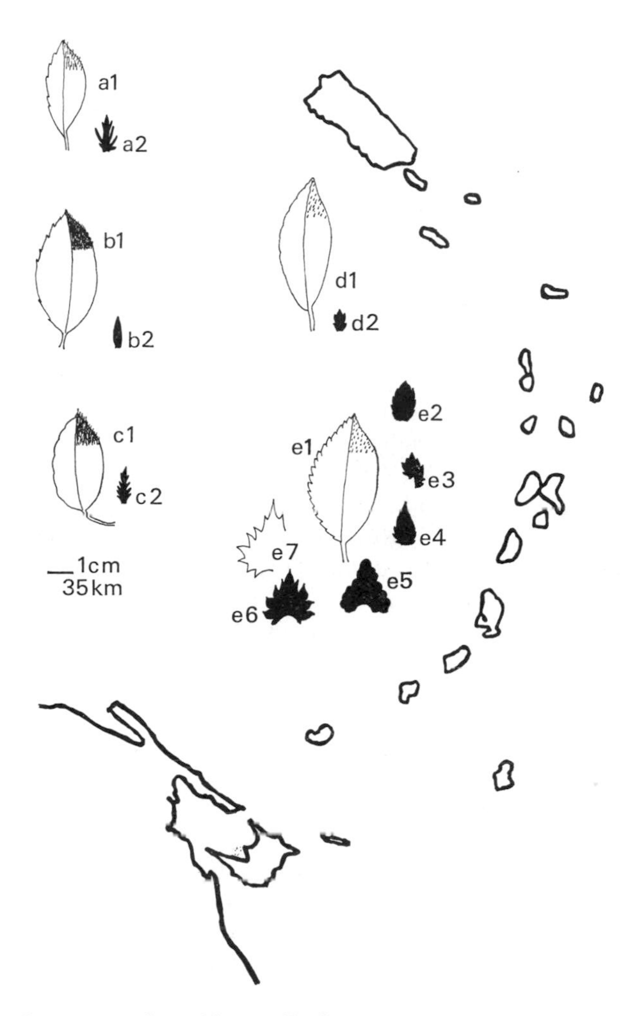

FIG. 5. The phenotypes of Caribbean Alloplecti. a. *Alloplectus domingensis* Urb.; 1 leaf shape, margin and adaxial vestiture (Ekman 1129, Dominican Republic), 2 sepal (Howard 12291 Dominican Republic). b. *A. pubescens* (Griseb.) Fawcett; 1 leaf shape, margin and adaxial vestiture (Wilson Type Jamaica), 2 sepal (Webster 5569 Jamaica). c. *A. grisebachianus* (Kuntze) Urb.; 1 leaf shape, margin and adaxial vestiture (Harris 12766 Jamaica), 2 sepal (Proctor 16539 Jamaica). d. *A. ambiguus* Urb.; 1 leaf shape, margin and adaxial vestiture, 2 sepal—both in cult. DBN. e. *A. cristatus* (L.) Mart.; 1 leaf shape, margin and adaxial vestiture (Herb. Gay 161 Martinique), 2 sepal var. *brevicalyx* Mort. (Fairchild s.n. St. Kitts), 3 sepal var. *brevicalyx* (Proctor 20041 Guadeloupe), 4 sepal var. *brevicalyx* (Hodge 825 Dominica), 5 sepal var. *crenatus* Mort. (Howard 11174 St. Vincent), 6 sepal var. *cristatus* (Proctor 17592 St. Lucia), 7 sepal var. *cristatus* (Herb. Gay 161 Martinique). All dimensions and vestiture density to scale.

isolation has created island populations with different modal values for morphological characters. The gene-pools of the two species, despite a distribution of several hundreds of miles along the Antilles, are envisaged as still being intact, as a result of the activity and mobility of pollen and/or seed dispersal vectors. Distributional data suggest that the Lesser Antillean species are derived from South American ancestors and are younger than the Greater Antillean species, which have a more remote Central American origin.

ACKNOWLEDGEMENTS

I should like to thank Dr Reginald P. Briggs (U.S. Geological Survey, Carnegie, Pennsylvania), Dr W. P. Woodring (Washington), Dr E. Robinson (Jamaica) and Dr John F. Tomblin (U.W.I. Seismic Research Unit, St. Augustine, Trinidad) for geological advice. I should also like to thank Dr R. A. Howard and the authorities of the following herbaria for the use of dried specimens: British Museum (Natural History); Royal Botanic Gardens, Kew; Institute of Jamaica; University of the West Indies at St. Augustine & Mona; Botanical Museum, Copenhagen; Instituto Botienico, Caracas; Missouri Botanical Garden; Gray Herbarium; Smithsonian Institution.

REFERENCES

Bond, J. (1960). "Birds of the West Indies". Collins, London.

Dietz, R. S. and Holden, J. C. (1970). *Scient. Am.* **223(4)**, 30–41.

Graham, A. and Jarzen, D. M. (1969). *Ann. Mo. bot. Gdn* **56**, 308–357.

Howard, R. A. (1970). (Personal communication).

Leeuwenberg, A. J. M. (1958). *Acta bot. neerl.* **7**, 291–444.

Mayr, E. (1963). "Animal Species and Evolution". Harvard University Press, Cambridge, Massachusetts, U.S.A.

Morley, B. D. (1968). "A biosystematic study of the genus *Columnea* L. in Jamaica". Unpublished thesis, University of the West Indies.

Morley, B. D. (1971). *Bot. J. Linn. Soc.* **64**, 81–96.

Morley, B. D. (1972). *Bot. J. Linn. Soc.* **65**, 25–36.

Morton, C. V. (1944). *Contr. U.S. natn. Herb.* **29**, 1–19.

Morton, C. V. (1953). *Fieldiana* **28**, 520–534.

Morton, C. V. (1971). *Phytologia* **21(3)**, 165–195.

Stearn, W. T. (1969). *Bull. Br. Mus. nat. Hist. (Bot.)* **4**, 181–236.

Stebbins, G. L. (1971). *Taxon* **20**, 3–16.

Wiehler, H. (1970a). "The distribution of the Gesneriaceae in the Caribbean islands". Unpublished manuscript, Cornell, New York.

WIEHLER, H. (1970b). "Studies in the morphology of leaf epidermis, in vasculature of node and petiole, and in intergeneric hybridization in the Gesneriaceae–Gesnerioideae". Unpublished thesis, Cornell, New York.
WOODRING, W. P. (1966). *Proc. Am. phil. Soc.* **110**, 425–433.

16 | Evolution and Endemism in the New Zealand Species of *Epilobium*

PETER H. RAVEN

Missouri Botanical Garden and Washington University, St. Louis, Missouri, U.S.A.

In all the world, there are some 200 species of the genus *Epilobium*. They are more or less evenly distributed at relatively high latitudes and altitudes, being restricted to the mountains in the tropics. In New Zealand alone, there are 37 native species in an area considerably smaller than the British Isles. These species, which are morphologically and ecologically diverse, are mostly endemic and all but one confined to the Australasias. In the entire continent of Eurasia, there are approximately 80 species of the genus; in North America about 45. Why then should there be so many species in New Zealand, whence have they come, and what has been their history in the region?

We have attempted to answer these questions in a comprehensive study of the genus throughout New Zealand and a portion of Australia (P. H. and T. E. Raven, 1972).

DERIVATION OF THE AUSTRALASIAN SPECIES

Epilobium is well represented on every continent except Antarctica. In considering the Australasian species, therefore, one's attention is immediately directed to two questions: (1) where did they come from; and (2) how many times did the group colonize Australasia? The formulation of these questions assumes that the genus did not originate within Australasia, a contention that will be documented by evidence to be presented on the following pages. In evaluating the derivation and relationships of the very diverse Australasian species of *Epilobium*, three lines of evidence have been brought into play: morphology, cytogenetics, and crossing relationships. These will now be discussed in turn.

MORPHOLOGY

Considering the genus *Epilobium* as a whole, the area of greatest diversity is the western United States. Here are found truly xerophytic species; the only annual species in the genus; the only species in the genus with aneuploid chromosome numbers; and both of the closely related genera, *Boisduvalia* and *Zauschneria*. *Chamaenerion*, a very distinctive group within *Epilobium*, has all of its six species in western Eurasia, but two extend to North America. Although the Australasian species of *Epilobium* are superficially diverse, they do not differ as fundamentally from one another as do species found in the Northern Hemisphere, as can be demonstrated in several ways.

In northern areas, for example, there are a great variety of strategies for overwintering amongst the species of *Epilobium*. Some have thread-like stolons from the base, terminating in solid corms; others have tight underground buds ("turions") covered with fleshy cataphylls; leafy rosettes from the base; rhizomes; leafy stolons; buds ("gemmae") in the leaf axils; and other sorts of perennation. In contrast, all of the Australasian species propagate themselves from buds near the surface of the ground, and the different growth forms amongst the species of this region can be seen as variations upon this common plan. The original condition seems to have been one in which the plants occurred in moist habitats and had leafy runners from the base or leafless ones just below the ground level. From this has evolved the tightly clumped habit characteristic of most Australasian species that grow in relatively xeric habitats. Another trend, toward increasing emphasis of the stolons and suppression of the erect shoots, has resulted in the evolution of the Australasian species that have a creeping habit. *Epilobium curtisiae* Raven is intermediate in this respect, being decumbent but still having definite inflorescences. The South American *Epilobium conjungens* Skottsb. resembles it in habit, but clearly had an independent origin, being different in virtually every other respect. In at least six evolutionary lines the trend has resulted in species in which all of the leaves are opposite—they normally become alternate in the inflorescence in *Epilobium*—and the flowers are borne individually in their axils. The stems continue to grow beyond the point where the flowers are borne. Why this peculiar plant form, unknown elsewhere, should have evolved repeatedly in Australasia, is unknown.

Another significant key to the relationships of the Australasian species is provided by their seed morphology. Although there is much variation in shape, most species have seeds that are broadly obovoid and the remainder can be understood as derivatives of this basic form. Studies with the scanning electron microscope have revealed that the pattern of the seed coats is basically similar in all Australasian species of *Epilobium*, in contrast to the much wider range of variability among the Northern Hemisphere species.

It has been suggested (Raven, 1967) that some of the distinctive groups of *Epilobium* in Australasia might have been derived from South America. The seed characteristics just discussed allow a definitive evaluation of this hypothesis. No species of the New World has broadly obovoid seeds except for *E. hirtigerum* A. Cunn. (*E. brasiliense* Hausskn.), almost certainly an early introduction from Australia into South America. All of the South American species have narrowly obovoid seeds that are attenuate at the ends, a common seed type in North America. On the other hand, broadly obovoid seeds of the sort that are characteristic of the Australasian species are well represented in Eurasia. On this evidence alone, it appears certain that the Australasian species of *Epilobium* were derived from Eurasia and the South American ones from North America, and that there has been no interchange, except for that involving the weedy *E. hirtigerum*, between the two areas.

CYTOLOGY AND CYTOGENETICS

All 45 native species of *Epilobium* in Australasia and each of their subspecies has the same chromosome number, $n = 18$, as determined by Hair for from one to several dozen strains of each species (J. B. Hair, unpublished). The same chromosome number is found in nearly all other species found elsewhere in the world, and therefore provides no clue to the origin or coherence of the Australasian species.

On the other hand, the cytogenetics of hybrids within the genus is most helpful in evaluating these questions. The most frequent chromosomal arrangement among the Northern Hemisphere species of the genus is the one that we shall term the AA arrangement (Raven and Moore, 1964; Thakur, 1965; Mosquin, 1968). It is characteristic of most Eurasian species. Another group of species, including *E. lactiflorum* Hausskn. and *E. hornemannii* Reichenb., has a chromosome arrangement that differs from the AA arrangement by two reciprocal translocations. When plants with this arrangement, which we shall term the CC arrangement, are crossed with ones having the AA arrangement, the hybrids form 15 bivalents and a ring of 6 chromosomes at meiotic metaphase I.

The Australasian species, on the other hand, have a chromosome arrangement that differs from the AA arrangement by a single reciprocal translocation (J. B. Hair *et al.*, unpublished). This arrangement, which we shall term BB, is known definitely to be characteristic of all 22 of the 45 Australasian species, representing a comprehensive sample of the diversity, that have been tested. A single plant of *Epilobium microphyllum* A. Rich. from the Rainy River, Nelson, New Zealand, had a chromosome arrangement that differed from BB by an additional reciprocal translocation, unlike three other populations of the species, which had the BB arrangement. We presume that the chromosome rearrangement which resulted in the genome of the "unusual" plant took place *in situ*.

Thus the Australasian species of *Epilobium* have their own characteristic chromosome arrangement, unknown elsewhere in the world. This strongly suggests that they have been derived from a common ancestor within Australasia. A hybrid between *E. pubens* A. Rich. from New Zealand and *E. denticulatum* Ruiz & Pav. from Argentina had 16 bivalents plus a ring of 4 chromosomes, suggesting that the South American species will prove to have the AA genome, and re-emphasizing their independent derivation and complete separation from the Australasian species.

CROSSING RELATIONSHIPS

Results from natural hybridization among the Australasian species will now be compared with those obtained earlier, mostly involving Northern Hemisphere species. These will be considered in four categories: (1) failure of seed set; (2) abnormal development of embryos or young plants; (3) "cytoplasmic effects" and (4) hybrid fertility.

(1) Failure of seed set. Thakur (1965), Brockie (1970), and others have reported that seeds are not normally set in crosses between members of sect. *Chamaenerion* and sect. *Epilobium*, owing to abnormal development of the embryos. When pollen from species with short styles and small flowers is placed on the stigmas of species with long styles and larger flowers, fertilization sometimes does not take place; this is a frequent phenomenon in flowering plants generally. With this exception, we have had little difficulty in obtaining seeds from any combination of species of sect. *Epilobium*. The few failures reported by Brockie (1966) in crosses between New Zealand species should be discounted. Brockie attempted most pollinations only once or twice, and he now considers mechanical difficulties to have been responsible in most if not all of these instances. There is no evidence for failure of seed set as a barrier to

hybridization between Australasian species of *Epilobium*, and no definite evidence for its operation between any two species of sect. *Epilobium* from anywhere in the world.

(2) Abnormal development of embryos or young plants. The production of non-viable or weak hybrids is very commonly observed in *Epilobium* (e.g., Michaelis, 1954, 1965a, b, 1966a, b; Thakur, 1965; Brockie, 1966, 1970). It should, however, be mentioned specifically that there is no evidence for failure of seed germination as a barrier to hybridization in the genus, nor for the phenomenon of plastid incompatibility which is so frequent in the genus *Oenothera*. Results from some 900 crosses made by Brockie and involving New Zealand species of *Epilobium* (W. B. Brockie, 1966 and unpublished) are summarized in Fig. 1 according to the taxonomy we shall employ in our forthcoming monograph of the Australasian species of the genus. Thanks to Brockie's efforts, the New Zealand species of *Epilobium* are probably better known, in terms of their crossing relationships, than any groups of plants of comparable size anywhere in the

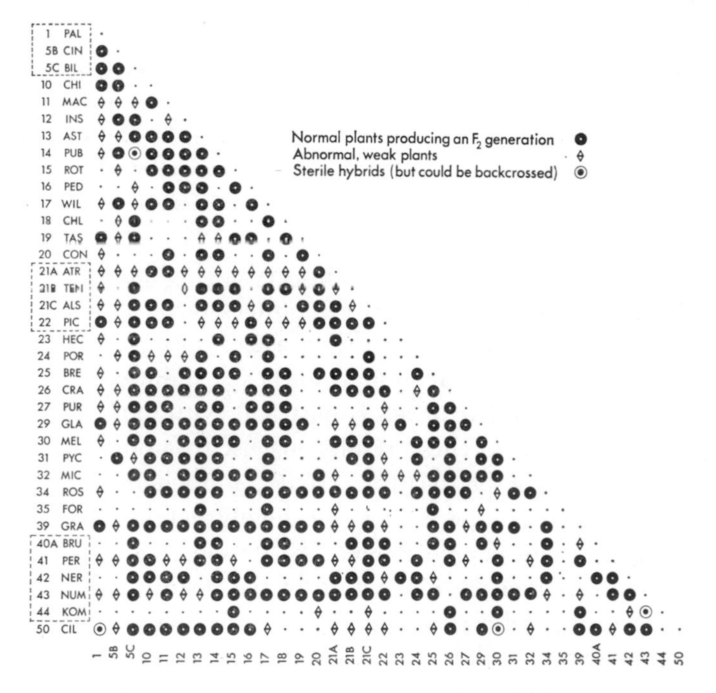

FIG. 1. Crossing relationships between some New Zealand species of *Epilobium*. The North American *Epilobium ciliatum* (50) is also included. Data from Brockie (1966 and unpublished); acronyms and numbers from Raven and Raven (1972).

world. Crossing results involving the North American *E. ciliatum* Raf. (*E. adenocaulon* Hausskn.) are likewise included in this figure.

The hybrids reported in this figure fall readily into one of three categories: (1) normal, vigorous plants that produce an F_2 generation upon selfing; (2) abnormal, slender, or matted plants which rarely produce any flowers; and (3) normal, vigorous plants that are sterile but produce a BC_1 generation on crossing with one or both parents. No reciprocal differences were noted in the approximately 85 instances in which the cross was made in both directions. It can readily be seen in Fig. 1 that certain groups of species stand out from the rest in terms of their tendency to produce abnormal hybrids, and the differentiation of these groups is postulated to have taken place within Australasia.

(3) Cytoplasmic effects. Cytoplasmic differences are very frequent among Northern Hemisphere species of *Epilobium* and have been discussed by many authors, but especially studied by Michaelis (e.g. 1954, 1965a, b, 1966a, b). In general, the reciprocal differences between hybrids fall into three categories: (1) crippled or abnormal plants obtained when the cross is made in one direction, normal ones in the other; (2) pollen-sterile plants obtained in one direction, more or less fertile ones in the other; and (3) small morphological differences between the reciprocal hybrids. In general, these phenomena seem to be controlled by the same sorts of factors; thus, crippled plants may also be pollen-sterile.

It appears likely that the reciprocal differences observed in *Epilobium* may be attributed to the differentiation of mitochondria. Pollen-sterile lines of wheat have mitochondria with very high oxidative and phosphorylative activity, and hybrids between pollen-sterile and pollen-fertile lines have an even higher activity level (Srivastava *et al.*, 1969). A mixture of mitochondria from parental strains of wheat exhibited mitochondrial complementation and activity levels comparable with those characteristic of the pollen-sterile line. These results suggest that both in respect of their oxidative and phosphorylative activity, mitochondria in different races of individual plant species may be highly heterogeneous and the results of mixing them unpredictable. The development of a functional individual depends upon the interaction of the nuclear genetic material with the cytoplasm, and certain combinations are dysfunctional.

In *Epilobium*, little cytoplasm and thus few mitochondria are contributed by the staminate parent. Abnormal hybrids between Australasian species of the genus are similar in respect of their morphology and fertility regardless of the direction of the cross. When species from other portions of the world, and even individual strains of some species, are intercrossed, reciprocal differences are often noted. The almost complete lack of reciprocal differences between Australasian species of *Epilobium* strongly suggests that all of these species have a closely similar cytoplasm. This stands in sharp contrast to the situation in Europe, where there is so much differentiation that Michaelis (1954) claimed that each strain of *Epilobium* should be thought of as having its own distinctive cytoplasm as well as its own genotype. It is very strong evidence for the derivation of all Australasian species from a common ancestor.

(4) Hybrid fertility. A high degree of sterility is characteristic of many hybrid combinations between Northern Hemisphere species. The information available for hybrids involving Australasian species is presented in Fig. 2. From this figure, it can readily be seen that there is no consistent pattern or delimitation of groups comparable with that shown in Fig. 1. Individuals have been found to vary widely in this respect when grown under different environmental conditions, and hybrid sterility does not appear to be an important factor in differentiating Australasian species of

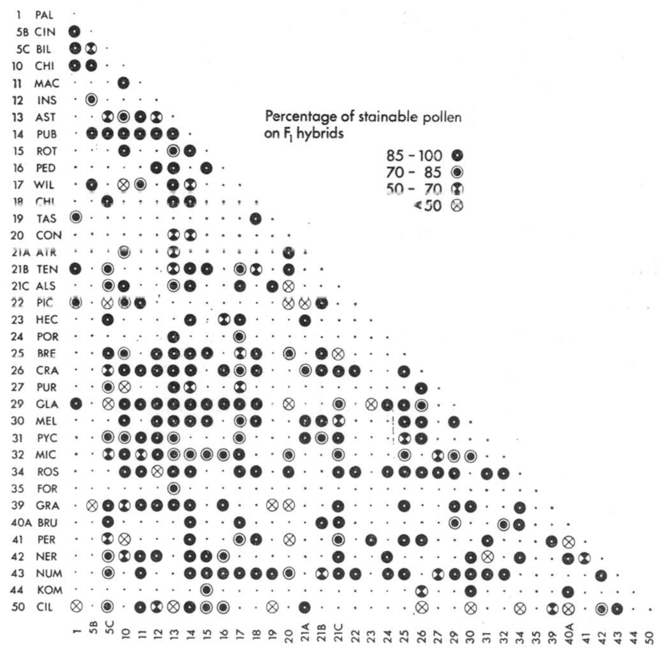

FIG. 2. Fertility of artificial hybrids between New Zealand species of *Epilobium*. The North American *Epilobium ciliatum* (50) is also included. Data from Raven and Raven (1972).

Epilobium. On the other hand, it can be seen from Fig. 2 that hybrids involving the North American *E. ciliatum* are often highly sterile, and that reciprocal differences are also found in hybrids with this species. The evidence summarized in Fig. 2 likewise suggests that the closely related species *Epilobium astonii* (Allan) Raven & Engelhorn and *E. pubens* A. Rich. may be differentiated from the rest in respect of the factors controlling hybrid fertility. One would assume that these species had become differentiated in this respect within New Zealand.

PATTERN OF DERIVATION

On the basis of the evidence just presented, the native Australasian species of *Epilobium* are a monophyletic group that was derived from Asia. They are homogenous and more or less distinctive in their morphology, cytology, and crossing relationships, although some differentiation and the evolution of 45 species has taken place subsequent to their arrival in Australasia. The facts that the Australasian species of *Epilobium* have their own unique chromosome arrangement and cytoplasm which are common to all members of the group; that they are to all intents and purposes interfertile; and that they are morphologically variations on a common theme—these facts, in the context of the genus, can only be interpreted to mean that they have been derived from a common ancestor.

HYBRIDIZATION AMONG THE AUSTRALASIAN SPECIES

On the evidence just presented, internal (or postzygotic) factors do not play an important role in the separation of the Australasian species of *Epilobium*. In an effort to understand why there are so many species in the area, and particularly in New Zealand, we must therefore turn to a consideration of external (or prezygotic) factors. These fall into two classes: (1) persistent autogamy; and (2) ecological differentiation.

AUTOGAMY

Perhaps 90% of the 200 species of *Epilobium* sect. *Epilobium* are predominantly autogamous, and there is no region of the world in which outcrossing species are more common than inbreeding ones. Many authors (e.g. Valentine, 1951; Lewis and Moore, 1962; Thakur, 1965) have interpreted autogamy as an important factor in limiting the occurrence of hybrids between species in this genus. Thakur (1965) has pointed out that, for the species of Europe, most of the common hybrids involve species in which at least some degree of cross-pollination is the rule, and those involving autogamous species are much less frequent. Four of the nine Australian species of *Epilobium* have at least some large-flowered populations in which outcrossing by bees and other insects is the rule, and this also seems to be the case in three of the 37 native species found in New Zealand, together with some populations of a fourth. Turning the statement around, in 33 of the 37 New Zealand species and nearly all populations of another, autogamy is virtually the only mode of reproduction. Insects are very rarely seen at the flowers of any species of *Epilobium* in New Zealand.

In the light of these observations, we believe that Brockie (1959) was fully justified in stating that for nearly all New Zealand species of *Epilobium*, the amount of gene exchange *even within* natural populations is very small. We would add that the possibility of gene exchange between populations is even less, and certainly far below the levels at which it would limit the effects of natural selection. In general, each population (and for all practical purposes, nearly each individual) of *Epilobium* in New Zealand is responding to selection completely independently of every other population.

The representation of outcrossing species of *Epilobium* in Australia is about the same as it is in Europe or North America, whereas in New Zealand it is much less. Of the nine species of the genus found in Australia, five are also found in New Zealand. Four of these have some highly outcrossing populations in Australia but are strictly autogamous in New Zealand; the fifth is strictly autogamous in both countries. These relationships are suggestive of an important role for autogamy in the production and maintenance of the many species of the genus found in New Zealand, where flower-visiting insects are much less frequent than in Australia.

ECOLOGICAL RELATIONSHIPS

In New Guinea and Australia, the relatively few species of *Epilobium* (13 in all) occur in habitats much like those in which species of the genus grow elsewhere in the world: moist places and along streams, drying flats, wet meadows, lakeshores, and open grassland. In New Zealand, the 37 native species occupy a very wide range of habitats; no other area has such an ecologically diverse array of species of *Epilobium*, and none has so many. The direct relationship between the large number of species and their ecological diversity is inescapable.

In our forthcoming monograph of the Australasian species of *Epilobium*, we have divided the habitats of the New Zealand species into 13 distinct categories: (1) lowland, moist, open habitats; (2) lowland, moist forest habitats; (3) lowland, dry habitats; (4) very wet habitats in areas of high rainfall; (5) rock cliffs and crevices, including limestone; (6) dry, shingly riverbeds at low elevations; (7) tussock grassland; (8) subantarctic moist tussock grassland; (9) wet habitats at high elevations; (10) streambeds and stony slopes at high elevations; (11) alpine fellfield and herbfield; (12) fine stable rock debris at high elevations; and (13) loose, unstable scree slopes. Within these broad categories, each of the 37 species of *Epilobium* can be distinguished, in most cases very sharply, from all others. This is by no means true of the species found in Europe, as noted by Valentine (1951), or those of North America.

SYMPATRIC OCCURRENCE

Thus far, we have seen that the New Zealand species of *Epilobium* are essentially interfertile. In comparison with the species found elsewhere in the world, they are more highly autogamous and much more highly differentiated ecologically. Despite their ecological distinctiveness, many of the New Zealand species of *Epilobium* occur sympatrically, as shown in Fig. 3. The hybrids that have been observed in nature are also reported in this table, as are a number of combinations in which the hybrids would be extremely difficult if not impossible to separate from one or both parents.

Despite the importance of autogamy and ecological differentiation as factors limiting hybridization between the New Zealand species, hybrids are relatively frequent in nature, with 38 definite and 10 additional suspected combinations being reported in Fig. 3. In several localities, hybrids have been found fairly frequently in disturbed areas, as along roadbanks. This has, for example, been discussed for the area around Porters Pass, Canterbury, by Brockie (1966).

Locally, we have found stable populations that were obviously of hybrid origin. For example, at Tiraumea Saddle in Nelson Lakes National Park, small boggy openings occur in what otherwise would be a continuous forest of *Nothofagus*. In one of the drier of these, *Epilobium komarovianum* H. Lév. was frequent, together with an almost continuous cover of hybrids between this species and *E. insulare* Hausskn. In the immediate vicinity, *E. insulare* was infrequent, occurring very sparsely in the same clearing and slightly more frequently in wetter openings just below. This would

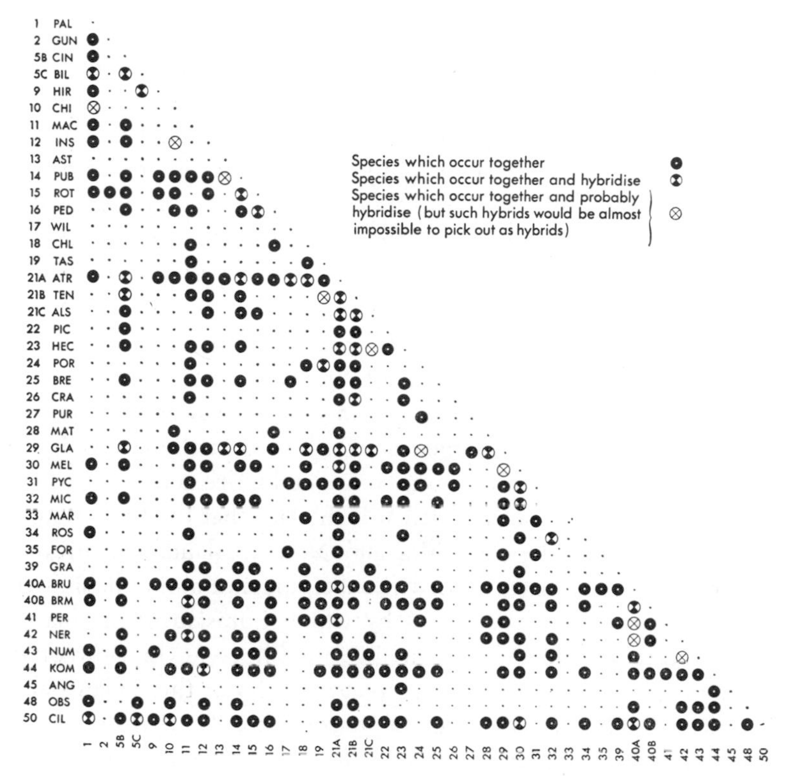

FIG. 3. Sympatric occurrences of native and naturalized species of *Epilobium* in New Zealand; from Raven and Raven (1972).

seem to be a clear example of a fully fertile hybrid that is more successful than either parent in this particular ecological setting.

Many species of *Epilobium* in New Zealand resemble the F_1 hybrids between two other existing species, and it seems likely that some have had a hybrid origin. Certainly the sort of genetic system that is operative in the New Zealand species of this genus is a highly flexible one. Autogamy and ecological differentiation together can maintain species—favorable genetic combinations—indefinitely, but hybridization can produce new, fertile lines that may be well adapted to particular habitats and then spread to become the sorts of major units that we recognize as species.

THE PATTERN OF EVOLUTION IN AUSTRALASIA

In common with the other families of Myrtales except for Myrtaceae itself, the Onagraceae seem to be of Tertiary origin (Muller, 1970, p.

440). At a time when more or less direct migration between Australasia and South America was possible across Antarctica—the middle Eocene (Raven and Axelrod, 1972)—*Epilobium* might not have even originated. The lack of direct relationship between the Australasian and South American species of *Epilobium* certainly suggest that, in any case, it was not present in the Southern Hemisphere at that time. The first reliable records of Onagraceae from the Southern Hemisphere (*Fuchsia*) are those of Couper (1960) from the lower Miocene (not Oligocene) of New Zealand. When did *Epilobium* reach Australasia?

During the early Tertiary, most of Australia was south of 28°S lat, approximately the present position of Brisbane (review in Raven and Axelrod, 1972). Subsequently, it has moved northward about 15° of latitude, and New Guinea emerged above the sea in the late Oligocene. In the process Australasia has become more and more accessible to immigration from Asia. Particularly critical is the fact that the high mountains in New Guinea, Malesia, Australia and New Zealand are all of late Pliocene or Pleistocene age. Both the subalpine and alpine habits within Australasia and the likely pathways by which Eurasian plants may have reached the area are no older than the late Pliocene. In view of these facts, it appears likely that at least a great majority of these plants arrived in Australasia in the late Pliocene or more recently. *Epilobium* is fairly well represented in the fossil record in New Zealand from this time forward, but there is also a single record from the early Oligocene (Couper, 1960). Regardless of whether this record is confirmed or not, it is clear that most of the adaptive radiation of *Epilobium* in Australasia has taken place in the past several million years, and particularly in connection with the drastic climatic changes that occurred during the Pleistocene.

Australia and New Zealand lie in the path of the prevailing westerlies. Since *Epilobium* arrived in the region only in the late Tertiary, when these lands were in approximately their present positions, it appears certain that the genus spread from Australia to New Zealand and not in the opposite direction. Five of the nine species found in Australia are common to the two areas, and two of these must have reached New Zealand at least twice, judging from the morphological relationships.

Why then are there so many more species of *Epilobium* in New Zealand than in Australia? The answer would seem to be of considerable generality, for there are a number of other groups, such as *Celmisia*, *Hebe*, *Myosotis*, *Aciphylla*, *Anisotome*, *Carex*, and *Astelia*, that seem to exhibit analogous patterns, with far more species in New Zealand than in Australia. Hybrids are well known in all of these groups, which are important, large genera in the subalpine and alpine areas of New Zealand, each containing a large majority of endemic species. In general, the species are interfertile, and the chromosome numbers either invariate or at only two levels. All of these are characteristics that would be expected in plant groups that have been derived recently from common ancestors and which had radiated recently into diverse habitats, with the interspecific recombination of genetic material following hybridization playing a major role in differentiation. It is customary to think of these groups as having originated in New Zealand and then spread to Australia; but in view of the realities of dispersal within the Australasian region and the probable time that most of these genera arrived in the area, the opposite pathway of migration appears far more likely in all cases.

In every case, these plants inhabit the open, often subalpine or alpine habitats that developed in New Zealand for the first time following the late Pliocene. During the Pleistocene, glaciation and the attendant climatic changes, together with the full uplift of the majestic mountains of New Zealand, led to the establishment of a whole array of new diverse habitats. These were formed in a complex geomorphological and meteorological sequence that must have been accompanied by an equally complex pattern of migration and extinction amongst the plants of open habitats, such as *Epilobium*. The importance of interspecific recombination in adjusting to a rapidly and drastically changing habitat, particularly on an island where migration from the outside is strictly limited, has been stressed by Rattenbury (1962). The ability to preserve favorable genetic combinations through persistent autogamy would have been no less important, and is common to most, but not all, of the groups listed above.

During the late Tertiary, those plants that could grow in subalpine and alpine habitats in New Zealand were presented with vast open areas that developed on an isolated, forest-covered island where relatively few kinds of plants, and initially no alpine plants, were represented. Under the influence of Pleistocene climatic changes and the effect they had on the varied rock types of New Zealand in building soils, the subalpine and alpine microhabitats of New Zealand must have changed in a manner similar to the colored particles in a kaleidoscope. Like *Celmisia*, *Hebe*, *Poa*, *Carex*, *Myosotis*, *Aciphylla*, and a few other groups that we also believe came to New Zealand via Australia following the late Pliocene, *Epilobium* had the ability to colonize these open habitats and to utilize interspecific hybridization as a means of producing new kinds of populations in response to the changes. These, then, were the groups of plants that were able to keep pace with the turns of the kaleidoscope, and the results are seen today in the wonderful arrays of species that are such an impressive feature of the flora of New Zealand at the present day.

SUMMARY

The 45 native species of *Epilobium* in Australasia have their own unique chromosome arrangement and cytoplasm which are common to all members of the group. They are essentially interfertile. They are morphologically variations on a common theme, one which is well developed in Asia. In view of these facts, it is virtually certain that they were derived from a common ancestor of Asian origin within Australasia. The South American species, on the other hand, were clearly derived from North America, and there has been no exchange between South America and Australasia except for the early introduction of the weedy *E. hirtigerum* (*E. brasiliense*) from Australia into southern South America.

Within Australasia, migration has been into Australia and New Guinea, secondarily eastward to New Zealand which lies directly in the path of the prevailing westerlies. In New Zealand, 37 native species, 32 of them endemic, occur in a region considerably smaller than the British Isles. All but three of these species are highly autogamous, and they collectively exhibit a much greater ecological diversity than the species found anywhere else in the world. They are essentially interfertile, although there has been some differentiation in respect of the factors controlling interfertility. Natural hybridization is frequent, but is limited by autogamy and ecological specialization, both highly developed in New Zealand.

The mountains of New Zealand, Australia, New Guinea, and Malesia reached heights sufficient for the development of subalpine and alpine habitats only in the late Pliocene and subsequently. Australia has moved northward some 15° of latitude during the Tertiary. Thus opportunities for immigration of the plants of subalpine and alpine habitats from Asia into Australasia have existed only in the past several million years, although representatives of some of these groups might have been in the region earlier in locally favorable habitats. During the Pleistocene in New Zealand, large areas of open habitat appeared at middle and high elevations on an island that had hitherto been almost entirely forested. In the few plant groups that could grow in these areas, and especially those in which interspecific hybridization could rapidly recombine the genetic material into new adaptive combinations, adaptive radiation rapidly produced large clusters of species, as in *Celmisia*, *Hebe*, *Poa*, *Carex*, *Myosotis*, and *Aciphylla*. It is thus no coincidence that T. F. Cheeseman was able to write in 1906: "in *Celmisia* as in other large genera of the New Zealand flora, the species, such as they are, must be regarded as founded on an aggregation of several small prevalent characters rather than on conspicuous and important differences." It is further argued that all of the groups mentioned above reached New Zealand from Australia, despite the much larger concentrations of species seen at the present day in New Zealand.

ACKNOWLEDGEMENTS

This study was conducted while I held a New Zealand Department of Scientific and Industrial Research Senior Postdoctoral Fellowship in the year 1969–1970. Field work in Australia and subsequent studies in the United States were supported by a research grant from the U.S. National Science Foundation. Our activities in New Zealand were supported most generously by the D.S.I.R., and we are grateful to many botanists in New Zealand and Australia for their kind help in the field and laboratory and their discussion with us of the phytogeography of Australasia. We are especially indebted to the Director of the D.S.I.R. Botany Division, E. J. Godley, and to W. B. Brockie, J. B. Hair, and Keith West for their great kindness in many respects and unfailing support.

REFERENCES

Brockie, W. B. (1959). *Trans. R. Soc. N.Z.* **87**, 189–194.
Brockie, W. B. (1966). *N.Z. J. Bot.* **4**, 366–391.
Brockie, W. B. (1970). *N.Z. J. Bot.* **8**, 94–97.
Cheeseman, T. F. (1906). "Manual of the New Zealand Flora". N.Z. Board of Science and Art, Wellington.
Couper, R. A. (1960). *N.Z. Geol. Surv. Paleont. Bull.* **32**, 1–88, pls 1–12.
Dietz, R. S. and Holden, J. C. (1970). *J. Geophys. Res.* **75**, 4939–4956.
Lewis, H. and Moore, D. M. (1962). *Bull. Torrey Bot. Club* **89**, 365–370.
Michaelis, P. (1954). *Adv. Genet.* **6**, 287–401.
Michaelis, P. (1965a). *Nucleus* **8**, 83–92.
Michaelis, P. (1965b). *Nucleus* **8**, 93–108.
Michaelis, P. (1966a). *Nucleus* **9**, 1–16.
Michaelis, P. (1966b). *Nucleus* **9**, 103–118.
Mosquin, T. (1968). *Can. J. Genet. Cytol.* **10**, 794–798.
Muller, J. (1970). *Biol. Rev.* **45**, 417–450.
Rattenbury, J. (1962). *Evolution* **16**, 348–363.
Raven, P. H. (1967). *Blumea* **15**, 269–282.
Raven, P. H. and Moore, D. M. (1964). *Watsonia* **6**, 36–38.

RAVEN, P. H. and RAVEN, T. E. (1972). The Australasian species of *Epilobium* (Onagraceae). N.Z. Dept. Sci. Indust. Res. Bull. (in press).

RAVEN, P. H. and AXELROD, D. I. (1972). *Science, N.Y.* **176**. 1379-1386.

SRIVASTAVA, H. K., SARKISSIAN, I. V. and SHANDS, H. L. (1969). *Genetics* **63**, 611–618.

THAKUR, V. (1965). "Biosystematics of some species of *Epilobium*". Ph.D. Dissertation, Univ. of Durham.

VALENTINE, D. H. (1951). *In* "The Study of the Distribution of British Plants". (J. E. Lousley, ed.), pp. 82–90.

17 | *Nothofagus*, Key Genus to Plant Geography*

C. G. G. J. VAN STEENIS
Rijksherbarium, Leiden, Netherlands

INTRODUCTION

Southern beeches range in the southern half of the Pacific in South America from *ca* 33°–55°S and occur further in New Zealand, Tasmania, East Australia, New Caledonia, and New Guinea (including also d'Entrecasteaux Is. and New Britain) (Fig. 1).

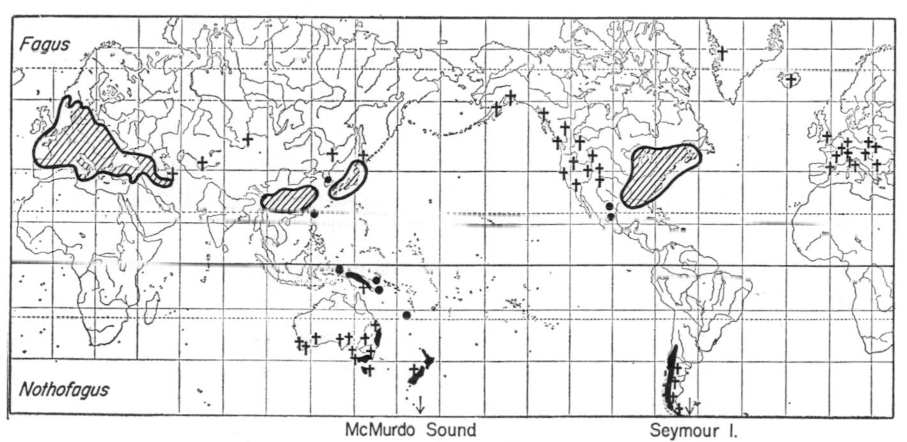

FIG. 1. Living and fossil record of *Fagus* and *Nothofagus*.

There are two sections in the genus. The leaf-shedding section *Nothofagus* (7–8 spp.) occurs in South America, with one species in Tasmania. The evergreen section *Calusparassus* (*ca* 28 spp.) consists of subsect. *Quadripartitae* (8 spp.) in South America and New Zealand, subsect. *Tripartitae* (1–2 spp.) in New Zealand, and subsect. *Bipartitae* (18 spp.) of which 13 are in New Guinea (and New Britain) and 5 in New Caledonia (van Steenis, 1953).

* A more detailed essay is published in *Blumea* **19**, 65–98 (1971).

Fagaceae in general prefer mesophytic conditions, but some grow under seasonally dry conditions. The ecology of the genera *Fagus* and *Nothofagus* is more restricted and is bound to a tropical to cool rain-forest climate, ranging from tropical lowland in New Caledonia and the submontane zone in New Guinea, to timberline in Fuegia and New Zealand; in New Guinea it is most common between (750–) 1000 and 3000 m.

As usual in Fagaceae *Nothofagus* displays gregariousness with frequent co-dominance (occasionally dominance) over large areas thereby producing a colossal biomass, possibly a good second after the conifers. The structure of the Lauro-Fagaceous mountain forest in New Guinea is to a large extent formed by Fagaceae. *Nothofagus* reaches here at maturity a height of some 40–50 m, with a columnar bole of 1–1½ (–2½) m diameter at breast height. These trees are several hundred years old.

Nothofagus shares various ecological features with most other Fagaceae. They are all monoecious and growth is seasonal; stands can easily be spotted by their red flush from the air. *Nothofagus* is not fire-resistant in the sense that it is in any way encouraged by fire; in fact the contrary is the case. *Nothofagus* is rather indifferent to soil, and it grows on various substrata, though more rarely on limestone. Its root system is closely associated with an ectotrophic mycorrhiza of the Agaricales. To this aspect Singer and Morello (1960) and Singer and Moser (1965) attach great significance, arguing that the ectotroph as a biological unit possesses high plasticity and high capacity and leads to dominance. This explanation sounds more likely than referring to allelopathic action.

Furthermore, *Nothofagus* shares with other Fagaceae a manifestly slow and very restricted capacity of dispersal. Germination power rapidly decreases with age, and there is no perennial seed reservoir in the soil. Holloway (1954) experimented with conifers and *Nothofagus*, and established that nuts cannot be dispersed by seawater. Preest (1963) observed dispersal by wind and put the maximum distance possible at 2–3 km. He also reviewed the situation concerning dispersal by birds, but the nut-eating birds are all endemic and sedentary. Epizoic dispersal can readily be discarded, and so can transport by icebergs. This restriction to short-distance dispersal leads to the view that the establishment of the geographical distribution of the genus during past geological epochs was extremely slow and furthermore bound to the presence of land. This is a principle which applies to the other Fagaceae as well as to many other plants, as Hooker, Engler, Diels, Skottsberg, Merrill, Florin, and Corner have all suggested.

This is one of the reasons that I have called *Nothofagus* a key genus for plant-geography. Other reasons are its full fossil record, its undoubted taxonomic status, the bi-hemispheric distribution of Fagoideae within Fagaceae, the fair knowledge of the ecology and morphology of its living species, and its adequately known geographical distribution, past and present.

THE FOSSIL RECORD OF *Nothofagus* AND ITS POLLEN

The full, reliable fossil record of *Nothofagus*, apart from macrofossils, is mainly known from fossil pollen which is unlike any other in the Fagaceae or elsewhere. The wide knowledge of this fossil pollen is largely due to its stratigraphic significance in New Zealand and Australia (Fig. 2).

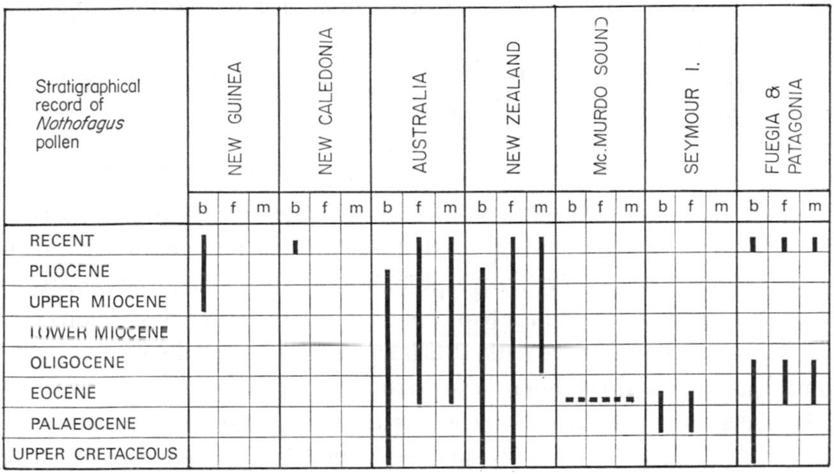

Stratigraphical record of *Nothofagus* pollen

NEW GUINEA | NEW CALEDONIA | AUSTRALIA | NEW ZEALAND | Mc.MURDO SOUND | SEYMOUR I. | FUEGIA & PATAGONIA

b f m | b f m | b f m | b f m | b f m | b f m | b f m

RECENT
PLIOCENE
UPPER MIOCENE
LOWER MIOCENE
OLIGOCENE
EOCENE
PALAEOCENE
UPPER CRETACEOUS

FIG. 2. Stratigraphical record of *Nothofagus* pollen.

In *Nothofagus* there are three pollen types, named after the specific epithets *brassii*, *fusca*, and *menziesii*. They do not coincide with taxonomic subdivisions of the genus; although the *brassii* type was until recently only known from the Papuan–New Caledonian living species of sect. *Calusparassus* subsect. *Bilamellatae*, it has now also become known from the South American *N. alessandrii* Espinosa (sect. *Nothofagus*), which is morphologically the most primitive living species of the genus in having seven nuts per cupule.

Figure 1 gives the complete living and fossil record of *Fagus* and *Nothofagus*. This shows that already in Upper Cretaceous times *Nothofagus* reigned in Fuegia, New Zealand and almost throughout Australia, with species belonging to two pollen types, *brassii* and *fusca*; the *menziesii* type dates from the Palaeocene. In Fuegia all three pollen types are still represented among the living species. Early Tertiary Antarctic localities are hitherto known from Seymour Island and McMurdo Sound. Some gaps are expected to be filled by future data, viz. the Tertiary record in Fuegia and older Tertiary finds in New Guinea, and fossils from New Caledonia.

It is not known how many species have inhabited this area; in most places pollen is abundant. In New Zealand several Tertiary species have become extinct; furthermore it appears from the living plants that one "pollen species" may be common to several taxonomic species. The complete development from the Cretaceous onwards runs, therefore, possibly to 100 species. The age of *Nothofagus* dates back, of course, to before the Upper Cretaceous because even at that period it occupied an immense range.

Records of *Nothofagus* pollen from other parts of the world appear to be based on erroneous identification.

POSITION OF *Nothofagus* WITHIN THE FAGACEAE

Nothofagus is an undoubted member of the Fagaceae and forms with *Fagus* the subfamily Fagoideae. It is quite distinct from *Fagus* in pollen and many morphological characters. Geographically the two genera are separated by the Formosa–New Guinea gap. *Fagus*, with *ca* 10 species, reigns over the northern hemisphere, its southernmost stations being the Sino-Himalayan area and Formosa.

The second subfamily of Fagaceae is the Quercoideae. This comprises the genus *Quercus* with *ca* 600 species, which is distributed all over the northern hemisphere (southernmost stations in the Malesian islands and Colombia), and the genus *Trigonobalanus*, which has 2 species, one in N. Thailand and one in Borneo and Malaya.

The third subfamily of Fagaceae is the Castanoideae which consists of three genera, *Castanea*, with *ca* 12 species in the northern hemisphere; *Castanopsis*, with *ca* 120 species in Malesia, S.E. and E. Asia and one species in S.W. North America; and *Lithocarpus*, with *ca* 300 species in Malesia, S.E. and E. Asia and one species in the southwestern part of the United States.

Fossil evidence of Fagaceae in the northern hemisphere is abundant from the Upper Cretaceous through the Tertiary, and it is found in many areas where the family no longer occurs today, as for example in South

Greenland and Iceland. This is similar to the situation of the southern beeches, which no longer grow on the Antarctic continent. Fossil evidence in the northern hemisphere is largely based on macrofossils, as the pollen of the other genera is not so characteristic as that of *Nothofagus*.

The present area of living Fagaceae at higher latitudes is of course what was left after the destruction of the glacial epoch. It may be that even the tropical montane genus *Trigonobalanus* was once more widely distributed; the fossil genus *Dryophyllum*, hitherto affiliated with *Quercus*, has recently been considered to be a fossil representative of *Trigonobalanus*.

To summarize, Fagaceae reigned in both hemispheres at high latitudes from the Upper Cretaceous; of the 7 genera, 6 are northern hemisphere. All 7 are represented in S.E. and E. Asia and one, *Nothofagus*, is southern hemisphere.

Geographically the interesting thing is that all genera border on or occur in an area extending roughly from Sino-Himalaya-Hainan-Formosa to New Guinea. The importance of this area is emphasized by taxonomic studies. Thus Dr Soepadmo, in his grand revision of Malesian Fagaceae for the Flora Malesiana (1972), says that in *Castanopsis* and *Lithocarpus*, for example, the most primitive species occur in Malesia and Indo China. Furthermore it appears that *Trigonobalanus verticillatus* Forman is an amphitetraploid, doubtless an ancient palaeoploid, with its unique, primitive characters and its affinity in some respects to both *Quercus* and Fagoideae.

The only reasonable conclusion seems to be to regard the Indo-Malesian area between Yunnan and North Queensland as the matrix area of the family, where the ancestral stock was tropical-montane and possibly slightly subtropical (van Steenis, 1971).

I have already several times, most recently in my Landbridge Theory (1962), emphasized that a considerable number of angiosperm families and alliances must stem from the tropics and I will elaborate this later below. This theory is basic to an understanding of angiosperm evolution and world plant geography; unfortunately it is not sufficiently recognized by chorologists working on temperate floras, but the idea is gradually gaining ground.

Recently, A. C. Smith (1970) has independently designated this same area as the matrix area of the Magnoliales and other primitive affiliated families of angiosperms, and regards it as the most important, primary centre of the origin of angiosperms. Again independently, Takhtajan (1969) has come to a similar conclusion, locating the origin of the angiosperms in the region between Assam and Fiji, and that of the typically northern hemisphere groups later, more particularly in the Yunnan area.

I do not want to go so far as to pin down the origin of *Nothofagus* to a special part of this matrix area, although it is reasonable to assume that it will have originated in the southern part and spread southwards; it may have been in New Guinea but also in northern Australia or Queensland. In the absence of fossil data on extinct matrix genera, I believe this roughly outlined location should suffice.

DISCUSSION OF FAGACEOUS EVOLUTION AND ITS COROLLARIES

In the foregoing I have given a brief synthesis of the status of *Nothofagus* and the fossil record of the genus and of other Fagaceae. This embraces the geographical distribution of the family from the Upper Cretaceous onwards, and the fossil evidence showing the effect of the Pleistocene Ice Age.

The Ice Age caused not only the retreat of the Lauro-Fagaceous forest in the northern hemisphere. It similarly removed *Nothofagus* from the Antarctic continent; and, by finally effecting the desiccation of Australia, it almost wiped out *Nothofagus* from the Australian continent, where it was one of the commonest trees under moister ameliorated conditions during most of the Tertiary, when Australia was partly covered by epeiric seas.

There are several important corollaries to our theory. A southern "origin" of *Nothofagus*—the word "cradle" was even used once—appears to be out of the question. Apart from the undeniable agreement of fossil and present distribution, this is taxonomically impossible, as austral Cupuliferous ancestors are completely absent.

In passing I may remark that several groups resemble *Nothofagus* in their geographical distribution, and this has induced several plant geographers to view them, rather thoughtlessly, as "austral groups", that is, having their cradle in the sub-Antarctic and having spread from there to lower latitudes. In many cases this is fictitious; their primary origin and centre was in the tropics or subtropics, whence they migrated to high latitudes in the South Pacific. There they evolved further, until the gradually deteriorating climate towards the end of the Tertiary took its toll. I will return to this subject below.

As the origin of *Nothofagus* is inevitably tied up with that of the other Fagaceae, it follows that it can only have reached Fuegia by way of

Australia via New Zealand and Antarctica. And as it was already in Fuegia in the Upper Cretaceous, with even two pollen types, it follows again that the matrix area must have already existed at an earlier time, mid-Cretaceous or even earlier, as *Nothofagus* is a notoriously slow disperser.

This panorama of Fagaceous development based on fossil evidence and recent distribution necessitates former land connections in the northern hemisphere; in my view, these were between Europe and North America and between Asia and North America, as suggested by the fossil evidence. In the South Pacific, too, there must have been land in early times where the ocean is now. New Caledonia, with five species of *Nothofagus* of the *brassii* pollen type widely spread through the island, lies at some 1500 km from New Guinea, where the nearest allied living *Nothofagus* species are found, and at an equal distance from Australia where fossil, possibly related species were living in the Tertiary. Of similar magnitude are the water expanses now found between Tasmania and New Zealand and between New Zealand and Antarctica, and they are less but still sizeable between Antarctica and Fuegia. They have obviously come into being by considerable subsidences. New Zealand and the surrounding islands as far as New Caledonia are, for example, vestiges of a huge drowned ridge, Tasmantis (Lord Howe Rise and Norfolk Ridge) according to the palaeontological record as produced by Fleming (1962). Between Antarctica and Australia are the huge Macquarie-Balleny and the South Tasmania Ridges.

Drift may also have occurred and helped the spread of plants with Nothofagoid ranges. But the theory here developed, on the basis of the interpretation of plant-geography and the fossil record, cannot be reconciled with continental drift in Wegener's sense, as it would never explain the existence of counterparts in these bi-hemispheric ranges. I have earlier already expressed this viewpoint of the "steady state" in my Landbridge Theory (1962) and it is agreeable to find support from Takhtajan (1969) and A. C. Smith (1970), who share the view that, after diversification, the primary spreading of the phanerogams occurred in a world of which the physiographic configuration was not vastly different from what it is today.

An important point must now be made, deduced from the axiom that plants travel together, migrate as communities, and are sorted out by their ecologies; this is that there must be other plant groups with a similar transtropical, Indo-Australian bi-hemispheric distribution, homologous with that of the Fagaceae.

INDO-AUSTRALIAN BI-HEMISPHERIC RANGES HOMOLOGOUS WITH THAT OF FAGACEAE

Ranges which extend over both hemispheres and which have obviously developed in this way by moving from south to north, or vice versa, crossing the tropics through Indo-Malesia, are common. They occur at various levels, at family, genus or even at specific rank, or they may involve super-family affinities.

Taxonomically, they display variety; sometimes the family is as a whole more or less evenly spread over both hemispheres; sometimes a genus or supra-generic taxon is distinctly either northern or southern; sometimes a genus itself has a uni-hemispheric satellite. It may also be that a family is distinctly split into a northern and a southern counterpart; and there are cases in which a family is mainly uni-hemispheric but one of its genera bi-hemispheric. It may even happen that a single genus retains its identity from the northern to the southern temperate zone.

To explain this, I must make a short digression on plant dynamism in general. This variation in types of geographical distribution is bound up with the course of evolutionary development in the past, with origin and extinction. Both evolution and extinction are erratic processes. There are no obvious reasons why certain groups evolved, why some families have few but large genera, others many genera, or why some, though proven to be very ancient, did not evolve to any extent but kept their identity intact. Extinction is a similarly erratic process, though in some cases one can attempt to correlate it roughly with major changes in land surface, or deterioration of climate (cold or drought) on a major scale.

Distribution, also a historic process, is much less erratic, because once a species or genus evolves, or arrives, it starts to spread, utilizing during this process the subsequently available territory and suitable ecological conditions. Other plants evolving or arriving at approximately the same time and place will, or at least can, avail themselves of the same environmental conditions which the physiography at that time provides. From then onwards they will travel together and will share the subsequent vicissitudes of the physiography. Some may not be able to keep pace where they meet unsuitable environments and, being unable to adapt, are lost during the primary migration; others may undergo change.

Keeping together will largely depend on agreement in adaptive and hence ecological response. It is clear that the closer this agreement, the more similar the evolving ranges will be in essential aspects, and homologous ranges will result. They are also sometimes called equiformal ranges; this term includes not only secondary "progressive" equiformal ranges, but also broken or very disjunct ranges, depending on the changes in physiography and climate which affect the evolution of the ranges.

From this view of range dynamics, one can infer that plants exhibiting homologous ranges must have had a common history, and have filled the common part of their ranges synchronously. This principle of historic plant geography has often been employed, but though it is an important tool, it has not been fully utilized for disentangling the chronological aspect of evolution and range history.

The application of the principle should be handled with caution, and several conditions must be satisfied before we can exclude coincidence. Thus, we must be aware that the common part of the range may result merely from a simultaneous migration caused by a shared major environmental event during part of the history, such as the exchange of arctic and alpine plants during the Pleistocene Ice Age, when the species had, at least in part, quite different distributional histories behind them. Such ranges are only partly homologous.

Returning to the Fagaceae, the essence of Fagaceous distribution, past and present, is twofold. It is bi-hemispheric through Indo-Malesia and it consists of two counterpart ranges which almost replace one another, (only overlapping in New Guinea); these are, a wide range in the northern hemisphere with several genera and many species, and a uni-generic southern range (*Nothofagus*), which forms a "tail" embracing the southern Pacific.

Of these two components, the bi-hemispheric range through Indo-Malesia is the more important, while the counterpart feature seems to me to be peculiar to the group, and to be a speciality which need not be characteristic of bi-hemispheric distributions in general.

Bi-hemispheric counterpart ranges are represented in several other groups, at both family and genus level, e.g. Staphyleaceae/Cunoniaceae, Magnoliaceae/Winteraceae, and *Poterium*/*Acaena*, *Dillenia*/*Hibbertia*. But such vicarious counterparts are not general among bi-hemispheric ranges. This is understandable if we think in terms of the theory here put forward, that many groups had their primary matrix area in the Indo-Australian tropics and subtropics and reached Fagaceous range size as early as the Upper Cretaceous; it would be the exception rather than the rule if all these ancestral groups separated into two distinct vicarious lineages, one northern and one southern. This would involve a rather rare north–south splitting of the original gene-pool.

Therefore it is not surprising that cases where groups are not sharply split into two separate taxonomic counterpart lineages are the more common. For example Ericaceae and Epacridaceae are rather neatly exclusive; but the distribution of the genus *Gaultheria* of the Ericaceae resembles closely that of Fagaceae, while *Pernettya* partly shares the range of *Nothofagus* from the neotropics to Tasmania. A similar situation is found in the pair *Carex–Uncinia*, in which *Carex* is worldwide but *Uncinia* almost exactly matches the range of *Nothofagus*. Another example is the *Veronica/Hebe* complex in which *Veronica* is worldwide and the *Hebe*-complex, even in detail, similar to the range and development of *Nothofagus* (with some satellite genera such as *Parahebe* and *Detzneria*).

There are also Nothofagoid ranges without distinct northern counterparts, e.g. *Libertia* (Iridaceae), *Oreomyrrhis* (Umbelliferae), and *Drapetes* (Thymelaeaceae). There are also genera or genus-complexes which even more than *Gaultheria* approach the total area of Fagaceae fossil and present, e.g. *Araucaria* and *Euphrasia*.

The recent southern range of *Araucaria* is almost an exact replica of that of *Nothofagus*, although it occurs also on the Brazilian shield where *Nothofagus* is not found (but might be expected in the fossil state); and the northern fossil range is very similar to that of the other Fagaceae. Recently Florin (1963) considered these fossils to belong not exactly to the recent genus *Araucaria*, but to extinct, closely associated genera, which would yield still more resemblance to Araucariaceae and Fagaceae in fossil and recent range.

Euphrasia also occupies a range which is homologous with that of the family Fagaceae (including its fossils) (Fig. 3). It is not known from the fossil state, and its habit is not well adapted to show up even if it were fossilized. This genus has clearly kept its identity through the ages; *Odontites* in the northern hemisphere and *Aganosperma* and *Siphonidium* in New Zealand are obvious derivatives, and perhaps some other genera. Besides, it has been able to maintain scarce but distinct transtropical footholds in a nice series representing a clear lineage.

It is remarkable that in New Guinea, New Zealand, and Patagonia these three genera, *Nothofagus*, *Araucaria* and *Euphrasia*, grow together. This is, in the light of the colossal physiographic changes since the Upper Cretaceous, an illustrative achievement, token of a marked stability of overall conditions, and sustaining the steady state principle. Why *Nothofagus* and *Araucaria* have not maintained footholds in Malesia west of New Guinea, where they are abundant today, is of course beyond firm reasoning. It

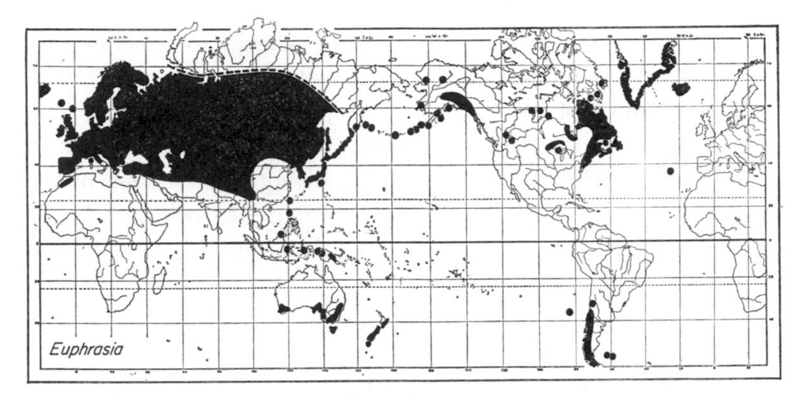

FIG. 3. Geographical range of the genus *Euphrasia*.

could be supposed that there is a marked difference in even short-distance dispersal capacity; both tropical-montane genera are forest dwellers and very slow and near-land migrants, which would be a serious drawback during the vicissitudes of the unstable physiography of the Malesian archipelago. The more microtherm *Euphrasia* of open alpine land can possibly migrate more easily from one mountain to another; but I had better refrain from such very hypothetical explanations.

I must add, though, that the idea of "maintaining" footholds in *Euphrasia* does not mean that I suggest that the present localities of *Euphrasia* in Malesia date back to the primary ancient migration. Mountain plants necessarily have to "jump" from eroding mountains to those emerging by orogenies in their vicinity and within reach of their capacity of normal short-distance dispersal. During this process new species, neo-endemics, may evolve. In a few cases palaeo-endemics may be maintained, as I assume from the primitive features shown by *Eryngium moluccanum* Steen. and *Ranunculus fasciculiflorus* H. Eichl. of Mt. Binaya in Ceram (Moluccas).

Bi-hemispheric trans-Indo-Australian ranges are fewer in woody than in herbaceous plants, but there are admittedly more herbaceous plants than woody ones. On the other hand the ranges of the woody plants are relatively seldom disjunct, but among the herbaceous ones there are many which are disjunct or almost so. *Oreomyrrhis* has only 2 stations in the Malesian islands, *Trisetum* 1, *Eryngium* 1, *Centunculus* 1, but the following have none: *Anemone*, *Angelica*, *Aphanes*, *Apium*, *Asperula*, *Barbarea*, *Caltha*, *Capsella*, *Centaurium*, *Convolvulus*, *Crepis*, *Daucus*, *Eritrichium*, *Erodium*, *Geum*, *Gratiola*, *Gypsophila*, *Lepidium*, *Limosella*, *Linum*, *Lotus*,

Lythrum, *Mentha*, *Myosurus*, *Papaver*, *Prunella*, *Samolus*, *Scleranthus*, *Thlaspi*, *Tillaea*, *Trigonella*, etc. Imagining the vicissitudes of the physiography of the Indo-Malesian tropics during the geological past, from the Cretaceous to the Recent, it is clear that only rather rarely could meso- and microtherm genera maintain themselves. They lived a precarious life; when mountains became eroded they had to disperse to others which were being uplifted. Hence, most will be older than the mountains where they are found today.

CHRONOLOGY OF THE BI-HEMISPHERIC FAGACEOUS RANGE AND ITS IMPLICATIONS

Accepting the facts which have been presented and the theory, based on these facts, that the sandglass-shaped bi-hemispheric trans-Indo-Malesian tropical ranges are indeed homologous (it is irrelevant whether they are disjunct or still more or less continuous), we conclude that they must have come into being by means of land areas along the western side of the Pacific, ranging from the East Siberian shield to Antarctica. This need not have been continuous land all the time, as some unimaginative opponents of the "landbridge", better "land-theory", have sometimes unreasonably deduced. On the contrary, land connections must have had a chequered history.

This is precisely the conclusive point on the homology of the plant ranges; the plants experienced the complicated physiographic history together, so far as the gaining of the ranges was concerned. From this it follows that they gained, or lost, the homologous part of their ranges synchronously. This view does not, of course, clarify their descent and ancestral history previous to the Upper Cretaceous. However, it stands to reason that for Fagaceae and other groups showing their range pattern, the (sub)tropical area between Yunnan and Queensland must have been a cradle for many groups, where their ancestral matrix led to evolution and served as a primary distribution centre.

If this conclusion is accepted, it leads necessarily to a critical revision of the relationship between the age and the morphological primitiveness of Cormophyta, two different aspects which are too often erroneously assumed to be identical. It may be shocking to some to find it advocated here that Araucariads, Magnoliaceae/Winteraceae, Fagaceae, and Scrophulariaceous complexes such as *Veronica/Hebe* and *Euphrasia* must have shared a common evolutionary and range history from or even before the Upper Cretaceous; yet, as shown by the fossil record, the Fagaceae at least already extended at that epoch from Greenland to Antarctica, and it must surely have taken them some time to gain this colossal distribution.

However, taken in conjunction with the fact that many angiosperm groups were already differentiated in the Early Cretaceous, and that most were present in the Middle Cretaceous, having descended from ancestral lines already started in the Upper Jurassic, the conclusion is less revolutionary than it may seem at first sight. One must of course get used to the idea that morphologically advanced groups of, for example, sympetalous orders can also be of high antiquity. The Fagaceae, which is generally not regarded as a primitive family, is proven to be among the earliest fossil phanerogams known.

It is highly desirable not to conclude that plants with primitive features are on that account manifestly older than those with derived characters. Naturally, plants with primitive features must be old, but those with derived features may be nearly as old. It could well be that the evolutionary differentiation of the phanerogams, soon after the emerging of the first ones to the phanerogamous level, led to an explosive radiation within a geologically very short period, as advocated by Němejc (1956). After this was accomplished, the lineages spread, evolved and changed, leaving extinctions in their wake, and so on. But a number of families and genera successfully stood the test of time and kept their identity until recent times. Success, measured by geological age, has nothing to do with so-called primitiveness in structure, *teste Ginkgo*, of which my Japanese colleagues tell me that among wayside trees in the city of Tokyo it withstands best the air-pollution from which several other trees suffer severely.

In my essay on plant geography (van Steenis, 1962), I have come to conclusions about the great age of still-living angiosperm groups, irrespective of their place in the taxonomic system and of their possession of more primitive or more derived morphological characters. From the large number of pantropical families showing this great age I deduced a tropical origin for the angiosperms which may possibly date back to the Early Cretaceous or even to matrices in the Upper Jurassic.

I feel satisfied that this opinion has been shared recently by other colleagues. I mention for example Hawkes and Smith (1965), who deduce from the taxonomy and distribution of *Bromus*, *Gossypium*, and *Solanum* that these genera must date from the Early Cretaceous. Still more recently Stearn (1971) has advocated a similar view on the age of the Acanthaceae.

In the palynological record which Muller (1970) surveyed and tabulated, the Acanthaceae are meagrely represented by one genus from the Upper Miocene and one from the upper part of the Lower Miocene, yet they are at present abundantly pantropical. In discussion, Muller is of the opinion that this cannot be due to the bad preservation capacity of the pollen. There remain two possibilities to explain the discrepancy between the assumed great age of the family and the too low minimum age of the pollen record. Either the record is incomplete, or the Acanthaceae developed early but produced pollen which differed so much from the present pollen types prevailing in the family that it cannot be recognized as Acanthaceous. Future research may reveal intermediate pollen types which link older unrecognized fossil pollen with more recent structures.

I may add that the same discrepancy between pollen record and age is found in several other families at the "tail" of Muller's tabulation, as for example in the Loranthaceae. This may be due to the two factors already mentioned, an incomplete record, and evolution of pollen types more or less independent of gross morphology.

REFERENCES

Fleming, C. A. (1962). *Tuatara* **10**, 53–108, maps 3–4.

Florin, R. (1963). *Acta Horti Bergiani* **20**, 121–312.

Hawkes, J. G. and Smith, P. (1965). *Nature, Lond.* **207**, 48–50.

Holloway, J. T. (1954). *Trans. R. Soc. N.Z.* **82**, 329–410.

Muller, J. (1970). *Biol. Rev.* **45**, 417–450.

Němejc, F. (1956). *Sb. nat. Mus. Praze* **12**, 65–143.

Preest, D. S. (1963). *In* "Pacific Basin Biogeography" (Gressitt, ed.), pp. 415–423.

Singer, R. and Morello, J. H. (1960). *Ecology* **41**, 549–551.

Singer, R. and Moser, M. (1965). *Mycopath. Mycol. appl.* **26**, 129–191.

Smith, A. C. (1970). The Pacific as a key to flowering plant history. Univ. Hawaii. Harold L. Lyon Arn. Lect. Vol. 1, pp. 1–28.

Soepadmo, E. (1972). Fagaceae. *Fl. Males.* **I**, 7 (in press).

Stearn, W. T. (1971). *Bull. Br. Mus. nat. Hist. (Bot.)* **4**, 261–323.

Takhtajan, A. L. (1969). "Flowering plants. Origin and dispersal". Oliver and Boyd, Edinburgh.

Van Steenis, C. G. G. J. (1953). *J. Arnold Arbor.* **34**, 301–374.

Van Steenis, C. G. G. J. (1962). *Blumea* **11**, 235–542.

Van Steenis, C. G. G. J. (1971). *Blumea* **19**, 65–98.

18 | Some Evolutionary and Phytogeographical Problems in the Aegean

ARNE STRID

Department of Plant Taxonomy,
University of Lund, Lund, Sweden

INTRODUCTION

Ever since the days of Darwin and Wallace islands have been a favourite playground for students of evolution. There are obvious reasons for this. Islands are closed communities; problems of isolation, dispersal, and differentiation can be studied in a more purified form than is generally possible in mainland areas. If the principal geological history of the region is known the rate of evolutionary change can sometimes be estimated.

The Aegean area with its multitudinous islands and very diversified topography is in many respects well suited to the study of evolutionary processes. The outlines of the palaeogeographical development of the Aegean have been established in the works of Creutzburg (1963, 1966), Pfannenstiel (1954), and others. The extensive floristic and phytogeographical studies of Rechinger (1943, 1950) have provided a firm basis for more detailed biosystematic investigations.

1. *Palaeogeography*

Until mid-Pliocene the present Aegean archipelago was part of a land mass connecting southern Greece with Asia Minor. This land bridge was bounded to the north by a lake covering large parts of the present North Aegean Sea and to the south by the Sea of Crete, separating it from the south Aegean island chain which is part of the Alpine fault system. In the late Pliocene and early Pleistocene, sea arms penetrated from the south-east, subsequently isolating a Cycladian land block. The present Aegean archipelago was formed in several stages interrupted by periods when contact was re-established between some of the islands. On the border of the Sea of Crete vertical movements due to local tectonic activity may have complicated the picture.

During the large Riss glaciation when the regression of the Aegean sea may have amounted to about 150 m, the islands in the northern and eastern Aegean certainly formed part of the mainland and apparently Crete was connected with Peloponnisos. Some evidence also indicates that the northern Cyclades may have been in contact with the Euboea–Attica area. During the most recent glaciation (Würm) the east Aegean islands as well as the small islands in the Gulf of Saronikas (i.e. Salamis, Aigina, etc.) were most probably connected with the mainland, whereas most islands in the Cyclades group remained isolated from one another.

The main features of the late geological history of the Aegean are reflected in Rechinger's phytogeographical subdivision of the area. There is general agreement among geologists, botanists and zoologists as to the early isolation of Karpathos. Botanical and zoological evidence suggest that the break between the Cyclades and the east Aegean islands must be of considerable antiquity, although some geologists maintain the idea of east–west connections in the Kalimnos–Amorgos area as late as the Riss glaciation. The palaeogeographical development of Ikaria is somewhat obscure. The distribution of several plant species suggests a relatively late land connection between the south-eastern Cyclades and eastern Crete, which, however, is contradictory to present palaeogeographical knowledge. The sequence of events in the break-down of the land bridge between Crete and Peloponnisos is also a matter for discussion.

2. *Phytogeographical Subdivision*

The area can thus be divided into four phytogeographical regions (Fig. 1; cf. Rechinger, 1950).

(1) The south Aegean island chain, characterized by a very high degree of endemism, especially on Crete and to some extent on Karpathos.

(2) The west Aegean region, a number of rather small islands off the coasts of Attica and Peloponnisos with a flora very similar to that of the Greek mainland.

(3) The east Aegean region with a strong Turkish element (excluded from Flora Europaea and referred to the Flora of Turkey area).

(4) The central Aegean region (the Cyclades in a phytogeographical sense) with a flora comparatively poor in species and mainly of a European character.

Recently the term Kardägäis ("Heart of the Aegean") has been coined by Greuter (1970, 1971) to cover the Cyclades and the central part of the south Aegean island chain (Andikithira to Karpathos).

FIG. 1. Phytogeographical subdivision of the south and central Aegean according to Rechinger (1950).

Over the past decade the Cyclades and Ikaria have been subject to a detailed floristic study by Runemark and co-workers, whereas Greuter has concentrated on Kithira, Crete, and Karpathos. Floristically, the south and central Aegean is now comparatively well known.

Parallel to this basic work some groups have also been subject to more penetrating biosystematic studies. One of these groups is the *Nigella arvensis* complex (Ranunculaceae).

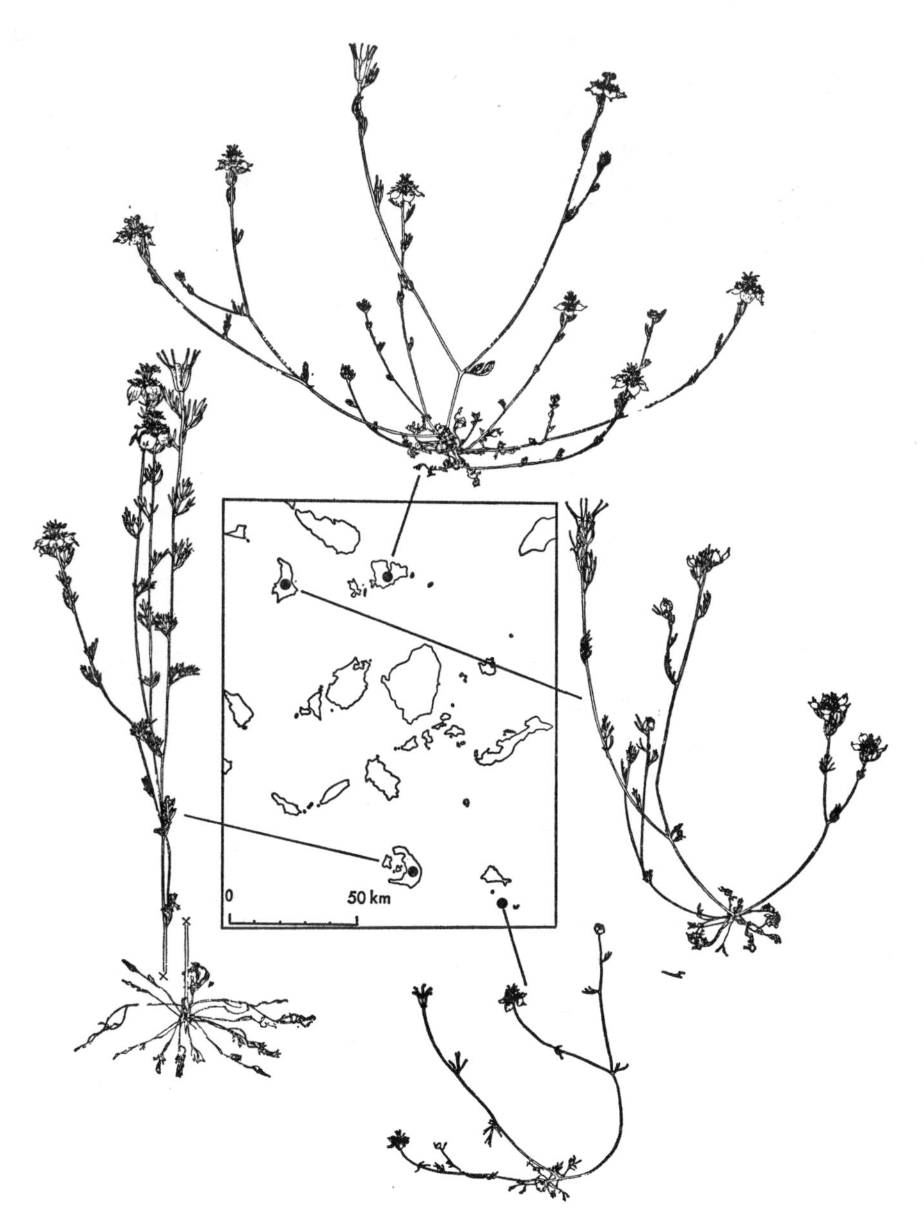

FIG. 2. Examples of morphological variation in *Nigella degenii*. Distinct local races are found on the different islands. Compiled from Strid (1970).

Nigella—A MODEL CASE

Nigella is a genus of diploid annuals with some 20–25 species in the Mediterranean and Near East. In the Aegean the members of the *N. arvensis* complex clearly fall into two different groups:

(1) the outbreeders (the "*N. arvensis* coenospecies") which present a particularly confusing picture of morphological variation.

(2) The selfers, two distinct and comparatively invariable species.

1. The Nigella arvensis *coenospecies*

The outbreeding *Nigella arvensis* coenospecies seemed to be represented by a discrete local race on nearly every major Aegean island (Fig. 2). In contrast to this seemingly quite irregular variation, gradual transitions with respect to morphological characters were found in the mainland populations.

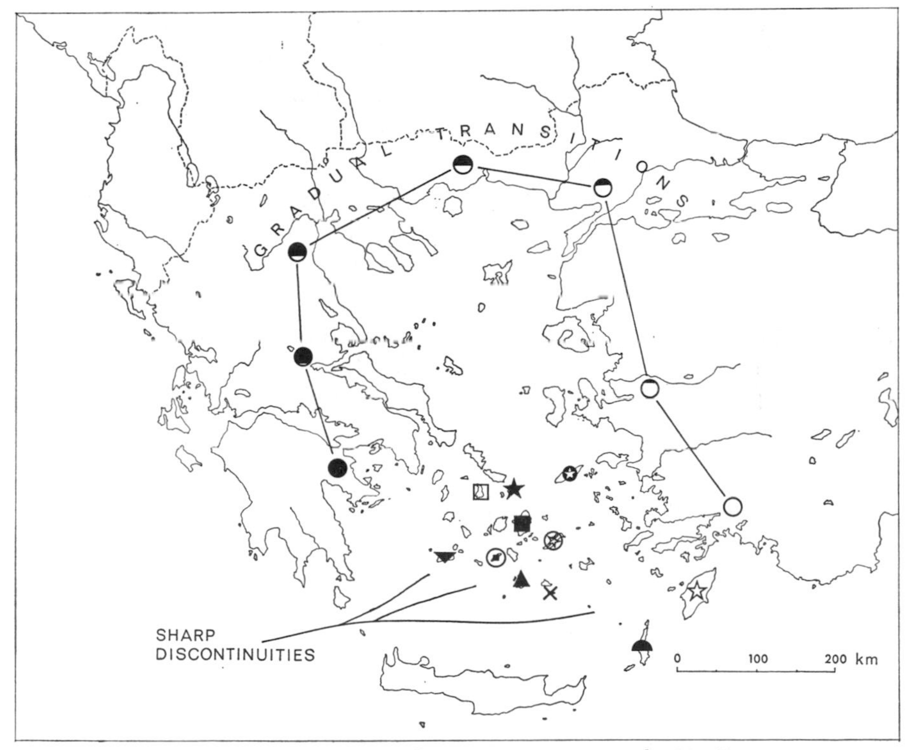

FIG. 3. Diagrammatic representation of variation patterns in the *Nigella arvensis* coenospecies. Sharp discontinuities between island populations stand out in contrast to cline-type variation in mainland areas.

It soon became obvious that extensive hybridization experiments as well as morphological analyses would be needed for a better understanding of this variation and its evolutionary and taxonomic significance. Fifteen populations, covering as much as possible of the morphological variation were selected for this work and crossed reciprocally in all possible combinations. A small separate crossing scheme was designed for the Turkish and east Aegean material. Development, pollen fertility (i.e. per cent stainable pollen), and in some cases also the meiosis and seed setting of the hybrids were observed. Reduced fertility values were recorded in a large number of cases. Plants with less than 45–50% stainable pollen generally do not produce seed.

Sterility barriers in the *N. arvensis* coenospecies appeared to be distributed according to a geographical pattern largely corresponding to the phytogeographical subdivision of the Aegean proposed by Rechinger (Fig. 4). Though often morphologically distinct, the Cycladian populations (*N. degenii* Vierh.) are generally interfertile, but form sterile or semi-sterile hybrids with populations from all surrounding areas. The Karpathos and Ikaria populations drop out both morphologically and genetically and have been recognized as separate species. The isolated phytogeographical position of Ikaria has only comparatively recently been appreciated, but is now supported by fairly strong evidence from a number of plant groups. The clearly abnormal distribution of *N. arvensis* L. ssp. *brevifolia* Strid (common on Rhodes and known from two nineteenth century collections on western Crete) may result from occasional long range dispersal. Kithira has a phytogeographically intermediate position between Peloponnisos and Crete as shown by the three *Nigella* species occurring in this area (Fig. 5; cf. Greuter and Rechinger, 1967).

On the mainland, the more or less continuous morphological variation of *N. arvensis* apparently corresponds to a cline with respect to genetic constitutional similarity. The end members of this chain of forms give rise to almost sterile hybrids when artificially crossed, while interjacent populations form more fertile hybrids the shorter the distance between them. From a taxonomic point of view this is a very awkward situation. The chain of forms has more or less arbitrarily been cut into three segments which have been given the rank of subspecies: ssp. *aristata* (Sibth. & Sm.) Nym. in southern Greece (approximately south of the 40th parallel), ssp. *arvensis* in northern Greece (as well as most of continental Europe), and ssp. *glauca* (Boiss.) Terracc. in western Turkey.

Analyses of meiosis in some F_1 hybrids indicate that sterility is mainly

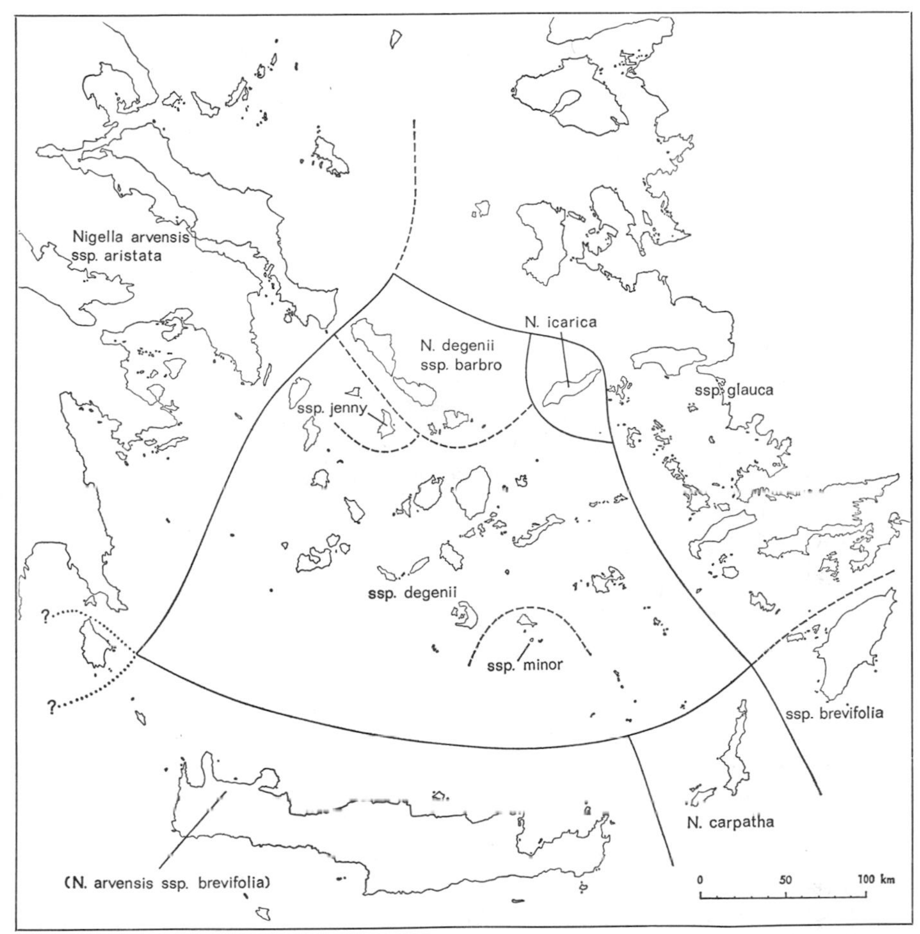

FIG. 4. Distribution of taxa in the *Nigella arvensis* coenospecies (cf. Fig. 1). Redrawn from Strid (1970).

due to genetic factors and not caused by major cytological disturbances. The distribution of fertility values in F_2 families obtained by artificial selfing of semi-sterile F_1 plants support the theory that fertility reductions in the primary hybrids are under multifactorial genetic control.

2. Nigella doerfleri

In addition to the nine outbreeding taxa shown in Fig. 4 there are two self-fertilizing members of the *Nigella arvensis* complex: *N. doerfleri* Vierh. which is distributed on the Cyclades, Crete and Andikithira, and *N. stricta*

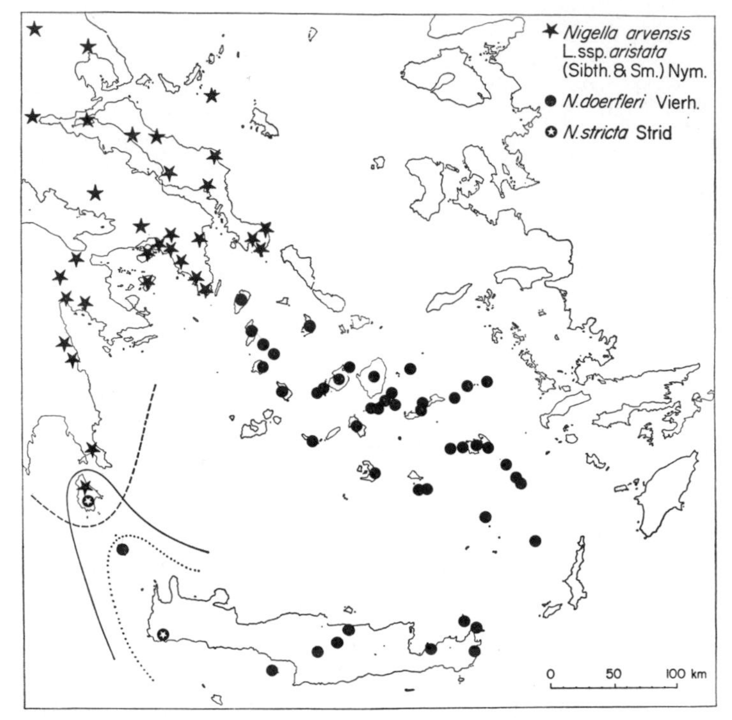

FIG. 5. Phytogeographically intermediate position of the island of Kithira is reflected in the distributions of three species of *Nigella*. For *N. doerfleri* and *N. stricta* the map gives the total distribution ranges.

Strid, a distinct, new species which was only collected for the first time in 1964 and is found in a few localities on Kithira and westernmost Crete (Fig. 5).

Nigella doerfleri has a similar distribution to the outbreeding *N. degenii*, but is a "rigid species" showing little morphological variation. A crossing scheme including 12 populations was designed to study the infraspecific genetic variation of this species.

A surprisingly high number of F_1 combinations (36%) were found to have average pollen fertility values below 90%, and were classified as semi-sterile. The populations differed greatly in their combining ability (0–10 fertile F_1 combinations out of 11). When pairs of mutually inter-fertile populations were compared with respect to their mode of reaction in crosses with every one of the remaining 10 populations, they could be classified as either (1) reacting in the same way, i.e. producing F_1 hybrids with pollen fertility values falling on the same side of the 90% limit, or (2) reacting differently. As a result of these comparisons the 12 populations were found to represent at least eight different constitutional types. In *Nigella doerfleri* a complicated pattern of genetic differentiation is thus concealed under a surface of relative morphological uniformity. It should be stressed that, contrary to the *N. arvensis* coenospecies, no geographical pattern could be observed in the distribution of fertility relationships. The *N. doerfleri* case can be taken as a warning to taxonomists against drawing too far-reaching conclusions from single cases of fertility reductions in crosses between pure lines of a self-fertilizer.

3. *Effects of the Breeding System*

The two selfers, *N. doerfleri* and *N. stricta*, clearly fall outside the *N. arvensis* coenospecies. They can only with great difficulty be crossed mutually or with members of the *arvensis* coenospecies, and the hybrids produced are completely sterile. A detailed comparison was made between some populations of *N. degenii* (outbreeder) and *N. doerfleri* (selfer)

TABLE I. Summary of differences between the two Aegean endemics *Nigella degenii* Vierh. and *N. doerfleri* Vierh. Simplified from Strid (1969 and 1970)

	N. degenii	*N. doerfleri*
Breeding system	Cross-fertilizing	Self-fertilizing
Morphology and habit	Very variable	Rather uniform
Diameter of flowers	(12–) 15–26 mm	9–15 mm
Petals	Large, conspicuously coloured	Small, inconspicuously coloured
Flowering period	June to end of July	April–May
Distribution	On large islands only	On both large and small islands
Pollen fertility	Often reduced	Rarely reduced
Seed setting	Slow, incomplete (empty capsules often found)	Rapid, complete
Taxonomic position	Closely related taxa found in surrounding areas; hybrids easily produced, semi-sterile	Distinct; hybrids with other taxa difficult to produce and completely sterile

which occur in the same area (Figs 4 and 5). The results strongly underline the fundamental importance of the breeding system for the entire differentiation pattern of plant species. Differences were found with respect to rate of development, variation pattern, distribution, fertility, etc. (Table I; for details see Strid, 1969).

Erysimum SECT. *Cheiranthus*

The phytogeographical subdivision of the Aegean suggested by the distribution of sterility barriers in the *Nigella arvensis* coenospecies is partly

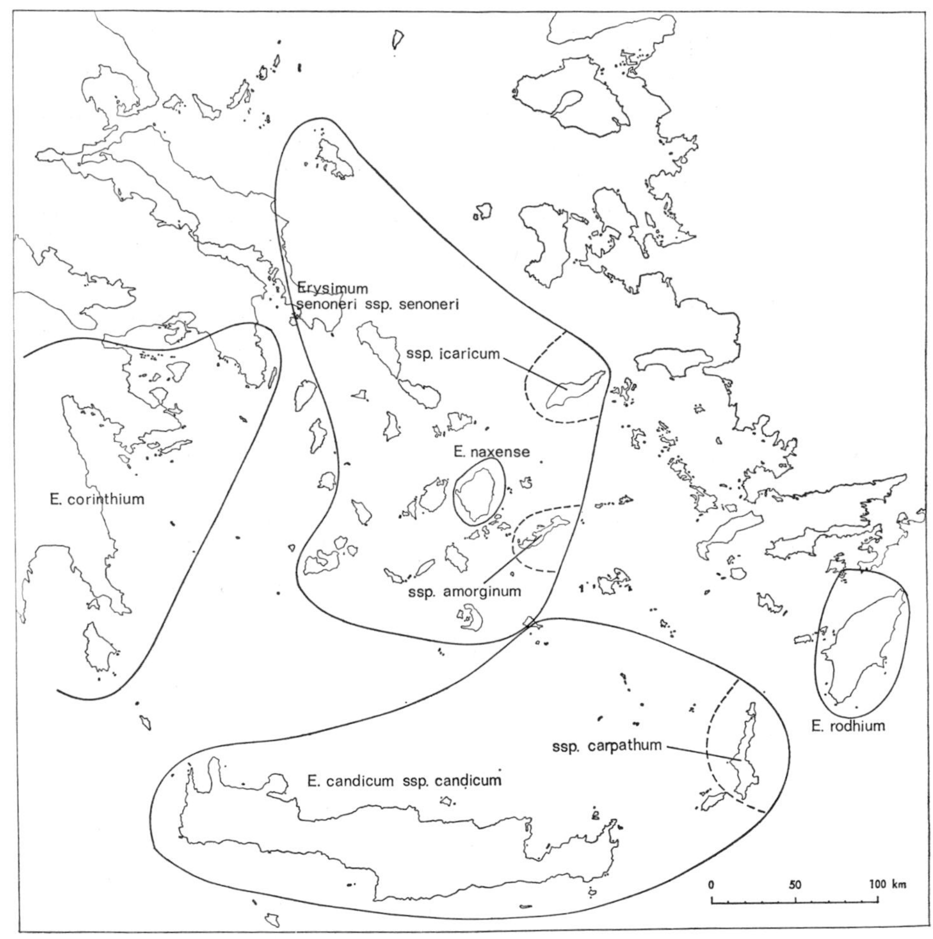

FIG. 6. Distribution of taxa in *Erysimum* sect. *Cheiranthus* (cf. Figs 1 and 4). Redrawn from Snogerup (1967b).

supported, partly contradicted by Snogerup's (1967a, b) results from a similar investigation of *Erysimum* sect. *Cheiranthus*, a group of perennial chasmophytes (Fig. 6). It can be observed, for instance, that the gap between the Cyclades and Euboea–Skiros is bridged by *E. senoneri* (Heldr. & Sart.) Wettst. ssp. *senoneri*, and that the Cretan taxon has an outpost on the island of Anafi. Similar to the situation in *Nigella*, separate taxa (but of different rank) have been recognized on Rhodes, Ikaria and Karpathos. The occurrence of an endemic species on Naxos is not paralleled in *Nigella*. The peculiar small-population system in *Erysimum* sect. *Cheiranthus*, discussed by Snogerup (1967b), strongly influences the local variation pattern of these taxa.

PROBLEMS OF SMALL PLANT COMMUNITIES

Members of the cliff communities often show very interesting patterns of variation and particular attention has been paid to them in Runemark's inventory work. Evolutionary and phytogeographical aspects of chasmophytic communities were discussed by Snogerup (1971). The species composition in cliff communities as well as in other small, isolated plant communities appears to be rather haphazard. Runemark (1969) applied the theories of genetic drift to random variation in the frequencies of individual species in such communities and coined the term "reproductive drift" for this phenomenon. In very small plant communities the species composition would thus be under influence of a random factor, reproductive drift, in the same way as very small populations are subject to genetic drift with respect to gene frequencies. In all probability similar processes—be they called drift, founder principle, or non-adaptive radiation—have played a significant role in the evolution of flora and vegetation on the Aegean islands.

SUMMARY

Differentiation patterns in (a) the *Nigella arvensis* complex (Ranunculaceae) and (b) *Erysimum* sect. *Cheiranthus* (Cruciferae) are presented as model cases for evolutionary diversification of plants in the Aegean area. The effects of the breeding system are discussed with reference to two endemic species of *Nigella*, *N. degenii* (outbreeder) and *N. doerfleri* (selfer). Evolutionary and phytogeographical aspects of very small plant communities are discussed.

REFERENCES

Creutzburg, N. (1963). *Kritika Kronika*, pp. 336–344. Iraklion. (In Greek with German summary.)

Creutzburg, N. (1966). *Erdkunde* **20**, 20–30.

Greuter, W. (1970). *Repert. Spec. Nov. Regni. Veg.* **81**, 233–242.

Greuter, W. (1971). *In* "Evolution in the Aegean" (A. Strid, ed.), pp. 49–64. Gleerup, Lund.

Greuter, W. and Rechinger, K. H. (1967). *Boissieria* **13**, 1–206.

Pfannenstiel, M. (1954). *Freiburger Universitätsreden, N.F.*, **18**.

Rechinger, K. H. (1943). "Flora Aegaea". *Denkschr. Akad. Wiss., Wien. (Math.-Nat. Kl.)* **105** (1).

Rechinger, K. H. (1950). *Vegetatio* **2**, 55–119, 239–308, 365–386.

Runemark, H. (1969). *Bot. Not.* **122**, 90–129.

Snogerup, S. (1967a). *Op. bot. Soc. bot. Lund* **13**, 1–70.

Snogerup, S. (1967b). *Op. bot. Soc. bot. Lund* **14**, 1–86.

Snogerup, S. (1971). *In* "Plant Life of South-West Asia" (P. H. Davis, P. C. Harper and I. C. Hedge, eds), pp. 157–170. University Press, Aberdeen.

Strid, A. (1969). *Bot. Not.* **122**, 380–397.

Strid, A. (1970). *Op. bot. Soc. bot. Lund* **28**, 1–169.

19 | Endemism in the Genus *Alchemilla* in Europe

S. M. WALTERS

Botany School, University of Cambridge, Cambridge, England

In any study of plant geography, it seems desirable to proceed in two steps; first we need to assemble the facts about the distribution of the groups we are interested in, and only then can we try to interpret those facts in the light of our understanding of the processes of vegetational change and of microevolution. We hope that our facts are not in dispute, and are adequate to bear the weight of speculative construction we like to put on them. In the case of most ordinary genera and species of the European flora, we do not think the facts are inadequate or disputed—although it is possible that we are sometimes too complacent about this! But when we come to the critical apomictic genera, we all realize that the situation may be different. For this reason I think it best to start this consideration of endemism in *Alchemilla* with some appraisal of the facts, and then proceed to speculate on their significance where the facts seem to be adequate.

What sort of things do we know about the distribution of *Alchemilla* species? We can usefully start with Linnaeus, because his very simple view of *Alchemilla* taxonomy still provides a very practical basis for the ordinary botanist two centuries later. There are in fact three Linnean species: *A. vulgaris* auct. (the Common Lady's Mantle), *A. alpina* L. (the Alpine Lady's Mantle) and *A. pentaphyllea* L., a small and very characteristic plant of snow-patch vegetation in the Alps of C. Europe. In N. and N.W. Europe generally we can usefully divide all *Alchemillas* into the two Linnean species, *A. vulgaris* and *A. alpina*; but the modern taxonomy divides each of these aggregates into a number of taxa to which binomials have been conveniently applied, so that we now have between two and three hundred species of *Alchemilla* described in Europe. Most of these taxa are endemic to Europe; relatively few are wide-ranging, extending into Asia or more rarely to Arctic N. America. Among the European endemics are many species known only from a limited area, and it is such species which are particularly relevant to our enquiry.

A good point to start in our assessment of the narrow endemic Alchemillas is with the one Linnaeus himself clearly recognized and described—*A. pentaphyllea*. It is significant that Linnaeus recognized it; there is very rarely any difficulty in identifying the plant, and the rare intermediates which do occur in the Alps between *A. pentaphyllea* and members of the *alpina* aggregate presumably were unknown to Linnaeus. Here there is a good endemic *Alchemilla* for which neither the taxonomy nor the distribution data are doubtful or inadequate; we shall return to it later.

No other *Alchemilla* endemic is quite so "respectable" as *A. pentaphyllea*. Some, however, are sufficiently clear-cut and well-known for us to feel some confidence in the available data. As an example we can take *A. faeroensis* (Lange) Buser, an attractive plant possessing characters of leaf-shape and flower structure intermediate between the "*alpina*" and the "*vulgaris*" Alchemillas, and restricted to the Faeroes and E. Iceland. *A. faeroensis* belongs to a group of species (Series *Splendentes*) possessing these intermediate characters, and in assessing its claim to be endemic we must be satisfied that it is distinct from all other *Splendentes*, which occur in the Alps and the Pyrenees. At this point we recognize a basic difficulty which actually applies to all groups, not just to so-called "critical" ones, namely what level of difference do we accept in any topodeme before we distinguish it by a name? We can grossly inflate the number of endemics by accepting a slight, statistically significant difference as a basis for taxonomic separation. In this way our taxonomic presuppositions or fashions determine our facts. There is no neat solution to this difficulty; but at least we should be aware of it. So far as *A. faeroensis* is concerned, you must take my word for it that it *is* satisfactorily separable from all other members of the *Splendentes* on characters of hair-distribution and leaf-shape. But two things should be said. Firstly, we know very little about some *Splendentes* species described from the Alps, and in particular we have an inadequate idea of their variation. Secondly, other *Splendentes* species could still be undescribed in the Alps—Rothmaler described an apparently distinct member of this group, *A. kerneri*, as recently as 1962—so that we cannot be confident that we know all the Continental material yet.

If this must be said of a relatively distinct endemic like *A. faeroensis*, we must make the point even more strongly in most other cases. The taxonomy of *Alchemilla* is well known in N. and W. Europe, but badly worked out in the Balkan mountains; this is a very familiar situation which applies to our knowledge of the flora in general, and is only accentuated in critical groups. In *Alchemilla*, however, there is a peculiar feature of our knowledge, arising from the fact that the specialist in the genus, Robert Buser, devoted many years of his life to painstaking study of Alchemillas in the Alps, and distinguished many species there which hardly any later workers have been able to recognize so confidently. How are we to assess the apparently high local endemism in the Alps consisting as it does almost entirely of Buser species? It is easy to be sceptical about such endemism and to dismiss it as the product of unequal specialist attention. I believe, however, that there is a real phenomenon here. For example, outside the Alps, in Europe, it is possible to accommodate nearly all the *alpina* aggregate into one of two species, viz. *A. alpina* L. *sensu stricto*, and *A. plicatula* Gand. The former is the only species of the aggregate in Scandinavia, for example, and the latter the only species on the Balkan mountains. Inside the main area of the Alps, however, there are described some 20 taxa of variably wide range and this accords with any superficial assessment one might make of the total variation within the *alpina* aggregate in different parts of Europe. This means that the fact of greater local variation within the *alpina* aggregate in the main European mountain ranges of the Alps is well-established, though any detailed conclusions about the endemism of *alpina* species would be more difficult to rely on.

Perhaps I should now assess the value of the only described endemic *Alchemilla* of the British Isles, since I was responsible for it—*A. minima* Walters, of which the *locus classicus* is the limestone of Ingleborough, in N.W. Yorks. This dwarf plant retains its habit in cultivation, a fact which led me to describe it as a distinct species in 1948. It is, however, obviously closely related to the widespread *A. filicaulis* Buser [including *A. vestita* (Buser) Raunk.], and it is an open question what is the appropriate rank for such a genetic dwarf taxon. Rothmaler reduced it to a variety of *A. filicaulis*. I have recently become very interested in genetic dwarf *Alchemillas* with the discovery (or more strictly re-discovery) of a Faeroese dwarf *Alchemilla* originally described as *A. faeroensis* var. *pumila* by Rostrup in 1880. This plant seems to bear a relationship to typical *A. faeroensis* analogous to that of *A. minima* to *A. filicaulis* (Walters, 1970). In these two cases we find local endemic dwarf taxa in areas where we know that there has been a long history of quite heavy sheep grazing; and we might speculate that such dwarf variants are the products of recent intense selection in heavily-grazed topodemes of *Alchemillas*.

This case has conveniently taken us from a consideration of the facts of endemism in *Alchemilla* to the realm of speculative interpretation. We can now consider more generally the problem of interpretation. Most *Alchemilla* species are apparently total apomicts; the single achene develops precociously in the young flower and the contents of the anthers degenerate before the pollen stage is reached. Such chromosome numbers as have been counted both in the *vulgaris* and the *alpina* aggregates are very high (more than 90, up to *ca* 220 in *A. faeroensis*). The single exception is the endemic Linnaeus recognized, *A. pentaphyllea*, which is apparently a normal sexual species with well-formed pollen and $2n = 64$. What then can we say about the endemic *A. pentaphyllea* and its relationship to the rest of the genus? Firstly, it does not seem possible on grounds of morphology to think of *A. pentaphyllea* as ancestral to the high polyploid apomictic complexes which represent the rest of the genus in Europe, and we are forced to assume that a number of ancestral diploid or low polyploid taxa must have become extinct during the evolution of the genus. Secondly the sexuality of *A. pentaphyllea* does explain the existence of a few rare, narrowly endemic taxa which Buser described and rightly guessed were hybrids between *A. pentaphyllea* and certain species of the *alpina* aggregate. It seems (though this still needs experimental verification) that some *alpina* species have a proportion of functional pollen and can therefore behave as male parent in a sexual cross. It is remarkable tribute to Buser's knowledge and interpretative insight that he called such plants hybrids on field evidence, knowing nothing of the cytogenetic situation.

Turning now from the unique case of *A. pentaphyllea* to the numerous cases of endemic apomicts, what might we guess has been the origin of such taxa? Here our knowledge of the microevolutionary situation in other genera may help us to interpret what we see in *Alchemilla*. A particularly useful parallel is found in the related genus *Potentilla*, where there are some Sections within which the ancestral diploid and low polyploid species still coexist in Europe with high polyploid partial or complete apomicts, plausibly derived from the ancestral taxa by a complex history of allopolyploid microevolution (cf. Smith *et al.*, 1971). Thus it seems very reasonable to derive the high-polyploid *A. faeroensis* from some ancient hybridization between a *vulgaris* and an *alpina* which were still partially sexual. Unlike *Potentilla*, however, we can no longer find possible sexual ancestors, and there is therefore no chance of experimental investigation.

An interesting question now arises as to the likelihood of polytopic origins for these apomictic taxa. The *Splendentes* are morphologically intermediate between "*vulgaris*" and "*alpina*"; are we to assume that each named taxon has had a separate, single origin in some ancient event of hybridization? Bound up with this question is the question of the age of the apomicts. The widespread Alchemillas of the *vulgaris* aggregate show distribution patterns which suggest that they are of an age comparable with many ordinary sexual species; there are, for example, arctic–alpine species (*A. glomerulans* Buser), northern montane species (*A. monticola* Opiz), etc. Some endemic species may therefore be relatively old "relict" endemics (*A. faeroensis* is likely to be in this category) whilst others are best interpreted as recent or neo-endemics. It seems reasonable to think of the recent local spread of a genetic dwarf endemic such as *A. minima* from an original single point of origin. Do we envisage a single origin for *A. faeroensis*—presumably in a very distant epoch? All this is obviously mere speculation. One feature of the ecology of Alchemillas would suggest that we should be very cautious about interpreting modern distribution patterns: this is the obviously anthropochoric nature of most species. Recent spread along railway lines, roads and around settlements is so obvious for some species that we might guess that the present-day distributions bear little relationship to any "native" patterns before human interference (cf. Bradshaw, 1962). Perhaps at this point we might conveniently bring our speculations to an end.

REFERENCES

Bradshaw, M. E. (1962). *J. Ecol.* **50,** 681–706.

Smith, G. L., Bozman, V. G. and Walters, S. M. (1971). *New Phytol.* **70,** 607–618.

Walters, S. M. (1970). *Fragm. flor. Geobot.* **16,** 91–98.

20 | Patterns of Distribution in the Combretaceae

A. W. EXELL
British Museum (Natural History), London, England

C. A. STACE
Botany Department, The University, Manchester, England

INTRODUCTION

The Combretaceae are a family of trees, shrubs and lianes found in a wide variety of habitats throughout the tropics. The classification we adopt together with an indication of the distribution and number of species in each genus is given in Table I. Details of the characters used in separating the various taxa are given by Exell and Stace (1966). The numbers of species given for each genus must in most cases be regarded as very approximate, depending upon our knowledge (in a few instances very fragmentary) of the group concerned.

The general pattern of distribution of the genera is a fairly familiar one. Of the two large genera *Terminalia* is pantropical, while *Combretum* is missing only from Australasia and the Pacific islands. One of the mangrove genera (*Lumnitzera*) is found in all three palaeotropical continents, and the other two (*Laguncularia* and *Conocarpus*) in both America and Africa (the latter just reaching Arabia). Two of the remaining 15 genera occur in Africa and Asia, but the rest are confined to one continent. Thus the strongest inter-continental relationships are between Africa and Asia and between America and Africa; there are virtually no amphi-Pacific connections.

Within the two widely-distributed genera the same sort of features recur. In *Combretum*, for instance, all the sections (perhaps about 50) are confined to one continent, except *Cacoucia* (Aubl.) Engl. & Diels and *Combretastrum* Eichl., which are found in both America and West Africa;

TABLE I. Classification of the Combretaceae according to Exell & Stace (1966), together with the geographical range (America, Africa, Asia, Australasia) and estimated number of species in each genus

Taxon	Am	Af	As	Au	Species
Subfam. STREPHONEMATOIDEAE					
Strephonema		Af			7
Subfam. COMBRETOIDEAE					
Tribe COMBRETEAE					
Subtribe COMBRETINAE					
Combretum	Am	Af	As		200
Quisqualis		Af	As		16
Calopyxis		Af			22
Meiostemon		Af			2
Thiloa	Am				3
Guiera		Af			1
Calycopteris			As		1
Subtribe PTELEOPSIDINAE					
Pteleopsis		Af			10
Subtribe TERMINALIINAE					
Terminalia	Am	Af	As	Au	150
Ramatuella	Am				6
Terminaliopsis		Af			1
Bucida	Am				9
Buchenavia	Am				22
Anogeissus		Af	As		14
Finetia			As		1
Conocarpus	Am	Af	(As)		2
Tribe LAGUNCULARIEAE					
Laguncularia	Am	Af			2
Lumnitzera		Af	As	Au	2
Macropteranthes				Au	4
20 genera					475 species

Trichopetala Engl. & Diels, which occurs from S.E. Asia to Africa (including India and Madagascar); and *Ciliatipetala* Engl. & Diels, which is basically African but just reaches Arabia.

A greater degree of inter-continental affinity is in general shown by rain-forest species than by those of savannas, presumably due to the fact that the latter have probably developed in the less distant past.

As might be expected, the area of distribution of each genus is very frequently proportional to the number of species, though, again not surprisingly, there are some exceptions to this. The three mangrove genera, in particular, are very widely dispersed (due to their habitat) but have only two species each. *Combretum* has over 100 species in Africa but only about 35 in America and roughly the same number in Asia. *Terminalia*, on the other hand, is relatively abundant in Malaysia (about 50 species) and Madagascar (about 35) compared with elsewhere (about 30 each in Africa and America). *Buchenavia* possesses 22 species in South America, the great majority in the Amazon basin, and *Ramatuella* consists of six species confined to a region of the Amazon about the size of the British Isles (Exell and Stace, 1963). Similarly *Calopyxis* is endemic to Madagascar yet Perrier de la Bâthie (1953) recognized as many as 22 species. In general one could say that the Amazon, Madagascar and the Malaysian islands have been areas of particularly active speciation, and this is a pattern repeated in many other families.

One can conclude from the preceding data, just as others have done in other tropical groups (e.g. van Steenis, 1969), that there is often *not* a close correlation between the size and the geographical range of a genus, and that small relatively homogeneous areas often support a number of closely related species. Whether the number of species in or the geographical area of such a genus is the better measure of its age is a matter of conjecture.

Thus these general characteristics of the family are for the greater part unremarkable. In a number of specific instances there are situations eminently worthy of further study and analysis, but in this paper we wish to pursue a somewhat different approach, and the outline above is mainly given as a background to this.

THE GEOGRAPHY OF CHARACTERS

Characters define taxa, so in one sense at least there must be a close correlation between the two. In addition one might expect similar sorts of characters to occur more or less sporadically throughout a wide area as parallel adaptations to similar environments. This is the case in a wide range of families in both tropical and non-tropical climates, and in the Combretaceae one encounters a good many admirable examples of this sort of situation. We have selected three for brief mention.

1. *Floral Characters in the Subtribe Combretinae*

Three of the seven genera exhibit reductions in floral structure: *Calopyxis*

(Madagascar) lacks petals (except in one species); *Meiostemon* (Madagascar and southern tropical Africa) lacks one whorl of stamens; and *Thiloa* (America) lacks both the above. Similar reductions are shown by two of the 200 or so species of *Combretum*: *C. apetalum* Wall. ex Kurz (Asia) lacks petals and *C. rupicola* Ridl. (Fernando Noronha, an island 350 km off the Brazilian coast) lacks petals and is, moreover, dioecious. These reductions are frequently compensated by the massing together of flowers to form an attractive unit, by the copious secretion of nectar, or by the increased development of the upper receptacle. In some members of the *Combretinae* the individual flowers have become particularly conspicuous (Fig. 1). The extreme example is found in *Quisqualis* (Africa and Asia),

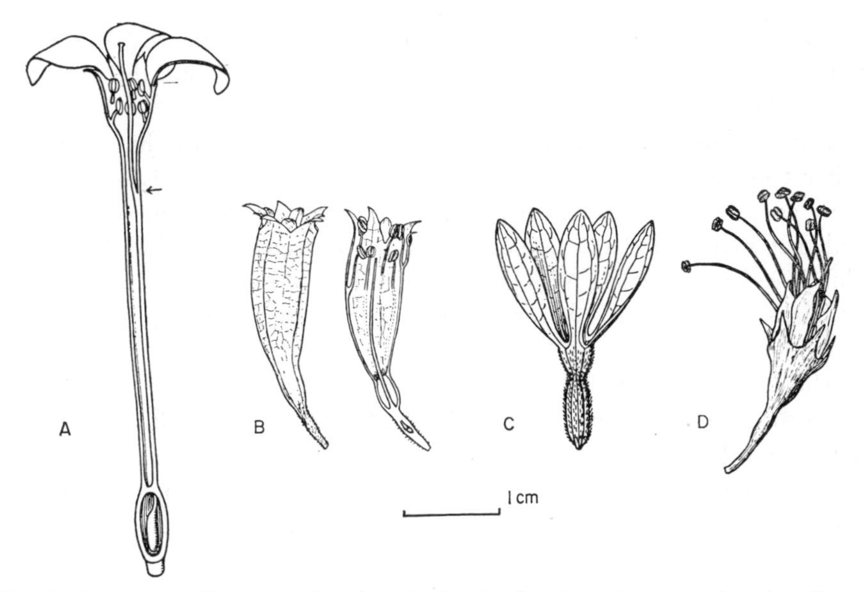

Fig. 1. Conspicuous flowers within the subtribe Combretinae. A. *Quisqualis indica* (from Exell, 1954). B. *Calopyxis grandidieri* (from Capuron, 1967). C. *Calycopteris floribunda* (from Exell, 1954). D. *Combretum coccineum* (from Capuron, 1967).

where the upper receptacle is greatly elongated, the petals are strongly developed and in one species at least the leaves around the inflorescence are brightly coloured (Exell, 1969). In *Calopyxis* (Madagascar) the upper receptacle is enlarged and corolla-like, and in *Calycopteris* (Asia) the calyx-lobes are similarly conspicuous. In Africa and America the more attractive *Combretinae* are mainly *Combretum* species with elongated upper receptacles and brightly-coloured petals (e.g. sections *Conniventia*

Engl. & Diels, *Grandiflora* Engl. & Diels and *Trichopetala*) or stamens (e.g. sections *Combretum* and *Megalanthera* Exell).

2. *Fruit Characters in* Combretum

The great majority of *Combretum* fruits are broadly winged, the tetramerous species usually having four and the pentamerous five wings. In most cases the fruits are wind-dispersed through the air, although, as we shall point out later, this may not always be the function of the wings. Wingless (or ridged or very narrowly winged) fruits are much rarer; this is thought to be a reductional adaptation to water dispersal. The occurrence of such fruits is apparently both geographically and taxonomically scattered. For instance the two amphi-atlantic sections (*Cacoucia* and *Combretastrum*) have both winged- and wingless-fruited species in both America and Africa, apart from the fact that *Cacoucia* has only a single (wingless) species in America. In Asia the single species in each of sections *Acuminata* Engl. & Diels and *Kaloedron* Miq. has wingless fruits, as do two of the four species of section *Tetragonocarpus* C. B. Clarke. In Africa this character is found in the monotypic section *Acuta* Keay, in one of the two species of *Lasiopetala* Engl. & Diels, and in all but about two of the large section *Chionanthoida* Engl. & Diels, among others.

3. *Glandular Trichomes in Combretinae*

All the species of this subtribe possess glandular trichomes which are of two types: clavate stalked glands; and peltate scales (Fig. 2) (Stace,

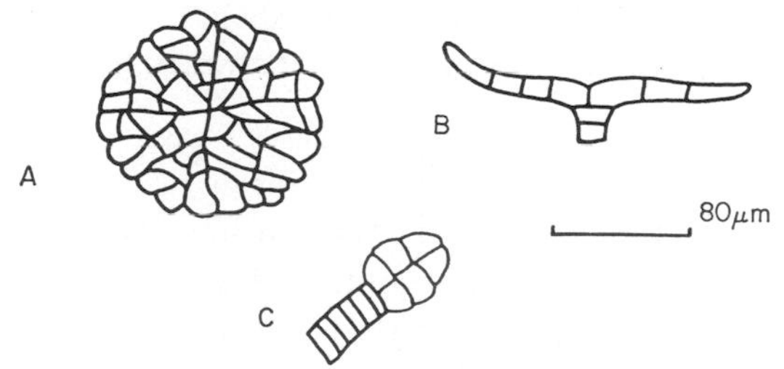

FIG. 2. The two major sorts of Combretaceous glandular trichomes. A. scale, top view. B. scale, vertical section. C. stalked gland, side view.

1965a). The distribution of these within the *Combretinae* is summarized in Table 2. Subgenera *Combretum* (scaly) and *Cacoucia* (Aubl.) Exell & Stace (glandular) of the genus *Combretum* both occur in America, Africa and Asia, while subgenus *Apetalanthum* Exell & Stace contains but a single

TABLE II. Distribution of glandular trichomes in Combretaceae: Combreteae: Combretinae

Combretum		
Subgenus *Combretum*		Scales
Subgenus *Cacoucia*	Stalked Glands	
Subgenus *Apetalanthum*	Stalked Glands	Scales
Meiostemon		Scales
Thiloa		Scales
Guiera		Scales
Calycopteris		Scales
Quisqualis	Stalked Glands	
Calopyxis	Stalked Glands	

Asian species. Hence *Combretum* possesses both glandular and scaly species in all three continents. Moreover the other six genera are so distributed that both sorts of trichome are again found in all continents, except for the absence of a glandular American genus. In addition *Calopyxis* (glandular) and *Meiostemon* (scaly) occur in Madagascar, which for many purposes is best treated as distinct from Africa.

These three sets of examples amply illustrate the point we wish to make: the tendencies or characteristics described are scattered fairly randomly over the total range of the taxon which possesses them, or, in other words, the characters and the taxa have a similar distribution pattern. Further examination of character-distribution in the Combretaceae shows, however, that this is not always the case, and we shall therefore now outline four examples of the opposite phenomenon.

4. *Fruit Characters in* Terminalia

The feature of the fruits we shall mention here is the same as that discussed previously under *Combretum*, although in *Terminalia* neither winged nor unwinged fruits are markedly the commoner. The winged fruits are mostly strongly zygomorphic, being flattened with two or three wings, and wingless (or very narrowly winged) ones are frequently large and fleshy with a thick stony endocarp (Fig. 3). In America all of the approximately 30 species have winged fruits. In Africa all but one of the 30 or so species have winged fruits, the single wingless species being confined to the east coast opposite Madagascar (Griffiths, 1959). In that island about 12 of the *ca* 35 species have winged fruits (Capuron, 1967), while in

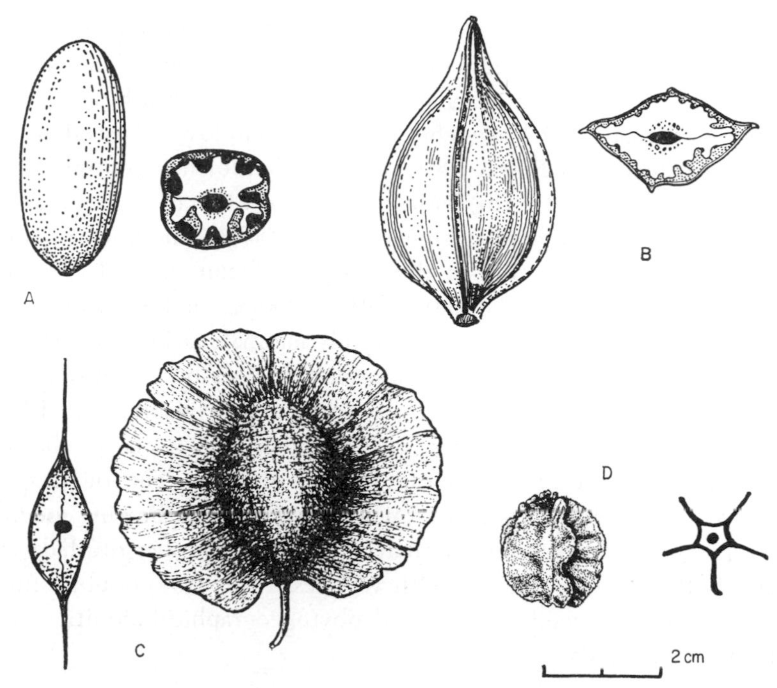

FIG. 3. Fruit-types in *Terminalia* and *Ramatuella*. A. *T. celebica*. B. *T. sepicana* (both from Exell, 1954). C. *T. neotaliala* (from Capuron, 1967). D. *R. crispialata* (from Exell and Stace, 1963).

Malaysia the figures are eight out of about 57 (Exell, 1954; Coode, 1969) Thus there is a definite progression towards fewer winged and more wingless fruits eastwards from America. The only exception to this generalization is that of central tropical Asia; up to date estimates have proved difficult to make, but Clarke (1878) intimated that half of the 12 Indian species were of each sort.

5. *Domatia in* Terminalia

These structures usually take the form of small pits or pockets in the vein-axils of the abaxial surfaces of leaves of woody dicotyledons, although they can take other forms and occur in other positions. Jacobs (1966) has reviewed some of the facts and opinions. Nothing is known of their function, but they are very largely constant in form and occurrence from species to species and can thus be usefully employed in taxonomy. In the Combretaceae there are two main sorts: pocket-shaped (marsupiform) and pit-shaped (lebetiform) (Stace, 1965b). In *Terminalia* there is a distinctive east–west progression in the occurrence and type of domatia similar to (but not correlated with) that shown by the fruits. The estimated percentages of domatium-bearing species, and in parentheses the percentages of domatia which are lebetiform, are as follows: Australasia 62% (86% lebetiform); Malaysia 58% (58% lebetiform); central tropical Asia 13% (75% lebetiform); Madagascar *ca* 40% (? % lebetiform); Africa 3% (0% lebetiform); America 23% (57% lebetiform). Thus in general there is a reduction in the number of domatia and also in the proportion of lebetiform domatia both eastwards and westwards from Australasia. Madagascar seems completely anomalous in this series; the figure of 40% is the only one not obtained directly by us but our source (Capuron, 1967) appears to us to be completely reliable. The 13% in Asia represents only four species; three of these (75%) have lebetiform domatia but these are all species found also in Malaysia. The fourth species is not found in Malaysia and, like the single domatium-bearing species in Africa (3%), has marsupiform domatia. The high number of Madagascan species both with domatia and with wingless fruits is no doubt a further confirmation of the well-documented phytogeographical affinities of that island with S.E. Asia.

6. *Scales in* Combretum

The anatomy of the scales of species of *Combretum* subgenus *Combretum* have proved of immense taxonomic value (Stace, 1969a, b), and many sorts can be recognized. They can be classed into three major groupings, between which intermediates are extremely rare (Fig. 4). It is not sug-

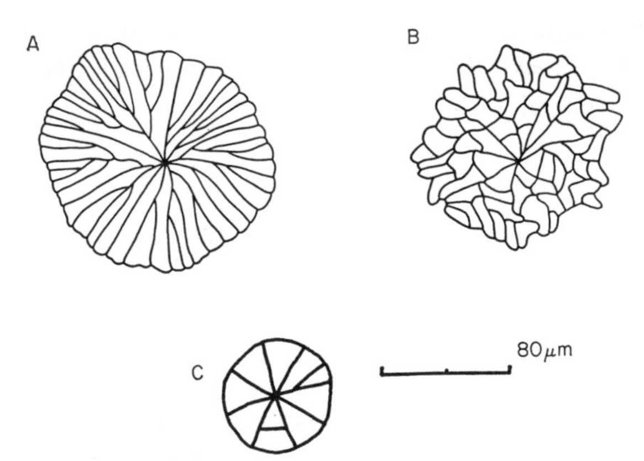

FIG. 4. The three main sorts of scales in *Combretum*, from top view. A. Large complex, all radial walls. B. Large complex, equal radial and tangential walls. C. Small simple.

gested that these scale-groups are of particularly fundamental phylogenetic significance (indeed, they all appear to pass through the same early stages of development), but the anatomy of the mature scales is extremely constant within each of almost all the sections of the genus recognized and so they are clearly of importance in assessing systematic and phylogenetic relationships. The three major scale-groups have differing frequencies in the three continents (Table III). The most prominent point

TABLE III. The number of sections of the genus *Combretum* possessing each of the three main scale-groups

Scale-group	Africa	Asia	America
Simple, small or medium (40–120 μm)	15	5	2
Complex and large (100–250 μm), nearly all radial walls	2	0	7
Complex and large (100–250 μm), equal radial and tangential walls	8	5	0

brought out in this summary is the great difference in the two complex scale-groups between the Old and the New Worlds. In America all the complex scales are of one sort, in Africa and Asia the vast majority are of the other. The two African sections with American-type complex scales are *Chionanthoida* and *Polyneuron* Exell; they do not seem to be closely related to any American sections. The large complex scales are more interesting than the simple ones in another way, for they are in general correlated with those floral characters which in this genus have usually been considered primitive. In particular the largest scales are usually found in species with relatively short, inconspicuous upper receptacles. The extreme examples are sections *Hypocrateropsis* Engl. & Diels and *Parviflora* Engl. & Diels, African and American respectively, which show a good many similarities yet have scales of completely different sorts. Two other such parallels are to be found: *Tomentosa* Engl. & Diels/*Elegantia* Engl. & Diels; and *Ciliatipetala–Glabripetala* Engl. & Diels–*Metallicum* Exell & Stace/*Combretum*. In these cases as well the scales of the African sections (given first) are quite different from those of the American ones. The American/African scale difference would be easy to understand if the sections in the two continents formed two separate taxonomic groups, but this is clearly not the case. The two scale-groups are found in their respective continents throughout a whole range of relatively little-related sections. Whatever the phylogenetic pathways of these sections it is clear that parallel evolution has taken place; either in flower and other features within each of the two scale-groups in their separate continents, or in scale characters within a whole range of different units of the genus. Perhaps the former is the more likely.

7. *Bush-fire Species of* Combretum

Fire is primarily a geographical factor conditioned in the first place almost entirely by the amount and incidence of the rainfall. Its effect on ecology, though probably not on speciation, has been greatly increased in the past by human action. In Africa the rainfall conditions necessary are approximately an annual precipitation of 25 to 150 cm and a dry period of five to seven months in which little or no rain falls. With a higher rainfall, rain-forest not susceptible to annual burning begins to appear and with an appreciably lower rainfall we begin to get semi-desert conditions with insufficient vegetation to burn. The region involved is the Soudano–Zambesian region extending in a semi-circle from Sénégal and Guinée eastwards to the Sudan, southwards through Kenya and Tanzania to Mozambique, and then westwards through Zambia, Rhodesia, the Transvaal and Botswana to Angola (Fig. 5). Owing to our limited knowledge and experience we shall virtually confine our discussion to the situation in Africa.

It is necessary to differentiate between spontaneous fires (mainly due to lightning) and fires started by man accidentally or deliberately for agricultural and other purposes, because the former have obviously occurred for a very much longer period of time and are thus more relevant to evolutionary studies. The occurrence of spontaneous fires other than in rare instances has often been doubted, but discussion by A.W.E. over a good many years with numerous experienced field workers has shown beyond doubt that fires have always occurred spontaneously and often very frequently as a result of lightning strikes both in Africa and elsewhere. Botanists who have provided such first-hand information include H. N. Ridley in Malaya, and F. White (Oxford), E. J. Mendes (Lisbon), and H. Wild, D. Mitchell, T. Müller and J. Gwyther (Rhodesia) in Africa. In parts of Africa special watches are kept during thunderstorms for strikes by lightning which are frequently accompanied by fires (at one station as many as 50 such fires were recorded in one season).

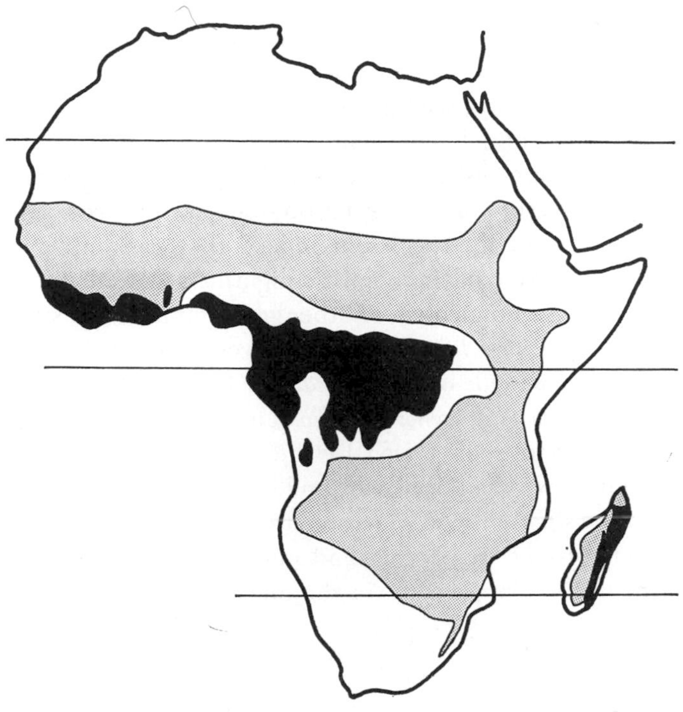

FIG. 5. Approximate extent of tropical rain-forest (black) and fire-prone savanna (stippled) in Africa.

Thus in Africa at least fire has been an important evolutionary factor for a long time and over vast areas of Africa only fire-resistant species can exist. Ecologically there is a complete range from dry open woodlands, often known as dry forest or *miombo*, to various types of grassland often with dwarf shrublets. Many perennial species persist underground and send up herbaceous or semi-woody shoots which are burnt off annually. Some are completely adapted to this habit and are variously known as nanopyrophytes or rhizomatous undershrubs, etc. Others persistently send up shoots until in favourable conditions, or, as one might say by good luck, they are able to produce enough woody growth to survive the annual conflagrations and to produce fire-resistant trees, sometimes with a thick layer of cork. D. B. Fanshawe (*In* Guerreiro, 1966), of the Zambia Forestry Dept., states that "the roots are often eight years older than the shoots and *Pterocarpus angolensis* (Leguminosae) may wait as much as 14 years for a successful growth of the aerial part". Some woody species produce enormous underground trunks making agriculture very difficult.

In such conditions plants usually flower immediately after the danger from fire has diminished, often at the end of the dry season and often before the leaves. The problem for a seedling is of course acute, and is made more so by intensive grazing; but fire can also be favourable in eliminating competition from non-resistant species and by eliminating some pests.

A well-known feature of the vegetation of fire-prone areas is the extraordinary variability of the species, if indeed they can be recognized as such, resulting in a kind of taxonomist's nightmare. Aubréville (in Guerreiro, 1966) once commented on the great number of synonyms in *Combretum* saying "*Après leur passage* [the fires] *il se produit une nouvelle feuillaison différente de celle des plantes adultes, et aussi formations de souches avec des feuilles distinctes des celles des plantes normalement développées*". This leads us naturally to a discussion of the effects of the fire factor on *Combretum*, which we shall deal with under three headings.

(*a*) *Habit*. *Combretum* produces pyrophytic species of the dwarf shrublet type with underground woody parts and annual herbaceous or semi-woody shoots burnt off by fire (Fig. 6). In the Sierra Leone–Dahomey–Nigeria

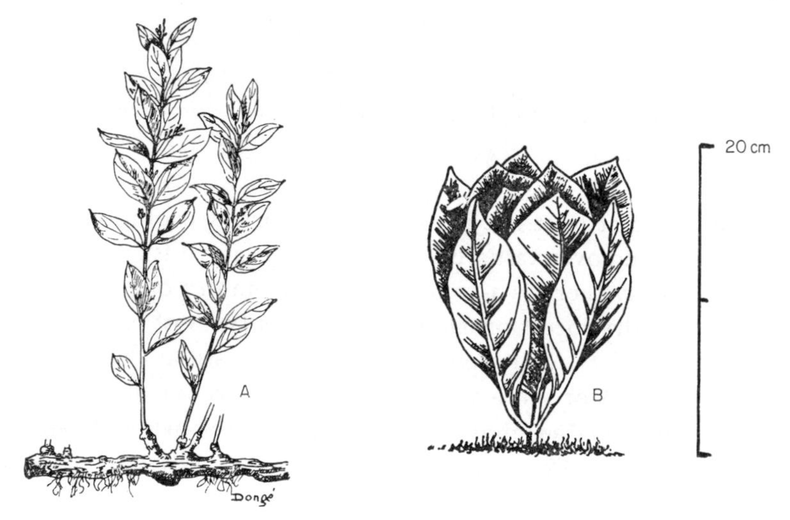

Fig. 6. Two pyrophytic species of *Combretum*. A. *C. simulans* (= *C. sericeum*) (from R. Portères, 1950, *Bull. bot. Soc. France* **97**, 180). B. *C. brassiciforme* (from Exell, 1963).

region we have six species, *C. bauchiense* Hutch. & Dalz., *C. harmsianum* Diels, *C. lineare* Keay, *C. relictum* Hutch., *C. brassiciforme* Exell and *C. sericeum* G. Don, all shrublets up to 1 m in height appearing in savanna soon after the grass fires and flowering within a few weeks (Keay, 1950,

1954). The first five belong to section *Glabripetala* of subgen. *Combretum* and the last one to section *Conniventia* Engl. & Diels of subgen. *Cacoucia*. *C. brassiciforme*, very closely related to two savanna trees, *C. schweinfurthii* Engl. & Diels and *C. gallabatense* Schweinf., is a cabbage-like species about 20 cm high with the inflorescences developing inside the "cabbage", an immensely different habit from that of its arboreal relatives yet almost identical as regards leaves, flowers and scales. Exell (1963) said of this species: "If these shrublets have evolved as a result of fire . . . the evolutionary pressure seems to have been in the direction of a rapid change in habit accompanied by but little change in the morphology of leaves, flowers and fruits. . . . It is as though change of habit has been the primary overriding necessity." Keay (1954) said that specimens of *C. sericeum* "have been collected which appear intermediate in habit between this species and *C. paniculatum* Vent., a scandent shrub or forest liane". Turning to the southern half of the Soudano–Zambesian fire-prone zone we find in the Angolan–Zambesian region a species, *C. platypetalum* Welw., of the same habit and ecology as *C. sericeum* and once more connected with the forest climber *C. paniculatum* by a series of intermediates. *C. platypetalum* is one of those excessively variable species (the variation in the size of the fruit, the shape of the leaves and in the indumentum is fantastic) which we think may owe their variability itself (both phenotypic and genotypic) to the effects of fire (see Fernandes, 1958). We treat it now as one very variable species but it could be divided specifically or infraspecifically into an indefinite number of taxa. There is no doubt in our minds that many of the dwarf shrublets so characteristic of these regions have evolved from rainforest climbers. There are two further shrublets in fire-prone regions of Angola: *C. viscosum* Exell (section *Ciliatipetala*), probably evolved from the closely related savanna tree *C. apiculatum* Sond.; and *C. argyrotrichum* Welw. ex Laws., the only species of section *Argyrotricha* Exell & Garcia. In northern India there is yet another of these dwarf species "burnt down annually by the forest fires" (Clarke, 1878), *C. nanum* Buch.-Ham. It belongs to section *Ovalifolia* Engl. & Diels and is clearly derived from the jungle climber *C. ovalifolium* Roxb.

(*b*) *Fruit*. As mentioned previously the fruits of *Combretum* are either wingless or winged, the latter generally being wind-dispersed through the air. Some winged fruits, however, have stouter, more rigid wings, which are thought to function as sails allowing the fruits to be blown along on the ground. Such fruits are found in several savanna species, particularly in *C. collinum* Fresen. (section *Metallicum*) and *C. zeyheri* Sond.

(section *Spathulipetala* Engl. & Diels). J. Gossweiler (personal communication to A.W.E.) observed at first hand fruits of these species being blown along the ground by the wind. This seems almost certainly an adaptation to fire which clears the ground and makes such a method effective. It does not exclude the possibility that there may also be some dispersal through the air.

(c) *Cotyledons*. This is a comparatively new idea* and it is unfortunate that there are still so few germinations recorded. There are four main types of germination known in *Combretum*: cotyledons remaining below ground; cotyledons arising from the hypocotyl above ground on short petioles; cotyledons arising below ground but coming above ground on long petioles; and cotyledons arising below ground but coming above ground and forming a usually peltate structure on a single long stalk formed by the connate petioles. The last type, which is very rare, concerns us here. The vegetative shoot comes up apparently independently and somewhat later at a distance of some inches. It has been recorded so far only in *C. collinum*, *C. fragrans* F. Hoffm. (section *Glabripetala*) and *C. zeyheri*, all successful savanna species in regions liable to fire. As yet this type of germination has not been recorded in any rainforest species in the whole family. We feel that there is a strong probability that this is connected with fire and guess that this rare and peculiar type of germination gives the seedling a second chance so to speak. These three species are trees or large shrubs and do not have their woody parts largely underground. Whether the seedlings have the capacity to shoot up again if burnt or eaten off we do not know. In any case circumstances demand that germination should take place about when the rains begin and that sufficient growth should be produced as rapidly as possible to be able to withstand the next season's fires. The dangers to the seedling are late fires and the intensive grazing which is itself correlated with the fire factor. The large peltate structure formed by the connate cotyledons can at once start photosynthesis when it comes above ground. If it is burnt or eaten off the independent vegetative shoot (that is independent above ground level) seems to give the seedling a second chance. Hypothesis of course, but it may be noted and may not be irrelevant, that out of about 100 to 120 African species of *Combretum*, two out of the three species we have noted as having this peculiar type of germination hypothetically adapted to fire, *C. collinum* and *C. zeyheri*, are the same as the two species already recorded as having fruits hypothetically adapted to fire. The exact

* See, however, Jackson (1968), ref. which just came to our notice, and has been added in proof.

advantage of the single stout cotyledon over two smaller ones each on a separate stalk is admittedly obscure. Apart from *Combretum*, T. Müller (in litt. to A.W.E.) has found that germination in *Terminalia sericea* Sond. is stimulated by burning the fruit, an obvious advantage to the plant in delaying germination until after burning and then facilitating it as soon as possible.

DISCUSSION

We have surveyed the geographical variation of seven groups of characters; three show a similar distribution to the genus or other taxon in which they occur, and four show a very different sort of distribution. It is not difficult to visualize two distinct situations which might give rise (and indeed have done so) to the latter phenomenon.

(1) Certain characters might be particularly prevalent in an area because they have evolved in a parallel fashion or convergently in a whole range of possibly little-related taxa. This has occurred frequently, usually as a response to selection pressures imposed by the environment, and examples are well-known to all botanists from school age onwards. The example relevant to the present discussion is the series of characters which is found in a wide taxonomic spectrum of fire-resistant species.

(2) Characters will be unusually common in a certain area if they are predominant in or completely characteristic of a taxon which is represented there by a high number of species. The characters might be absent from an area merely because the taxon which bears them is absent. For instance petal-less Combretinae are relatively far more frequent in Madagascar than elsewhere, because on that island *Combretum* is represented by very few species, being largely replaced by the related yet apetalous *Calopyxis*.

Hence in cases of a non-random distribution of characters we have two criteria to apply: the characters might be tied to a particular taxon or group of related taxa; or they might be developed in response to environmental conditions. None of the four sets of characters discussed falls into the first category because we deliberately avoided such features. Besides the pyrophytic characteristics we might suppose that the unequal distribution of *Terminalia* fruit-types could be explained in terms of the environment, for winged fruits are dispersed by wind and wingless ones by water or animals. In the Old World the increase in winglessness from Africa to Madagascar to Malaysia might well be a reflection of the increase in the ratio of rainforest to savanna and of water to land. And if Clarke's (1878) Indian figures are representative they could be explained in the same way. But all the American species of *Terminalia* have winged fruits, and most of them are rainforest species growing in an area of abundant watercourses. The six species of the very closely related *Ramatuella* also all have winged fruits, yet the fruits of all 22 species of *Buchenavia*, fairly closely related to and growing in exactly the same situations as *Terminalia*, are all wingless (Exell and Stace, 1963). It seems as if all the wingless-fruited American Terminalias have become Buchenavias (it should be noted that the two genera are separated on the basis of floral characters), and perhaps a truer statistical picture would be gained if these genera (and *Ramatuella*) were united for this purpose.

On the other hand we have been totally unable to account for the distribution pattern shown by the other two characters. It has been suggested that domatia are commoner in wetter climates (i.e. tropical rainforest rather than savanna) and this might possibly account for the gradation in the Old World. But the single African domatium-bearing species (*T. phanerophlebia* Engl. & Diels) is a plant of the low veld, and the Madagascan species with domatia are certainly not mainly rainforest trees. In America *Buchenavia* and *Terminalia* again show highly significant differences both in the number and in the type of domatia present.

We suppose that the situations regarding the domatia and scale-types and similar cases might be explained in one of the following ways:
(1) A process broadly akin to orthogenesis has taken place, where a tendency to evolve along certain lines is built into a group and continues to develop after that group has been divided into separate lineages; or

(2) The characters are, in fact, subjected to some sort of selection pressure by the environment. We have no idea what selective advantage one type of scale or domatium might have over another, but if there were one then it might of course be no more surprising to find an abundance of radial-walled scales in America than one of pyrophytes in Africa.

While we can offer no conclusive explanation of these phenomena we do believe their study has an important bearing on one aspect of taxonomic principle. It is often considered that small characters of seemingly no selective value are particularly valuable guides to phylogenetic affinities, because of their random rather than channelled evolutionary pathways. Examples which may be quoted are characters of trichomes, stomata, xylem vessels, pollen grains and classes of chemicals such as alkaloids. Yet the data we have presented indicate, we suggest, that this is not the case; such characters do not warrant special consideration and should be treated in the same way as others.

ACKNOWLEDGEMENTS

We are grateful to the following for permission to use figures already published: Trustees of the British Museum (Natural History), Controller of H.M.S.O., the officers of La Societé Botanique de France, and the Flora Malesiana Organization.

REFERENCES

CAPURON, R. (1967). "Les Combretacées Arbustives ou Arborescentes de Madagascar." Centre Technique Forestier Tropical, République Malgache, Madagascar.

CLARKE, C. B. (1878). *In* "Flora of British India" (J. D. Hooker, ed.), Vol. 2, pp. 443–461.

COODE, M. J. E. (1969). "Manual of the Forest Trees of Papua and New Guinea". Pt. 1, revised, Combretaceae. Lae, New Guinea.

EXELL, A. W. (1954). *In* "Flora Malesiana" (C. G. G. J. van Steenis, ed.), Ser. I, Vol. 4, pp. 533–589.

EXELL, A. W. (1963). *Kew Bull.* **17**, 233–236.

EXELL, A. W. (1969). *Curtis's bot. Mag.* **177**, 539.

EXELL, A. W. and STACE, C. A. (1963). *Bull. Br. Mus. nat. Hist.* (*Bot.*) **3**, 3–46.

EXELL, A. W. and STACE, C. A. (1966). *Bolm. Soc. broteriana* Sér. 2, **40**, 5–25.

FERNANDES, A. (1958). *Mem. Soc. broteriana* **13**, 33–46.

GRIFFITHS, M. E. (1959). *J. Linn. Soc. Bot.* **55**, 818–907.

GUERREIRO, M. G. (1966). "A floresta africana e os factores bióticos. Primeiras observações de um ensaio em Mocambique", pp. 38–148. Inst. Invest. Cient. Angola, Luanda.

JACKSON, G. (1968). *J. W. Afr. Sci. Ass.* **11** (2), 215–222.

JACOBS, M. (1966). *Proc. Kon. Nederl. Akad. Wet.* Ser. C, **69**, 275–316.

KEAY, R. W. J. (1950). *Kew Bull.* **5**, 255–257.

KEAY, R. W. J. (1954). *In* "Flora of West Tropical Africa" (J. Hutchinson and J. M. Dalziel, eds), 2nd ed., Vol. 1, pp. 264–281.

PERRIER DE LA BÂTHIE, H. (1953). *Ann. Mus. colon. Marseille*, Sér. 7, **1**, 1–43.

STACE, C. A. (1965a). *J. Linn. Soc.*, *Bot.* **59**, 229–252.

STACE, C. A. (1965b), *Bull. Br. Mus. nat. Hist.* (*Bot.*) **4**, 3–78.

STACE, C. A. (1969a). *Bot. J. Linn. Soc.* **62**, 131–168.

STACE, C. A. (1969b). *Brittonia* **21**, 130–143.

VAN STEENIS, C. G. G. J. (1969). *Biol. J. Linn. Soc.* **1**, 97–133.

Section V

SPECIAL TOPICS

21 | Migrations of Weeds

HERBERT G. BAKER

University of California, Berkeley, California, U.S.A.

PROBLEMS OF LOCATING ORIGINS

Although the material dealt with in this paper has been accumulated during many years of contact with the plants that by common agreement we call weeds, the thoughts contained in it have been put in order whilst on a safari across central and western Australia. This may be particularly appropriate if one believes that to get an adequate perspective one should step back from the scene to view it. In parts of central Australia, moisture and soil-fertility conditions are often so extreme that even where human-induced disturbance has occurred, relatively few weeds have immigrated. But this is most unusual; although disturbance of nature by agricultural and urban man usually produces habitats that are more xeric than those of the undisturbed vegetational climax, conditions in these habitats are not usually so severe as to preclude the establishment of a weed flora.

In some parts of the world, perhaps most obviously in the Middle East, interference on a significant scale with what we call natural vegetation has gone on for more than ten thousand years. Such disturbance in any particular area has waxed and waned under the influences of changing climates, erosion, population explosions and the deaths of civilizations, yet all the time the boundaries of man's influence have been expanded continuously by exploration and exploitation. For some of the common weeds, this means that we now have little hope of locating the starting points of their migrations.

A case in point is provided by *Oxalis corniculata* L., the little yellow oxalis which is so widespread in gardens, in lawns and in waste places in the warmer regions of the world. It may be one species or a complex of species. European floras often say that it is native to the New World, but California floras, for example, insist that it came from Europe. Jepson

(1925) even inclined to the belief that a form with abundant vegetative anthocyanin was introduced from Europe while the less anthocyanous form is a native Californian. Nevertheless, some of the most convincing claims for it as a native come from the Southern Hemisphere (P. Michael, personal communication, for Australia; Laing and Blackwell, 1952, for New Zealand). Nowadays, the complex of largely self-pollinated forms that we call *Oxalis corniculata* has achieved a cosmopolitan distribution, and plants have become well-established on islands as remote as Tristan da Cunha.

Sometimes, a restricted perusal of the literature can give a false impression of certainty as to the history of a weed taxon. Suggestions by one author are copied as facts by others and myths become accepted. The case of the Pineapple Weed, *Matricaria matricarioides* (Less.) Porter, is instructive in this respect. This weed is extremely common in compacted soil of roadsides and trodden places throughout the Northern Hemisphere and in several places south of the Equator. The spread of this plant through Europe, following its introduction in the mid-nineteenth century has been described beautifully (Hegi, 1929; Salisbury, 1932, 1961)—but where did it come from?

It has a very close relative in California, *Matricaria occidentalis* Greene, which grows in the Central Valley in saline places subject to flooding each winter. *M. occidentalis* is not a weed by any stretch of the term's definition, but there are mixed populations containing it and the weedy *M. matricarioides* which suggest that the latter is also at home in western North America. One would hardly get this impression from the literature!

The type specimen of *Matricaria matricarioides* (as *Santolina suaveolens*, Pursh, 1814) comes from North America, having been collected "on the banks of the Kooskoosky" in Idaho on the Lewis and Clark expedition to the West which was in the field from 1804 to 1806 (Hegi, 1929, is in error in stating that the type locality is in California). Nevertheless, Jepson (1925) doubted that it is native in California (as did Howell, 1949) and, of the other early flora writers of western North America, only Peck (1941), describing the flora of Oregon, did not indicate some doubt. Hegi (1929) supposed the species to be a native of eastern Asia (with just a possibility of it also being native in North America). Presumably as a result of Peck's willingness to accept it, Salisbury (1953) referred to *M. matricarioides* as coming to Europe from Oregon, but Clapham (in Clapham *et al.*, 1952, 1962) gave a slight twist to the Hegi attribution and stated that it is "Probably native in N.E. Asia but established in North America".

Salisbury (1961, p. 72) then modified his comment to "supposed to have come to this country from Oregon, although probably an introduction, there and in other parts of North America, from North-East Asia" (despite, elsewhere in the same publication (p. 87), calling it "the Pineapple Weed from Oregon").

A graduate student at Berkeley, Mr L. Maynard Moe, is working with these two species of *Matricaria* at the present time, and we hope that his studies will throw light on the origin of *M. matricarioides*, producing something more satisfactory than its relegation to the wilds of north-east Asia.

Another puzzler, again a denizen of trampled ground (and looking vegetatively rather like *Matricaria matricarioides*), is *Coronopus squamatus* (Forsk.) Asch., of the Cruciferae. According to Clapham (in Clapham *et al.*, 1962), this species is said to be a European native but to be introduced into much of the rest of the world. On the other hand, *Coronopus didymus* (L.) Sm. is supposed by the same author to be native only in South America and introduced elsewhere, including Europe. This seems strange, especially as only ten species are acknowledged for the genus and some American floras (e.g. Howell, 1949; Munz, 1959) claim that *C. didymus* has come to their areas from Europe rather than from South America.

These remarks are not made to poke fun at any of the extremely competent writers of floras, without whose work we should be truly lost, but simply to indicate that some weeds have been extremely successful in getting around the world since the end of the fifteenth century (or earlier) and that we now have little chance of pin-pointing their places of origin.

The problem of deciding whether a species is a native, or whether it was introduced deliberately or accidentally by man into the areas where it now grows weedily, is one which is not by any means restricted to temperate regions. Pan-tropical weeds are notoriously difficult to pin down to palaeotropical or neotropical origins. There are difficulties even when the family to which a species belongs is acknowledged to be characteristic of one continent or another.

We may take the case of the epiphytic cactus *Rhipsalis baccifera* (J. Mill.) W. T. Stearn (= *R. cassutha* Gaertn.). Everybody knows that the Cactaceae are characteristic of the Americas, and the presence of a species of that family in Africa and Asia leads immediately to the suspicion that it is introduced in the latter continents, particularly when it occurs commonly as an epiphytic weed in cacao and other plantations. *Rhipsalis baccifera* is present on both sides of the Atlantic Ocean—in the palaeotropics it is in Africa, the Mascarene Islands and Ceylon.

However, *Rhipsalis baccifera* shows adaptations for non-human dispersal that cannot be ignored. Its berries contain an extremely sticky pulp that causes seeds to adhere to the beaks of birds who probe them. Natural dispersal is achieved by the bird wiping its beak on a branch of a tree and implanting the seed in the process. There are also possibilities of internal carriage of seeds by the bird. A bird blown out to sea from the east coast of South America might have nowhere to clean either end of its body until it reached the African continent.

Such an event need not have been a frequent one. An isolated plant of *Rhipsalis baccifera* can set abundant seed (Baker, unpublished) and start a new population. In addition, growing as an epiphyte or on rocks, it scarcely comes into competition with native vegetation. Establishment after natural long-distance dispersal seems quite feasible for this species, at least as feasible as an alternative theory that it was introduced by sailors who hung it in their ships as a substitute for mistletoe.

The problems increase when, for various reasons, a species becomes more common in an area to which it has been introduced than it was in its native homeland. Freedom from pests and competitors, fortuitous discovery of more nearly optimal environmental conditions, genetic changes in the weed itself; all may produce this result. An example is provided by *Erodium obtusiplicatum* (Maire, Weiller & Wilcz.) J. T. Howell.

In 1906, Philipp Brumhard validated with a description a name that he had given (Brumhard, 1905) to a plant collected in Amador County, California. He called it *Erodium botrys* (Cav.) Bertol. forma *montanum*. Thirty-seven years later, Wagnon and Biswell (1943) reported that plants identifiable with this taxon were extremely common on the rangelands of California. They indicated its morphological peculiarities and some of the physiological differences between it and *E. botrys*, and it may be added that there are clear ecological differences, too (Baker, unpublished). At present, it is not only extremely abundant in the grazed, annual grasslands of California but is also common in similar habitats in Chile. In 1947, J. T. Howell raised this taxon to the level of a separate species under the name *Erodium obtusiplicatum*.

E. obtusiplicatum has also been recorded as a weed in Bedfordshire, England, growing in market-garden fields that had been fertilized with "shoddy" derived from the cleaning of fleeces in woollen mills (Dony, 1953). In view of the probable origin of this wool from Australia, its presence in that continent was to be suspected, and this is confirmed by collections from sheep-rearing regions of Queensland, New South Wales, Victoria and South Australia (Carolin, 1967, and specimens in the National Herbarium of New South Wales). Despite this abundance in situations to which it has been carried by man, there appear to be relatively few extant collections of this species from what must be its area of origin in the Mediterranean region (Wagnon and Biswell, 1943; Howell, 1947).

GENETIC CHANGES ASSOCIATED WITH WEEDINESS

Have colonists like *Erodium obtusiplicatum* changed genetically, or found unoccupied (because recently created) ecological niches, or do they represent cases of successful escape from the inhibitory attentions of pests? For these species we do not know the answers because the necessary research has not been done.

In the case of *Hypericum perforatum* L. (St. John's Wort, or Klamath Weed in California) we do know some of the answers because this is a pernicious weed and has been given the necessary attention, and it appears that all three factors have made a contribution to its success. Pritchard (1960, and unpub.) made a study in which he cultivated together, in a garden, samples of this species from several areas of the world. He showed that weedy material from Australia and California grew to a greater height, produced larger numbers of stolons and had a greater seed-output than non-weedy material from Europe. In addition, we know that freedom from pests has contributed to its success as a weed outside Europe because it is now controlled in Australia and California by the introduction of European chrysomelid beetles (Huffaker and Kennett, 1959). Finally, there is no question but that the conversion of forested land to grassland in northern California presented *Hypericum perforatum* with a niche to fill—which it did.

PATTERNS OF WEED-SPREAD

Since the first analysis of the spread of weeds in the British Isles made by Salisbury (1932), we are familiar with the pattern that is usually followed when a new weed arrives in a country. Following the original introduction, a number of disjunct populations belatedly make their appearance and, from these, there may be a filling-in of the intervening spaces. In California, we are following the spread of *Lythrum tribracteatum* Salzm. ex Ten., which was introduced to the state from the Mediterranean region, some time before 1931. It has produced massive populations in vernal pool situations in restricted areas of six counties, only two of which are adjacent. Physiological differences and patterns of isozymes show distinctions between the populations which suggest that each one has been started from a very small number of seeds, possibly carried in by waterfowl.

Even though some "weedy genera" exist (such as *Lepidium*, *Chenopodium* and *Amaranthus*), often it appears that only one species or a very few species of a genus manage to hit the jackpot and take advantage of man's aid in dispersal to spread far away from their places of origin. This may be particularly the case in the tropics where an example may be taken from the genus *Tridax*. This genus contains about 26 species (Powell, 1965) restricted to Central America and adjacent territories, except for *T. procumbens* L. which has become a prominent weed throughout the tropics. We are studying it to find out the reasons for its ability to spread and establish itself when its congeners are so restricted.

Even when a single species is involved in a weed invasion, only a part of the variability existent in that species is likely to be transferred to the new area of occupation. To some extent, this results from the self-compatibility (even regular autogamy) of many weeds (Baker, 1955a) reducing heterozygosity, together with the likelihood of a "founder effect" operating when the invasion is begun by one or a few seeds. Also, only part of a polyploid complex may be transported. Thus, tetraploid, pentaploid and probably hexaploid cytotypes of *Oxalis pes-caprae* L. exist in South Africa where it is native. Only the pentaploid has become established as a weed in California (Baker, 1965). Even at the level of sexual dimorphism, healthy weed populations are being established in at least two areas of northern California by the aquatic plant *Myriophyllum brasiliense* Camb. where only a pistillate individual was introduced originally and reproduction has been entirely vegetative.

When weeds do enter a new area, they usually come from one with approximately comparable climatic conditions. Thus, the California weed flora tends to be drawn from other areas of the world where at least the temperature regime approximates to that of the "Mediterranean" type. A great many annuals, perhaps most obviously grasses of the genera *Avena* and *Bromus*, can illustrate the contribution from the Mediterranean region itself, while *Madia sativa* Mol. comes from Chile and Argentina. *Cotula coronopifolia* L. is a contribution from South Africa and *Senecio minimus* Poir. [known in many floras as *Erechtites prenanthoides* (A. Rich.) DC.] is a relatively recent arrival from Australasia which is spreading quite rapidly in coastal California.

Most weeds appear to have had their origins in areas of lengthy agricultural and urban development and to have become successfully established in others where agriculture and settlement are newly instituted. Thus, western Europe has contributed much to the "new" flora of New Zealand and, of course, the Mediterranean region has provided the bulk of the plants presently living in California (at least below about 6000 feet above sea-level).

Occasionally, this immigration brings conspecific taxa into relatively close proximity. Thus, in California, *Prunella vulgaris* L. (Labiatae) exists in grasslands and the margins of coniferous forests as a native plant (*P. vulgaris* ssp. *lanceolata* (Bart.) Hultén) but distinctive forms of the same species (*P. vulgaris* ssp. *vulgaris*) have been introduced from Europe and constitute the weedy populations growing in lawns (Nelson, 1965). Hybridization is rare but where it occurs it has been followed by introgression into ssp. *lanceolata* (Nelson, 1964). On the other hand, *Veronica serpyllifolia* L. var. *serpyllifolia* (Scrophulariaceae), which also invades lawns in lowland areas of California, following its introduction from Europe, has not yet made contact with the native variety *humifusa* (Dickson) Vahl which is found in moist places in the coniferous forest zone of the mountains. In this case, it is in the British Isles that these varieties have come together and introgressed in some mountainous areas (Gilmour and Walters, 1959) and similar hybridization may be expected ultimately in California.

Sometimes the inflow of a new set of weeds may be related to a change in agricultural practices. A striking example of this is to be seen in the American tropics where, after thousands of years of shifting cultivation, pastoral activities are being introduced as the forests are cut. Parsons (1970) has analysed with great care the spread through the entire region of African grasses that, unlike tropical American grasses, had co-evolved with herds of grazing herbivores. Often, the cattle which graze them now also include African breeds in their ancestry. The "Africanization of the American tropics" is an interesting reversal of the flow of economic plants and weeds from tropical America to Africa which took place soon after contact was established between the continents nearly 500 years ago.

In the Aboriginal Reserves of central Australia, grazing by stock is not permitted and the usual weeds of overgrazed grasslands are absent. Nevertheless, Russian Thistle (*Salsola kali* L.) is present along moister roadsides and in creek beds, quite ready to move in should pastures be created.

RESTRICTIONS ON WEED DISTRIBUTION

Some analysis is possible of the restrictions imposed on the establishment of introduced plants in a new area even though, as will be emphasized later, successful weeds are often remarkably tolerant of environmental variation. Most obvious among the restrictions are the climatic ones.

On the Berkeley campus of the University of California, the lawns have been invaded by two grasses of tropical origin, *Cynodon dactylon* (L.) Pers. and *Pennisetum clandestinum* Hochst. ex Chiov. Above-ground portions of both are badly browned whenever a frost strikes the area. The grasses regenerate readily from their rhizomes and stolons and their invasion continues, but it is clear that the northern limit to their spread in California will be set by this susceptibility to freezing damage. Similarly, in California, the tropical water weed *Eichhornia crassipes* (Mart.) Solms (Pontederiaceae) reaches its northern limit in the delta of the San Joaquin and Sacramento rivers. Every winter, most of the material is killed by frost but occasional apical buds survive and its fantastic productivity and rate of vegetative reproduction by offsets, which Bock (1969) has shown to be just as high in a Californian summer as they are in the tropics, enable it to block the waterways that it infests anew each year.

Sometimes there is a surprise, however. *Myriophyllum brasiliense*, of comparable provenience to *Eichhornia crassipes*, is similarly exposed to the winter cold but appears to be unaffected by even the hardest frost to hit the same delta area.

Less dramatically than with the grasses and *Eichhornia*, other species sort themselves out into habitats according to the geographical regions from which they came. *Geranium robertianum* L. comes from European areas where summer moisture is available, being replaced by *G. purpureum* Vill. in the Mediterranean region (Baker, 1955b). *Geranium robertianum* is a minor weed in the San Francisco Bay region but it is restricted to the margins of year-round streams or else finds summer moisture in watered gardens or under the coast redwoods. Sweet Alison, *Lobularia maritima* (L.) Desv. (Cruciferae), has a coastal distribution in Europe (Clapham, in Clapham *et al.*, 1962); in the San Francisco Bay region it is a successful weed only within sight of the ocean. Even more striking is the case of *Brassica oleracea* L., the cabbage. Before it became domesticated, this was a sea-cliff plant in Europe; escaping from cultivation it has returned to its old cliff-face habitat on the northern side of San Francisco's Golden Gate.

Such sorting as this can also take place at the subspecific level. For example, in the White Clover, *Trifolium repens* L., Daday (1954a, b) has shown the existence of a "ratio cline" in proportions of plants having a potentiality for cyanide production on bruising of the leaves. The proportion is at its highest in populations growing in southwestern Europe and falls away to the north-east. It is not yet certain to what extent this cline is based on the temperature tolerances of the cyanogenic and acyanogenic plants and to what extent on predator pressure, but it appears that a similar cline has become established in North America where the species has been introduced.

Restrictions of an edaphic nature on the distributions of introduced plants are sometimes very clear. Thus, the foxglove, *Digitalis purpurea* L. (Scrophulariaceae) lives up to its European reputation as a "calcifuge" by only escaping from cultivation and becoming weedy in areas of acid soil in California. *Chrysanthemum segetum* L., the Corn Marigold, which prefers sandy soils in Europe, repeats that preference in its plentiful establishment on parts of the Mendocino County coast of California where the soils are derived from stabilized sand-dunes.

The nature of the agricultural use of land may have a profound effect upon the success of weeds from other areas. In California, the percentage of weed species which are California natives is almost zero in the most highly artificial environments of lawns and golf-course greens (because lowland, summer-green terrestrial communities, which these man-made situations mimic, are rare in California). The proportion of California natives rises in less drastically modified weed community types, and is over 60% (83% native to North America as a whole) in the rice-fields that have replaced the comparably wet tule swamps of the Central Valley (Baker, 1962).

So, for most of the minor weeds in California, there are real restrictions upon their successful establishment. Most of them fit into moulds that have already fashioned the native species. I have found this clearly demonstrated in my studies of seed weight in relation to habitat conditions in California (Baker, 1973). Large seeds are selected for in native plants growing in communities where seedlings are exposed to shade, to the threat of drought or to intensive competition; where these stresses are not imposed there is an opposing selection for large numbers of necessarily smaller seeds. The various types of natural community may be arranged in sequence according to the intensity with which these factors, taken together, operate. A similar sequence may be set up for weed community types. In both sequences, there is a strong positive correlation between mean seed weight for a community type and the intensity of the stress.

ADAPTATIONS IN VERY SUCCESSFUL WEEDS

Such restrictions as those quoted in the previous section of this paper are obvious only in their application to the merely moderately successful weeds. They are minimally operative in the cases of the highly successful weed species, one of whose outstanding characteristics is their extreme tolerance of environmental variation as long as they are not subjected to severe interspecific competition from natives (Baker, 1965, 1967).

A good example of a highly successful pantropical weed is *Ageratum conyzoides* L. (Compositae). It may be compared with its close relative *A. microcarpum* (Benth.) Hemsl.,* which is only slightly weedy in its home on the Central American plateaux. Table I shows the characteristics of plants of these species demonstrated by their growth under controlled conditions at the University of California, Berkeley. The much greater independence of environmental controls shown by individual plants of *A. conyzoides* is clear. It should be emphasized that this is a different strategy from the selection of a series of locally-adapted races through which a species may tolerate a range of environments even though each individual is narrowly adapted. Instead, the highly successful weeds have been selected for a "general purpose genotype" (or series of genotypes) which provide each individual plant with pre-adaptation to a range of environments, although presumably reducing the exactness of its adaptation to any one (Baker, 1965, 1967). Such pre-adaptation, especially when allied with a capacity for self-fertilization or apomixis, enables the weed to be an "opportunist", building up a more or less adapted population rapidly without the loss of time that would be involved in the selection of a

* My (unpublished) biosystematic studies show that generic distinction between *Alomia* (no pappus) and *Ageratum* (with pappus) cannot be maintained. Consequently, *Ageratum microcarpum* (Benth.) Hemsl. is the oldest name for this species which contains what Johnson (1971) has separated as 3 species, *Ageratum oerstedii* Robins., *A. petiolatum* Hemsl. and *Alomia microcarpa* (Benth.) Robins.

TABLE I. Comparative features of *Ageratum microcarpum* and *A. conyzoides* revealed by controlled environment experiments

Ageratum microcarpum (scarcely weedy)	*Ageratum conyzoides* (widespread weed)
Light requirement for germination	No light requirement for germination
Perennial	Life span 1 year
Flowers in second season of growth	Germination to flowering in 6–8 weeks
Flowering inhibited by high night temperatures	Flowering at low (10°C) or high (27°C) night temperatures
Flowering better with long (12 hour) nights	No photoperiodic control of flowering
Mesophyte	Tolerates waterlogging, drought
Self-incompatible	Self-compatible (largely self-fertilized)
Not very phenotypically plastic	Phenotypically plastic
$n = 10$	$n = 20$

special, locally-adapted variant. Also, with selfing or apomixis there need be no loss through genetic recombination of such general purpose genotypes once they have been produced and selected. Most weeds are capable of self-pollination or are apomictic.

Obviously, if human disturbance of an area continues and the weed remains long enough in the area, selection for closer local adaptation will come into play. I have offered evidence of the "general purpose genotype" (g.p.g.) strategy in the dandelion, *Taraxacum officinale* Wiggers (Baker, 1965) but, even in this apomict, Solbrig (1971) has provided evidence of genotypic variation within well-established local populations giving the possibility of microenvironmental adaptation.

The replacement of general purpose genotypes by those giving close local adaptation when a weed lingers in an area may be expected to be quicker in amphimictic species than in apomicts. Y. B. Linhart (1971), working with me at the University of California, has demonstrated that intra-population differentiation has taken place within Californian populations of the rather weedy native *Veronica peregrina* L., despite the nearly cleistogamous flowers (restricting genetic recombination through crossing) and the very tiny seeds (that might be expected to be widely dispersed at each reproductive act). The sub-populations of this species, which occur in and around vernal pools and ditches, are slightly differentiated morphologically but, more significantly, are differentiated in their competitive ability and resistance to physiological damage from flooding.

Thus, the general purpose genotype strategy appears to be of greatest importance during the colonizing phase of a weed's progress. It can be seen, for one more example, in the wide range of soil types inhabited by the pampas grass that has escaped abundantly along the coast of California. These apomictic plants seem best referred to *Cortaderia atacamensis* Pilger from Ecuador and Chile, rather than to the more frequently cultivated Argentinian species, *C. selloana* (Schult.) Asch. & Graeb., a dioecious taxon which Munz (1959) erroneously cites as a frequent escape. Mrs Martha Costas-Lippmann is studying this interesting situation in Berkeley.

Some native members of the California flora exhibit weedy tendencies even though the history of cultivation in this state covers less than 200 years. In these cases, the pre-adaptation probably results from a similarity in the natural habitats of the species and the waste places, agricultural land and roadsides provided at the present day by man. The genus *Amsinckia*, in the Boraginaceae, and *A. intermedia* F. & M., in particular, have had some success as annual roadside and arable land weeds. *A. intermedia* has even become a weed in eastern England (Salisbury, 1961) although on a restricted scale so that it is not mentioned by Clapham *et al.*, 1962). It is also to be found as a weed in Japan (H. Hara, personal communication). However, even this species has not yet become perfected as a weed, for it still needs an appropriate combination of temperatures and rainfall ("an *Amsinckia* year") for it to form sheets of plants in northern California. By contrast, the old reliable introduced mustards, *Brassica nigra* (L.) Koch and *B. geniculata* (Desf.) J. Ball, "do their thing" almost regardless of the climatic pattern every spring and summer, respectively. They are highly successful ("g.p.g.") weeds.

Another California native, naturally occurring in dry flats and fields, is the skunk weed, *Navarretia squarrosa* (Esch.) H. & A. (Polemoniaceae). This species has been able to show some success as a colonizer of not-too-frequently trodden trails, where the flexibility of its largely prostrate shoots and its finely divided leaves help to preserve it from damage. It has even attained the standing of a minor weed in Australia, where it is fairly common in some of the wheat-growing districts of New South Wales (Whittet, 1958, p. 404). It will be interesting to study its genetics.

We do not know the function, if any, of the mephitic odour of *Navarettia squarrosa*, but the possession of unusual chemical substances may be among the armoury of useful weapons possessed by some successful weeds. The Rosy Dock, *Rumex vesicarius* L. (*Acetosa vesicara* (L.) Löve), a native of Egypt and western Asia, has taken over vast areas of the Flinders Ranges, in South Australia, where it seems to be thoroughly unpalatable to the sheep which graze there, presumably on account of its oxalic acid content. Ironically, in the early years of its establishment (near Blinman), Black (1948) wrote of it (under the name *Rumex roseus* L.) as being "regarded favourably as sheep-feed". If it were preferentially grazed it could never have got out of hand as it has done.

In California, some tree species of *Eucalyptus*, especially *E. globulus* Labill., tend to behave weedily. It is unusual for many of the European immigrant and California native plants which did not co-evolve with these Australian eucalypts to be able to germinate and establish themselves under the shade of these trees; and it may be that, in addition to the exploitation of the soil by the root-systems of the trees and their blanketing of the soil-surface with their decay-resistant leaves, these eucalypts are able to discourage potential competitors allelochemically by the soluble phenols and volatile terpenes that are contained in and released from their leaves (Baker, 1966; Del Moral and Muller, 1969). However, it must be pointed out against this explanation of their success that competition from herbs and shrubs is only likely to be a serious matter for the eucalypts in the seedling stage, whereas most of their reproduction in California is by sprouting. Seedlings are usually only to be seen on new mineral soil rather than underneath existing trees. Nevertheless, the role played by allelochemicals in the establishment and spread of weeds is worthy of further attention; some examples involving herbaceous weeds appear to be incontrovertible (Anon., 1971; Whittaker and Feeny, 1971).

HYBRIDIZATION AND WEEDINESS

Hybridization with another species may help a potential weed to become more completely adapted to this way of life. A full spectrum of interspecific hybridization can be seen in the world's weed flora, from the formation of sterile F_1 hybrids to the production of vigorous, fertile amphidiploids or significant introgression.

A very successful weed in Australia (which is also spreading widely within the Americas) is the shrubby *Nicotiana glauca* Grah. from Argentina. Even in the desert regions of Australia it is to be found in washes and occasionally along roadsides. It has met and formed sterile hybrids with at least one native member of the genus. Thus, in Victoria, natural hybrids with *N. suaveolens* Lehm. have been recorded (Nicholls, 1936). This case is of particular interest because it involves two separate sub-genera in the classification by Goodspeed (1954), with quite different chromosome numbers ($n = 12$ in *N. glauca*; $n = 16$ in *N. suaveolens*) and different pollinators (largely hummingbirds for *N. glauca* in its native South America, while *N. suaveolens* is pollinated by moths). Should amphidiploids be formed, however, a new weed might be born and, in any case, the example sounds a warning that possibilities of hybridization cannot be ignored even when distantly related species are brought together.

Classic cases of hybridization between pairs of introduced weeds, producing new amphidiploid weeds are provided by Ownbey (1950, etc.) in his studies of *Tragopogon* in Idaho and Washington states. These have been reviewed recently by Stebbins (1971, pp. 157–9). Another presumptive amphiploid weed has begun to take over my garden in Berkeley. This is a vigorous, ground-covering white-flowered violet with forty chromosomes in its nuclei. It produces chasmogamous flowers from January through March and then follows with abundant cleistogamous flowers as the days get longer and warmer. It is enormously prolific in vegetative reproduction and its seeds germinate massively every February. It can be keyed out to *Viola alba* Besser ssp. *scotophylla* (Jord.) Nym. but, as that taxon is reported to have $2n = 20$ (Schmidt, 1961; Valentine *et al.*, 1968), the tendency of the weedy white violet towards *V. odorata* L. in some characters suggests that hybridization with that species (also with $2n = 20$) was followed by doubling of the chromosome number to produce our weedy white violet. The hybrid between *V. odorata* and *V. alba* is known from nature and has been called *V. multicaulis* Jord. *V. alba* is not mentioned as being in cultivation by Valentine *et al.* and Schmidt (1968) but is recommended as a ground-cover by Thomas (1970).

The weedy white violet is becoming very common in and around gardens in the San Francisco Bay region, but, of particular interest, is some further evolution which has gone on. In my garden a ''sport'' has appeared in which the violet flower colour has been restored and a number of other morphological changes are apparent. These changes have resulted from the loss of six chromosomes, so that the violet-flowered plant has $2n = 34$. Meiosis is upset in this ''mutant'', resulting in seed-sterility but, vegetatively, it appears to be even more vigorous than its white-flowered parent and it is spreading by stolons through the cover already established by the parent.

Sometimes the hybridization is between a cultivated species and an existing weed. Thus, *Raphanus sativus*, the cultivated radish, has been helped in becoming a weed in California by introgressive hybridization with *R. raphanistrum* L. which is already a weed (Panetsos and Baker, 1968). Without the introgression, but also by hybridization, pernicious weeds in another family, the Gramineae, have been introduced directly into the largely weed-free agricultural fields of California. These cases concern the genus *Sorghum*.

"Hybrid" sorghum (*Sorghum bicolor* (L.) Moench) with high yield and great uniformity of dwarf habit (very suitable for machine harvesting) is produced in the U.S.A. by the "male-sterility" method in which two inbred lines, one of which produces no pollen, are interplanted and allowed to cross-pollinate. Seed is then harvested from the male-sterile plants and is sent to the farmer to sow in his field to produce the annual "hybrid" crop. Unfortunately, in adjacent fields or along the roadsides and irrigation ditches around the seed-producing fields, other taxa of *Sorghum* may occur including the annual Sudan grass, *S. sudanense* (Piper) Stapf and the vigorous weedy perennials *S. almum* and *S. halepense* (L.) Pers. (Johnson grass). Sometimes it is the pollen of these non-grain sorghums which blows onto the stigmas of the male-sterile *S. bicolor* plants, producing hybrid seed of different sorts than those desired (Quimby *et al.*, 1958, pp. 32–3). All forms of *S. bicolor* and *S. sudanense* are diploid; *S. almum* and *S. halepense* are tetraploid. The interspecific hybrids are diploid, triploid or tetraploid (some of the triploids and all the tetraploids resulting from the involvement of unreduced female gametes, which male sterile *S. bicolor* produces rather frequently). Unknowingly the farmer sows the seed of these "off-types" in his field along with the desirable seed. The hybrids with Sudan grass are tall, with multiple culms and are seed-fertile. Being missed by the combine harvester, which is set to handle short-stemmed plants, they provide weed seed for future years. The hybrids with *S. halepense* (and the rarer ones with *S. almum*) inherit a perennial habit from the pollen parent, have considerable vegetative vigour and are more or less sterile, consequently wasting little energy on seed-production. Weeds such as these cannot be easy to eradicate by spraying because they and the crop plants amongst which they grow belong to the same genus, and they persist, after the crop has been harvested, by seed-reproduction or perennation.

RECENT AND CONTEMPORARY SPREADING OF WEEDS

Obviously, human influences have been crucial in bringing the taxa together in the previously described cases of new weed evolution. Even indirectly, however, the spread of weeds is being assisted by man, despite the existence of deterrents such as pure seed laws.

Ornamental plants are still being distributed, and some of the weeds of the world have resulted from the escape of this kind of plant from cultivation. A very striking example in coastal Western Australia is provided by the Cape Daisy, *Cryptostemma calendulaceum* R. Br. (=*Arctotheca calendula* (L.) Levyns). This immigrant from South Africa fills vast areas of pasture and cultivated land with its lemon-yellow flower-heads. But ornamentals, particularly perennials, are not frequently potential weeds, at least partly because it is considered undesirable for many kinds of ornamental plant to set seed. Seed pods may detract from the appearance of the plant, the formation of the pods may reduce further flowering, or the projection of seeds in an uncontrolled manner around a garden may be deprecated by its owner. As a consequence of this (and, sometimes, because to a gardener's taste, "double" flowers are preferable to normal flowers), many ornamental plants have been deliberately deprived of their ability to reproduce by seed. Clearly, seed-sterility reduces (but, as we have seen in the case of *Oxalis pes-caprae*, does not necessarily eliminate) their chances of naturalization as weeds.

Where annual ornamentals are involved, however, seed-reproduction is essential, even in cultivation, and these species retain a greater potentiality for successful escape. Thus, in Queensland, for example, the little *Ageratum* used as an edging plant in herbaceous borders (*A. houstonianum* Mill.) has managed to get away from captivity and become a weed. It has also established itself elsewhere in the tropics and, on the Cameroons Mountain in West Africa, it has met and hybridized with the pan-tropical weed *Ageratum conyzoides* (J. K. Morton, personal communication). Of all the seed-reproducing economic plants, those used as forage plants (which must often re-seed themselves naturally) are some of the most likely weeds and the frequent establishment of several species of *Trifolium*, *Medicago* and *Melilotus* testifies to this.

In addition, existing weeds are being spread as new agricultural lands are opened up to attempt the feeding of our burgeoning human population. Contributing very largely to their stocking with weeds is the construction of roads into previously virgin areas. The function of road-building in spreading weeds through California has been reviewed very thoroughly recently by Frenkel (1970). Lousley (1970) has also treated the subject for the British Isles. One western North American example may be added, however; that of the Turkey Mullein, *Eremocarpus setigerus* (Hook.) Benth., the only species in its genus and an inhabitant of bare ground in inland areas from California to Washington. Apparently intolerant of grass-competition, this species has taken advantage of the bare soil of the soft shoulders of roads built through the area to increase its numbers enormously. It is unusual among California annuals in growing and flowering during the summer months rather than during the fall, winter and spring. Because of this, it misses being sprayed with herbicides by road-crews which spray during the winter months. Also, by reason of its densely hairy covering, it is resistant to the effects of sprays that may hit it. The sprays, in fact, remove most of the potential competitors which *Eremocarpus* finds intolerable. Consequently, *Eremocarpus setigerus* has benefitted markedly from the activities of roadbuilder man.

Railroad-building must have had a comparable influence in the past, and one clear case of assistance in the spread of a weed may be quoted. This involves the building of the Costa Rica Northern Railway during the years 1880 to 1890. The railway was built largely by Jamaican labourers, brought from their own island to the Caribbean coast of Costa Rica. With them they brought *Ageratum houstonianum*, a common weed in Jamaica. It spread with the railway through the lowland tropical coastal area on the Caribbean side of the country and then, when the railway was completed up to the central plateau, *A. houstonianum* was put in contact with *A. microcarpum* which is indigenous on that plateau. Continuity between the populations of the two species has produced a cline of hybridization along the railway line and has presumably increased the weediness of *A. microcarpum* (Baker, unpub.).

In olden times in the United States, cattle drovers' trails through the Great Plains were the avenues along which weeds were dispersed. Mesquite (*Prosopis glandulosa* Torr.) was spread from Texas to Colorado and Kansas (and beyond) by the cattle which fed on its nutritious pods, carried the seeds for miles, and then dropped them in a fertilized seed-bed, with the potentially competitive grass grazed down for good measure (see discussion in Harris, 1966). Valuable grazing land has been lost to the mesquite jungles that have developed. In the tropics and sub-tropics, other woody leguminous weeds, such as *Acacia farnesiana* (L.) Willd. and *Lucaena glauca* (L.) Benth. owe some of their success to similar processes.

In California, the relatively rapid recent spread of *Carduus pycnocephalus* L. and *C. tenuiflorus* Curt. along tracks and roadsides and to form pure stands under the crowns of oak trees in the savanna areas of the Inner

Coast Ranges, Central Valley and Sierra Nevada foothills may be attributed to the superb growth of the thistles in conditions of improved nitrification, brought about by cattle passing along the tracks and congregating under the trees.

Release of land from agricultural use for incorporation into parks is resulting in the initiation of secondary successions and thereby changing the weed floras in the areas concerned, for some weedy species that are disfavoured in agricultural systems are pioneers in secondary successions. In the San Francisco Bay region, Poison Oak, *Rhus diversiloba* T. & G. (Anacardiaceae), an allergenic shrub or vine with excellent vegetative reproduction as well as capability for establishment from bird-distributed seed, has become much more common than historical records would indicate to have been the case a century ago (A. Clarke, pers. comm.).

Of course, there can be complications in the weed stories. Deductions from historical records are not always to be trusted. A common belief among Californians is that their hill slopes and valleys were covered with sheets of the state flower, the California Poppy, *Eschscholzia californica* Cham., from centuries past until recently. But the California poppy is a weedy plant, and has shown itself in that light when introduced into other countries, like Chile, that have an appropriate Mediterranean type climate. Its natural habitats include eroding cliffs, the rocky beds of rivers and creeks that have irregular flow, as well as stabilized sand-dunes and grasslands in which the cover is not complete. Human disturbance has increased its opportunities for growth.

It is probable that the enormous "sheets" of California poppies in the state at the turn of the century owed much to sequences of overgrazing and then undergrazing by cattle and sheep brought to California by Europeans.

And the complications continue to be created. Now that *Hypericum perforatum* is biologically controlled by chrysomelid beetles in northern California, it has not become extinct. It persists as scattered plants under the shade of trees (where the beetles are less effective) rather than in the open grasslands where it was such a pest (Huffaker, 1962) and it could well be that selection for greater shade-tolerance of a genetically controlled sort will now take place in *Hypericum perforatum* in California.

It has also been suggested that the Italian Thistle, *Carduus pycnocephalus* and the Medusa-Head Grass, *Taeniatherum asperum* (Simonkai) Nevski (*Elymus caput-medusae* L.) may be moving into the "vacuum" left in northern California grasslands by the biological control of *Hypericum perforatum*. If this be the case, it will be evidence that the removal of a particular weed is not enough; it must be followed by re-seeding with desirable species or else the niche will be filled by other weeds which are ready and waiting.

I hope it is clear that, despite some complications, weeds are some of the best material available for the study of the ecological features of plant migration and the strategies involved in adaptation to the environmental variations to be encountered in such migrations. Both the minor weeds, with their subjection to environmental restrictions, and the major weeds with their apparent insouciance are useful for these purposes. Despite the application of herbicidal sprays and other sophisticated means of weed-control, they are likely to be available for study for many years to come, and I recommend them to you for greater ecological attention.

ACKNOWLEDGEMENTS

Original work reported in this paper has been greatly aided by research grants nos. G-21821 and GB-8593 from the National Science Foundation, Washington, D.C.

REFERENCES

ANONYMOUS (1971). "Biochemical Interactions among Plants". Natn. Acad. Sci., Washington, D.C., U.S.A.

BAKER, H. G. (1955a). *Evolution* **9**, 347–348.

BAKER, H. G. (1955b). *Watsonia* **3**, 160–167.

BAKER, H. G. (1962). *J. Calif. hort. Soc.* **23**, 97–104.

BAKER, H. G. (1965). *In* "The Genetics of Colonizing Species" (H. G. Baker and G. L. Stebbins, eds), pp. 147–168. Academic Press, New York and London.

BAKER, H. G. (1966). *Madroño* **18**, 207–210.

BAKER, H. G. (1967). *Taxon* **16**, 293–300.

BAKER, H. G. (1973). *Ecology* (In press).

BLACK, J. M. (1948). "Flora of South Australia" (Second Edition), Part II. Government Printer, Adelaide, South Australia.

BOCK, JANE H. (1969). *Ecology* **50**, 460–464.

BRUMHARD, P. (1905). "Monographische Übersicht der Gattung Erodium". Inaug.-Dissert. Botan. Gart. Univ. Breslau. 59 pp. Univ. Breslau, Breslau.

BRUMHARD, P. (1906). *Repert. Spec. Nov. Regni. Veg.* **2**, 116–119.

CAROLIN, R. C. (1967). *Contr. N.S.W. natn. Herb.* **102**, 1–23.

CLAPHAM, A. R., TUTIN, T. G. and WARBURG, E. F. (1952). "Flora of the British Isles". Cambridge University Press, Cambridge.

CLAPHAM, A. R., TUTIN, T. G. and WARBURG, E. F. (1962). "Flora of the British Isles". (Second Edition). Cambridge Univ. Press, Cambridge.

DADAY, H. (1954a). *Heredity, Lond.* **8**, 61–78.

Daday, H. (1954b). *Heredity, Lond.* **8**, 377–384.

Del Moral, R. and Muller, C. H. (1969). *Bull. Torrey bot. Club* **96**, 467–475.

Dony, J. G. (1953). *In* "The Changing Flora of Britain" (J. E. Lousley, ed.), pp. 160–163. Botanical Society of the British Isles, Oxford.

Frenkel, R. E. (1970). "Ruderal Vegetation along Some California Roadsides". Univ. of California Publs. in Geography, Vol. 20. 163 pp. Univ. of California Press, Berkeley, California, U.S.A.

Gilmour, J. and Walters, M. (1959). "Wild Flowers—Botanizing in Britain". Collins, London.

Goodspeed, T. H. (1954). "The Genus Nicotiana". Chronica Botanica Co., Waltham, Mass.

Harris, D. R. (1966). *Ann. Ass. Am. Geog.* **56**, 408–422.

Hegi, G. (1929). "Illustrierte Flora von Mittel-Europa", Vol. VI, Part 2. J. F. Lehmanns Verlag, München.

Howell, J. T. (1947). *Leafl. west. Bot.* **5**, 67–68.

Howell, J. T. (1949). "Marin Flora". Univ. of California Press, Berkeley.

Huffaker, C. B. (1962). *Can. Ent.* **94**, 507–514.

Huffaker, C. B. and Kennett, C. E. (1959). *J. Range Mgmt.* **12**, 69–82.

Jepson, W. L. (1925). "A Manual of the Flowering Plants of California". Assoc. Students, Univ. of California, Berkeley.

Johnson, M. F. (1971). *Ann. Mo. bot. Gdn.* **56**, 6–88.

Laing, R. M. and Blackwell, E. W. (1952). "Plants of New Zealand" (Fifth Edition). Whitcombe and Tombs, Christchurch, New Zealand.

Linhart, Y. B. (1971). "Differentiation within Natural Populations of California Annual Plants". Ph.D. thesis (Genetics). Univ. of California, Berkeley, California, U.S.A.

Lousley, J. E. (1970). *In* "The Flora of a Changing Britain" (F. Perring, ed.), pp. 73–83. Botanical Society of the British Isles and E. W. Classey, Hampton, Mdx.

Munz, P. A. (1959). "A California Flora". Univ. of California Press, Berkeley, California, U.S.A.

Nelson, A. P. (1964). *Evolution* **18**, 487–499.

Nelson, A. P. (1965). *Brittonia* **17**, 160–174.

Nicholls, W. H. (1936). *Victorian Nat.* **53**, 64–65.

Ownbey, M. (1950). *Am. J. Bot.* **37**, 487–499.

Panetsos, C. and Baker, H. G. (1968). *Genetica* **38**, 243–274.

Parsons, J. (1970). *Tübinger. geogr. Stud.* **34**, 141–153.

Peck, M. E. (1941). "A Manual of the Higher Plants of Oregon". Binfords and Mort, Portland, Oregon.

Powell, A. M. (1965). *Brittonia* **17**, 47–96.

Pritchard, T. (1960). *In* "The Biology of Weeds" (J. L. Harper, ed.), pp. 61–66. Blackwell, Oxford.

PURSH, F. T. (1814). "Flora Americae Septentrionalis". White, Cochrane and Co., London.
QUIMBY, J. R., KRAMER, N. W., STEPHENS, J. C., LAHR, K. A. and KARPER, R. E. (1958). *Bull. Tex. agric. Exp. Stn.* no. 912, 1–35.
SALISBURY, E. J. (1932). "The East Anglian Flora". Norfolk and Norwich Nats. Soc., Norwich.
SALISBURY, E. J. (1953). *In* "The Changing Flora of Britain" (J. E. Lousley, ed.), pp. 130–137. Botanical Society of the British Isles, Oxford.
SALISBURY, E. J. (1961). "Weeds and Aliens". Collins, London.
SCHMIDT, A. (1961). *Öst. bot. Z.* **108**, 20–28.
SOLBRIG, O. T. (1971). *Am. Scient.* **59**, 686–694.
STEBBINS, G. L. (1971). "Chromosomal Evolution in Higher Plants". E. Arnold, London.
THOMAS, G. S. (1970). "Plants for Ground Cover". J. M. Dent, London.
VALENTINE, D. H., MERXMÜLLER, H. and SCHMIDT, A. (1968). *In* "Flora Europaea" (T. G. Tutin and others, eds), Vol. 2. Cambridge Univ. Press, Cambridge.
WAGNON, K. A. and BISWELL, H. H. (1943). *Madroño* **7**, 118–125.
WHITTAKER, R. H. and FEENY, P. P. (1971). *Science, N.Y.* **171**, 757–770.
WHITTET, J. N. (1958). "Weeds". The Farmer's Handbook Series, N.S.W. Dept. Agric., Sydney.

22 | Comments on the History and Ecology of Continental European Plants

ECKEHART J. JÄGER

Sektion Biowissenschafton, Martin-Luther-Universität, Halle-Wittenberg, DDR

Compared with oceanic distribution patterns, continental plant ranges have not often been discussed; the last comprehensive treatment was by Eilart (1963). Hitherto the investigators of continental plants have been concerned mainly with the classification of distribution types and with the history of postglacial distribution. This paper will comment on the origin of continental European species and on the explanation of their range limits.

What are continental plants? There has been a long and useless discussion about this question. The distribution of a species of broadleaved forests such as *Hepatica nobilis* Mill. (Fig. 1.) is called continental by

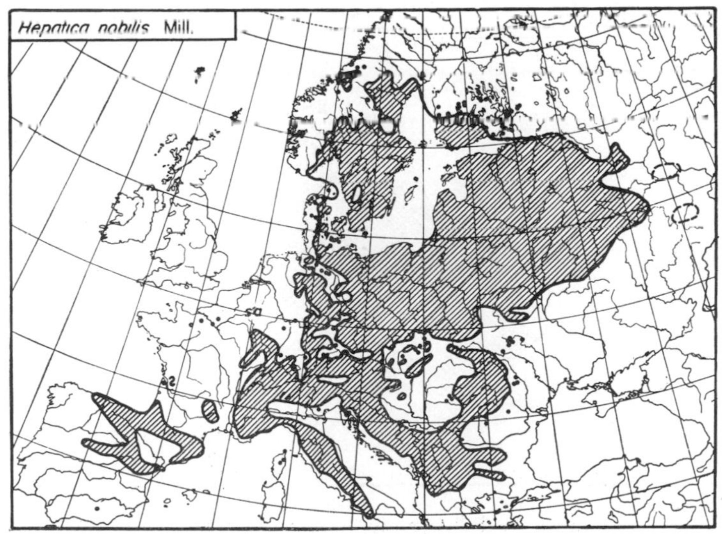

FIG. 1. Map showing the general distribution of *Hepatica nobilis* Mill.

Swedish, oceanic by Russian and intermediate by German plant geographers. *Arnoseris minima* (L.) Schw. & Kœrte (Fig. 2) is a continental plant according to Matthews (1937), but to a Polish botanist like Czeczott (1927), it is clearly oceanic. What about *Filago minima* (Sm.) Pers. (Fig. 3), which is widely distributed in England, but also in Poland; is this species more oceanic or more continental than *Arnoseris*? The reader will certainly agree that either is true. In the "Comparative Chorology of the Central European Flora" (Meusel *et al.*, 1965), which gives distribution maps of about 5000 plant species, the research team has tried to characterize distribution patterns from the point of view of a survey of the whole Eurasian continent. We therefore thought it best to give a range description covering the whole breadth (amplitude) of the oceanicity of the range. For such range descriptions in the "Comparative Chorology" a map of phytogeographical oceanicity was drawn (Fig. 4), compare Jäger, 1968. The two species shown in Figs 1 and 2 extend over the oceanicity-degrees (1)–2 and (1)–2(3) respectively in this map.

The map of the phytogeographical oceanicity was prepared by a comparison of many circumpolar plant ranges having a different degree of oceanicity. For example, the distribution of the *Carex remota* L. group (Meusel *et al.*, 1965, map 69a) extends over the whole Holarctic kingdom, but only to a certain distance from the sea. All the four parts of this

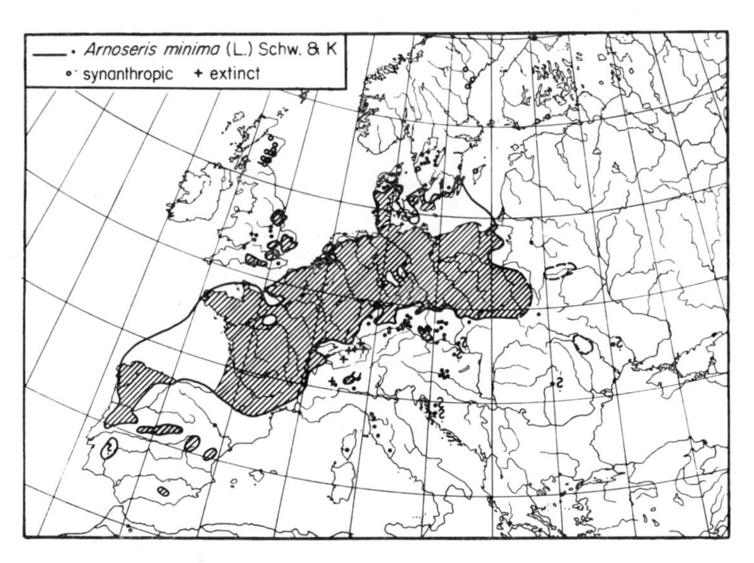

FIG. 2. Map showing the general distribution of *Arnoseris minima* (L.) Schw. et K. (after Meusel *et al.*, Vol. II in preparation, prepared by E. Jäger 1971).

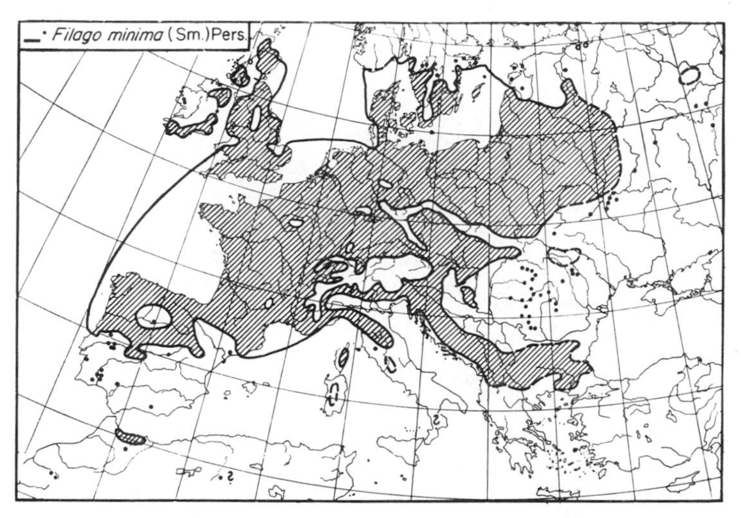

FIG. 3. Map showing the general distribution of *Filago minima* (Sm.) Pers. (after Meusel *et al.*, Vol. II in preparation, prepared by E. Jäger 1971).

range were considered to have a similar oceanicity-breadth. The same was done for continental distribution patterns like that of *Carex obtusata* Liljebl. (Meusel *et al.*, 1965, map 64d). In principle it would be possible to describe the amplitude of oceanicity according to this one scale, without any distinction between oceanic and continental ranges; but in the "Comparative Chorology" we hold tight to this distinction. Thus ranges, the main parts of which lie in degree three or four (C_2 or C_1) in the map (Fig. 4), are regarded as subcontinental or continental, compared with the oceanic and suboceanic ranges, the centres of which are in the first or second degree of oceanicity (OC_1 or OC_2). According to this treatment, the range of *Hepatica nobilis* (Fig. 1) is classified as a suboceanic distribution (*Asarum*-type; Jäger, 1970).

This oceanicity scale, which is one of the three dimensions of range description by the method of Meusel, 1943 (compare Meusel *et al.*, 1965), does not provide an ecological explanation of plant distribution. For this purpose we have to compare the distribution limits with certain climatic lines and geological limits. De Candolle used this method in 1855, and it still seems to be the only way to elucidate the factors which control actual plant ranges.

In this connection, it is important to distinguish clearly between the actual (synecological) range, which has developed under natural conditions, and the potential (autecological) range, which would be

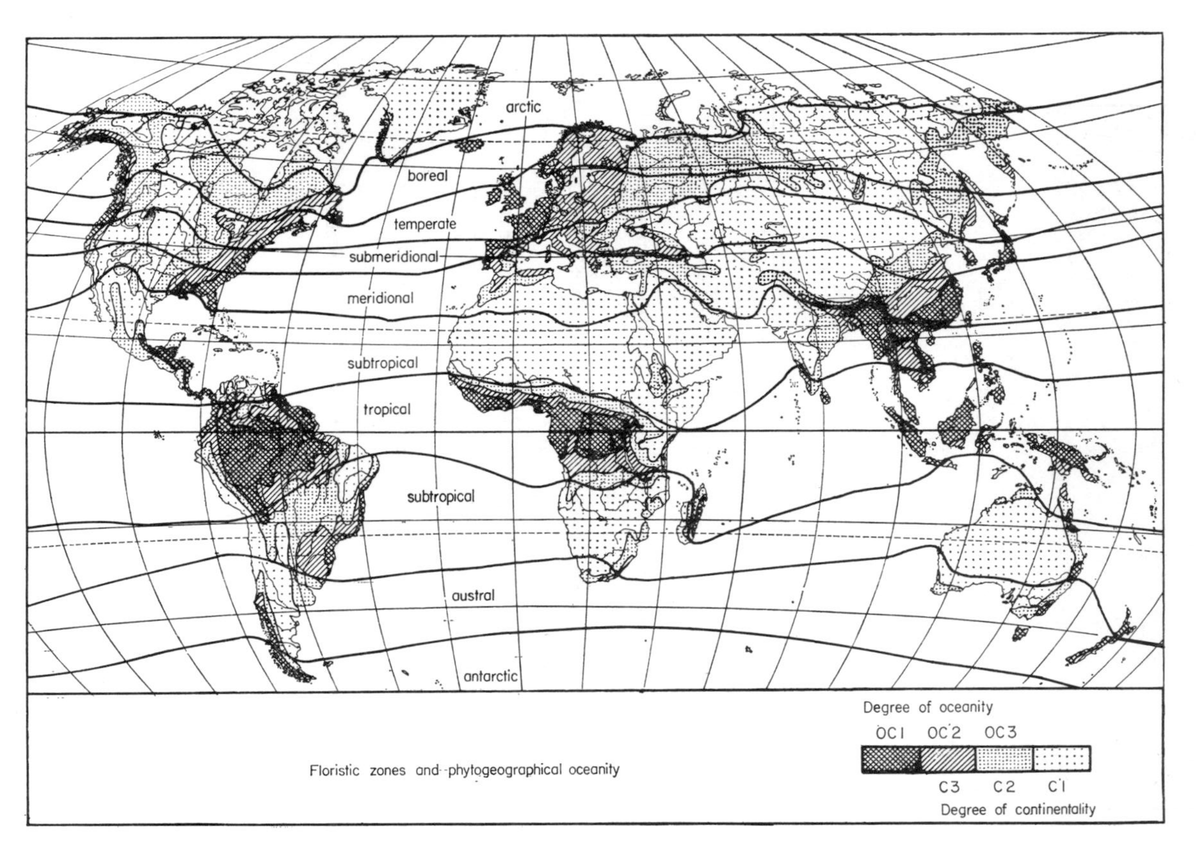

FIG. 4. Map of the world showing the degrees of phytogeographical oceanity (resp. continentality). After Jäger (1968).

developed if there were no competition from other plants. An essential difference between the actual and the potential distributions is often found, specially in continental plants. As a rule, in oceanic plants the region of best growth falls within the limits of the actual range. Judging from attempts, successful and unsuccessful, to cultivate them under open air conditions, the potential range is usually wider, but it is controlled by factors similar to those which control the actual distribution. On the other hand, in continental species the area of best growth may lie outside the actual range, and may be occupied by more vigorous oceanic plants; thus, in continental plants, the limits of the potential and the actual ranges are often controlled by very different factors. Hence, conclusions about controlling climatic factors which are made on the basis of a comparison of range limits with climatic boundaries, are a function of the synecological balance of competition rather than the autecology of the species concerned.

The most important continental European range patterns (in order from north to south) are the following.

(1) *Trifolium spadiceum-type:* "submed/mo-me/mo+sarm-scand-boreoross".* There are only a few European continental species in the boreal-zone, because most of the species of the boreal European flora have an Eurasian or circumpolar distribution. The ranges of the *Trifolium spadiceum*-type are not continental in the strict sense. To this type belong:

Picea abies (L.) Karst. (Meusel *et al.*, 1965, map 20d); *Alnus incana* (L.) Moench (Meusel *et al.*, 1965, map 120a); *Trifolium spadiceum* L. (Meusel *et al.*, 1965, map 236a); *Melampyrum sylvaticum* L.

The following have a similar distribution:

Trollius europaeus L. (Meusel *et al.*, 1965, map 156a); *Geranium bohemicum* L.; *Epilobium collinum* C.C.Gmelin; *Daphne mezereum* L.; *Campanula latifolia* L.

Characteristic features of this distribution pattern are the gap in the oceanic western part of Europe, and the disjunction between the north-eastern lowland area and the range in the mountains of south and middle Europe.

On the whole this pattern coincides very well with the European part of the boreal coniferous forest region. These species occur in regions with

* Abbreviations as in Meusel *et al.* (1965, map 258); the terms used and their abbreviations are: Submediterranean (submed), mountainous (mo), middle-European (me), Sarmatian (sarm), central-European (centraleur), Scandinavian (scand), Boreorossic, north Russian (boreoross), Pontic (pont), Pannonian (pann), Iberian (iber), Algerian (alger).

a short growing season and a high humidity during the growing season, and this is in agreement with their occurrence in the Caucasus, in eastern Turkey, and on the mountains of the Balkan peninsula, and with the gap in western Europe. The 12°C July isotherm coincides well with the northern limit of distribution, so this limit may be controlled by summer temperature. (The climatic conditions of this distribution pattern have been treated in detail by Jäger, 1969).

Most of the species included in this type are circumpolar or Eurasian in affinity; only a few, such as *Trifolium spadiceum*, are examples of the penetration of European taxa into the northern coniferous forest flora, which today seems to be mostly Pacific in origin and to have the character of a foreign invader in Europe.

(2) *Sempervivum soboliferum-type*: "pann-sarm-balt-eastscand". A second range-pattern comprises some psammophilous plants which extend from the temperate to the boreal zone:

Festuca polesica Zapal. (Meusel *et al.*, 1965, map 37b); *Koeleria grandis* Bess.; *Silene tatarica* (L.)Pers. (map Miniaev, 1965); *Dianthus fischeri* Sprengel (map Miniaev, 1965). *D. arenarius* L. (Fig. 5a). *Gypsophila fastigiata* L. (Fig. 5d). *Pulsatilla patens* (L.)Mill. (Meusel *et al.*, 1965, map 161b). *P. pratensis* (L.) Mill. (Meusel *et al.*, 1965, map 162b). *Cardaminopsis arenosa* (L.)Hay. (Fig. 5c). *Sempervivum soboliferum* Sims. (Fig. 5e). *Thymus serpyllum* L. (Fig. 5b).

These species are mostly related to Eurasian continental groups. However, some have a European relationship (*Sempervivum*, *Festuca cinerea* group). The connection with America is slight in this more southern type.

The characteristic shape of the distribution pattern, with its wide extension from south to north (often connected with a slight taxonomic differentiation in this direction, compare Fig. 5a, b, d, e) has up to now been explained mainly in terms of postglacial migration. But the recent range has been developed as a result not only of extension, but also of repeated restriction. During the late-glacial and preboreal periods of post-glacial times, these light-loving plants were able to spread rapidly, because the climate on the eastern side of the inland-ice anticyclone was continental, and closed forest communities were lacking. Afterwards, in the boreal period, their spread was stopped by the competition of the forests, their occurrence has been restricted and reduced, and they have become locally extinct.

We can see some fossil finds marked on the map (Fig. 5d), but there are probably many more. So we must not regard recent localities as indicators

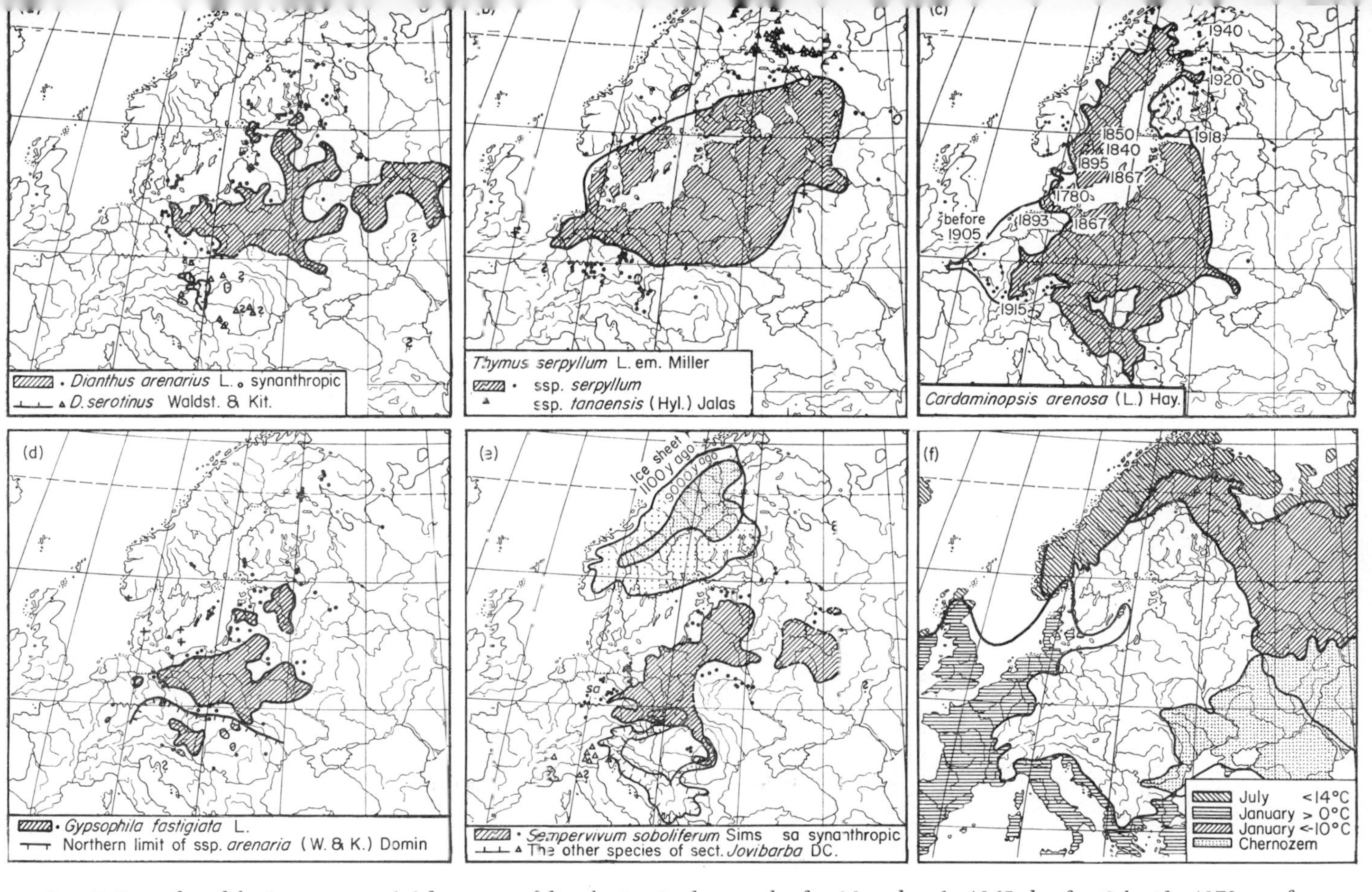

FIG. 5. Examples of the *Sempervivum soboliferum*-type of distribution (a, d, e, partly after Meusel *et al.*, 1965; b, after Schmidt, 1970; c, after Meusel *et al.*, 1965, changed) and climatic and edaphic lines for the interpretation of this type of distribution (f, after Atlas mira, 1964).

of migration routes, but rather as sites that were suitable for preserving these plants during the last few thousand years.

In regressive plant ranges like these it is sometimes difficult to find a reasonable explanation in terms of climatic factors. The reason is that plants with regressive ranges are able to maintain old colonies for a long time under recent climatic conditions, though these conditions are not suitable for the establishment of new colonies. Hence, the boundaries are fixed, and indicate the gaps in the previous range, in our case the gap caused by the ice sheet of the late quaternary glaciation, which is drawn on the map of *Sempervivum soboliferum* (Fig. 5e). Greuter (1970) has pointed out the distinction between the two main range groups, one being well correlated with climatic factors, the other to be explained in historical terms. This is the difference between progressive and regressive plant ranges.

The *Sempervivum soboliferum* range-pattern can be explained (except the gap in Sweden and some northern outposts) by the winter temperature or the length of the growing season in the northeast; by drought and the soil in the southeast (the limit of the black soil, Fig. 5f, follows the lines of annual precipitation, Fig. 6d); by the winter temperature, which restricts the competing oceanic psammophilous plants such as *Corynephorus canescens* (L.) Beauvin in the west; and by summer temperature in the northwest (Fig. 5f). It is characteristic of a progressive range such as that of *Arabis arenosa* (L.)Scop. (Fig. 5c), that the species occurs, as expected, in the Swedish part of the climatic range.

(3) *Galium triandrum-type*: "sarm-centraleur". Examples are:
Cerasus fruticosa (Pall.)Woron. (Meusel *et al.*, 1965, map 227a). *Cytisus ratisbonensis* Schaeff. (Meusel *et al.*, 1965, map 228c). *Galium triandrum* Hyl. (Fig. 6a). *Campanula bononiensis* L.; *Inula hirta* L. (Fig. 6c). *Aster amellus* L. (map, Jäger in Hegi 1964).

(4) *Astragalus asper-type*: "(eastsubmed)-pont-sarm-centraleur". Examples are:
Melica picta C.Koch (Meusel *et al.*, 1965, map 39d). *Iris pumila* L. (Meusel *et al.*, 1965, map 101d). *Astragalus asper* Jacq. (Meusel *et al.*, 1965, map 243d); *Linum flavum* L.; *Inula germanica* L. (Fig. 6b). *I. ensifolia* L. (map, Jäger in Hegi 1966).

In the temperate zone, but mainly in the two southern zones of the Holarctic kingdom, the submeridional and the meridional (Fig. 4), water is a far more important controlling factor than temperature. The range of the Sarmatian and the Pontic-Pannonian plants is not only in the north, but also in the west and south, very comparable with the values of annual precipitation (Fig. 6). Some parts of the southern distribution limit are apparently controlled by the seasonal distribution of rainfall, because submeridional steppe plants grow only in regions with summer rain. In the map (Fig. 6d), the 500mm and 250mm lines of average annual precipitation are shown. They run parallel to the limits of Pontic-Pannonian ranges (Fig. 6b), but in cooler central Europe, and in eastern Europe with its prevailing summer rain, the 450mm-line would be in even better agreement. There the 500mm-line rather corresponds with the northern limits of the Sarmatian-central European species (*Galium triandrum*-type, Fig. 6a, c) of the forest-steppe. Only the last-mentioned type occurs in the lower Wisla valley, whereas the Pontic-Pannonian species are lacking there. This is probably connected with some temperature factor rather than with migration routes. The eastern limit is apparently controlled by winter temperature. A great many of the European steppe and forest-steppe plants have originated in regions with mild winters, and so they are probably not adapted to withstand very hard frost.

In its affinities and origins the European steppe flora is very heterogeneous. [I think it is a little too simple to assume an ancient mediterranean origin for the whole flora as Lavrenko (1969) does.] We find old isolated taxa such as *Dictamnus*, which, judging from its range and relationship, is a pre-Miocene element of the dry open forests of Eurasia. Other steppe species have arisen from the belts of European summer-green and evergreen broad-leaved forests. Examples of this are the Pontic-Pannonian *Melica picta*, closely related to *Melica nutans* L., and *Vinca herbacea* W. & K. Other species of *Vinca* have apparently been elements of evergreen forests. Certainly, many Pontic-Pannonian (and also Pontic-south Siberian) species are derived from the steppes of the mountains of the Orient, e.g. *Linum flavum* (sect. *Syllinum*). Finally, there are European steppe plants belonging to continental Eurasian taxa, the most primitive species of which are often of middle Asian, sometimes also of east Asian origin, an example of the last being *Dracocephalum*.

The connections with north America are rather slight in the submeridional Eurasian steppe flora. Only a few taxa, the areas of which reach into the boreal zone, could have used the Bering-bridge to get into the New World (e.g. *Carex supina* Wahlb.—*C. spaniocarpa* Steud., *Astragalus danicus* Retz.*s.l.* and some others).

(5) *Nigella-type*: "med-pont-middle-eur". Secondarily the Pontic steppe flora has been enriched by an influx of many plants adapted to an even

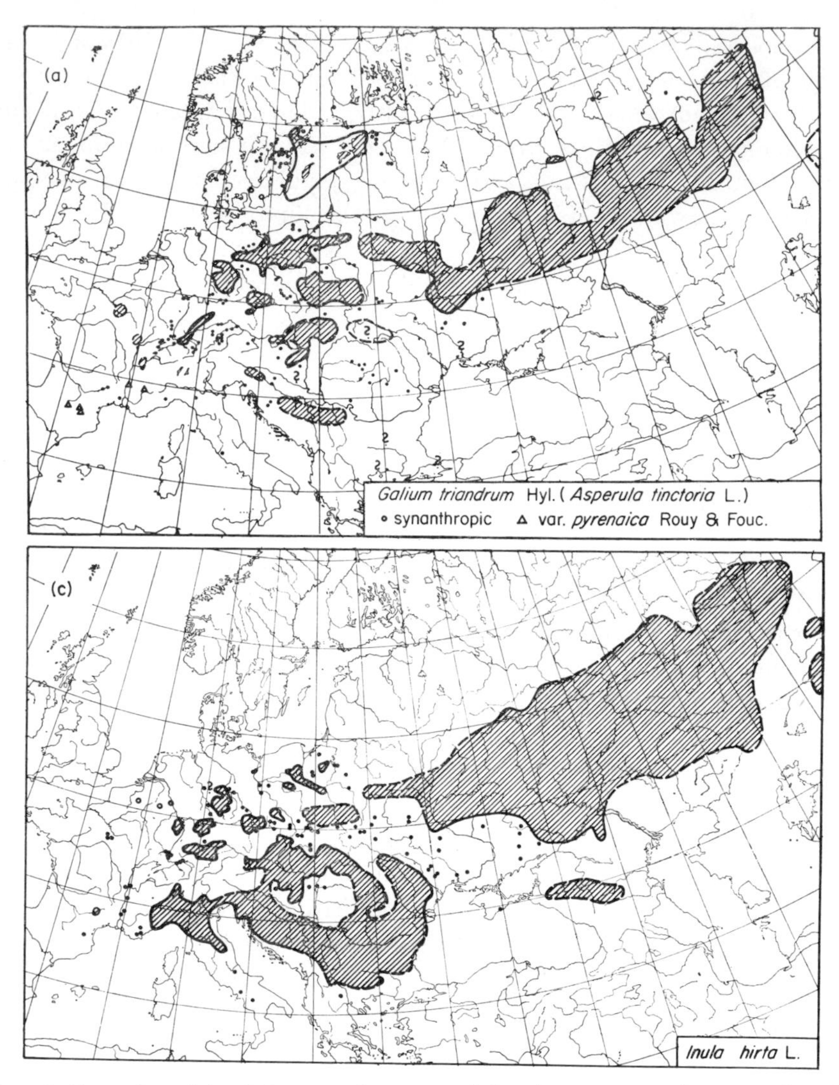

FIG. 6. Examples of the *Galium triandrum*-type of distribution (a, c, after Meusel *et al.*, Vol. II in preparation, prepared by E. Jäger 1971),

more arid climate. These annual or geophytic species, typified by *Nigella*, are derived mainly from Turkey and Iran too, only a few being of western Mediterranean origin. But their main development took place apparently later than that of the submeridional steppe plants and was connected with the drying out of the southwestern part of Eurasia during the Miocene. There are some examples of more primitive perennial species of the sub-

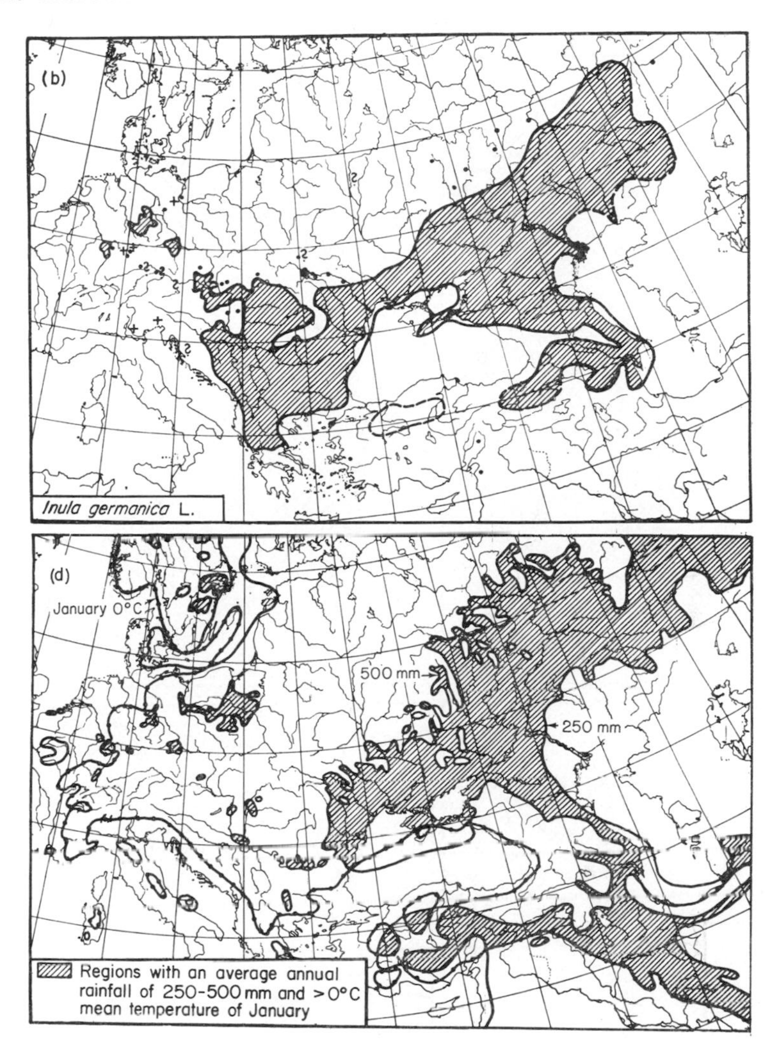

FIG. 6 (*continued*) the Astragalus asper-type (b, after Meusel *et al.*, Vol. II in preparation, prepared by E. Jäger 1971), and climatic lines for the interpretation of these types of distribution (d, after Atlas mira, 1964).

meridional steppes and more advanced annual species with a mainly meridional distribution, e.g. in *Crambe*, *Rapistrum* and *Adonis* (sect. *Adonis* and sect. *Consiligo*).

The progressive ranges of the *Nigella*-type are easily interpreted. There is a very close coincidence of annual rainfall with the southern limit, of spring temperature with the northeastern and eastern limits; and of

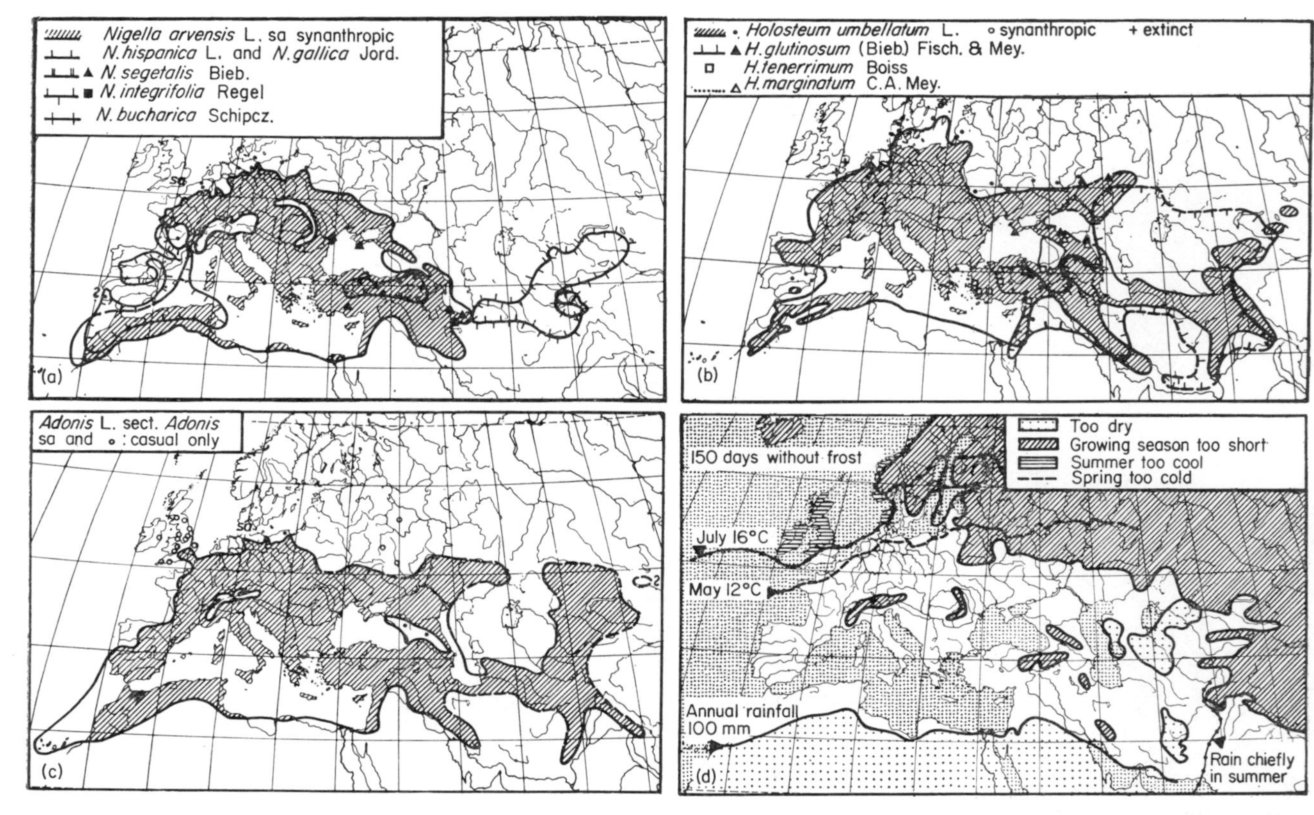

FIG. 7. Examples of the Nigella-type of distribution (a, b, c, after Meusel *et al.*, 1965, changed) and climatic lines for the interpretation of this type of distribution (d, after Atlas mira, 1964).

summer temperature with the northwestern limit of distribution (Fig. 7d). All these factors define the growing season of these mostly summer-annual species. Examples of taxa belonging to this type are:

Muscari (Meusel *et al.*, 1965, map 97c). *Colchicum* (Meusel *et al.*, 1965, map 90a). *Holosteum* (Fig. 7b). *Nigella* (Fig. 7a). *Adonis* sect. *Adonis* (Fig. 7c). *Fumaria vaillantii* Lois. (Meusel *et al.*, 1965, map 174b). *Rapistrum rugosum* (L.)All. (Meusel *et al.*, 1965, map 176d). *Conringia orientalis* (L.)Dum. (Meusel *et al.*, 1965, map 177c). *Thlaspi perfoliatum* L. (Meusel *et al.*, 1965, map 179d). *Eryngium campestre* L.; *Heliotropium europaeum* L.; *Odontites lutea* Rchb.; *Valerianella carinata* Lois.

Some of these taxa are distantly related to plants of southern Africa or of the dry Mexican-Californian region. On the other hand, we can find hardly any connection with the temperate broad-leaved forest.

(6) *Ononis tridentata-type*: "iber-(alger)". In southern Europe, the Ibero-Algerian dry region is certainly the most important centre of development of continental or xeric plants, and this is not surprising because the Iberian peninsula was the only large, semiarid landmass of Europe in early Tertiary times. At present in this region there are some very dry regions with less than 400mm annual rainfall, enclosed by high mountain chains. Here the Eurasian steppe plants are not able to grow, and they are restricted to the foothills of the north-eastern part of the peninsula.

The xerophytic taxa occurring in this Ibero-Algerian area, e.g.:

Gypsophilla struthium Loefl.	*Ononis tridentata* L.
Herniaria fruticosa L.	*Linum suffruticosum* L.
Lepidium subulatum L.	*Helianthemum squamatum* (L.) Pers.
Reseda stricta Pers.	*Atractylis humilis* L.

have been treated in detail by Jäger (1971). They are not so clearly connected with older types as the Pontic-Pannonian and Mediterranean-Pontic-middle European are. On the other hand, they themselves seem to be old and isolated and often the most primitive of a large number of Mediterranean species. Not only their high species number in the Iberian peninsula, but also the primitive character of the Ibero-Algerian species, suggest that this was an ancient centre of development for xerophytic plants such as *Ononis*, *Reseda*, *Herniaria*, *Atractylis* and *Helianthemum*. Only in more distant relationships are there interesting connections with other dry regions such as Mexico-California and southern Africa, and especially the rich Oriental floristic region (primary Irano-Turanian flora).

REFERENCES

Atlas mira, Fisiko-geografičeski. (1964). *Akad. nauk. SSSR, Moskva.*

Czeczott, H. (1927). *Bull. int. Acad. pol. Sci. Lett.* [*Cl. d.* (*math. nat.*)] *Ser. B.* (*Sci. nat.*) **1926**, 361–407.

De Candolle, A. (1855). "Géographie botanique raisonée". Paris.

Eilart, J. (1963). Pontiline ja pontosarmaatiline element flooras. *Akad. nauk. Est. SSR, Tartu.*

Greuter, W. (1970). *Repert. Spec. Nov. Regni. Veg.* **81**, 91.

Jäger, E. J. (1968). *Repert. Spec. Nov. Regni. Veg.* **79**, 157–335.

Jäger, E. J. (1969). *Ber. dt. bot. Ges.* **81**, 397–408.

Jäger, E. J. (1970). *Repert. Spec. Nov. Regni. Veg.* **81**, 67–92.

Jäger, E. J. (1971). *Flora Jena* **160**, 217–256.

Lavrenko, E. M. (1969). *Vegetatio* **10**, 11–20.

Matthews, J. R. (1937). *J. Ecol.* **25**, 1–90.

Meusel, H. (1943). "Vergleichende Arealkunde". Bornträger, Berlin.

Meusel, H., Jäger, E. and Weinert, E. (1965). "Vergleichende Chorologie der zentraleuropäischen Flora". Vol. I (Vol. II in preparation). Fischer, Jena.

Miniaev, N. A. (1965). *Vest. leningr. gos. Univ.* **21**, ser. biol., 44–56.

Schmidt, P. (1970). *Wiss. Z. Martin-Luther Univ. Halle-Wittenb. math. nat.* **18**, 810–818.

23 | Floristic Connections between Southeast England and North France

FRANCIS ROSE

King's College, London, England

INTRODUCTION

The Southeast of England and the North of France lend themselves particularly well to a phytogeographical comparison. They are in effect two regions very close together geographically, formed in geological time as originally a single region by the same processes of geomorphology, but separated in the early Quaternary by the formation of a shallow arm of the sea. Pinchemelle (1954) has drawn attention to their close geomorphological connections. During phases of the Pleistocene glaciations, the channel between them became dry, providing a land connection across the short distance between them. This land connection was not closed until quite recently, geologically speaking;—round about 5500 B.C. in the Post-glacial period (Pennington, 1969). This means that from the onset of temperate conditions at the opening of the Post-glacial, about 8300 B.C. until 5500 B.C., temperate species of plants could migrate freely across the dry channel and North Sea into Southeast England.

It has been frequently postulated that the position of the present Straits of Dover was the final point of breach of the land link between Britain and France in the Post-glacial. There is no real direct evidence of this, but it has been suggested that comparative studies of the flora on both sides of the narrow part of the Channel could provide useful information on this matter. If this were the position of the longest existing Post-glacial land connection, one might expect to find floristic evidence of this in a richer temperate flora in Kent than in other English counties.

In spite of the closeness of the two regions and their common geological and geomorphological background, there are however differences of importance in the environments in the two regions. Great Britain is an

Island, and this fact reflects itself in a greater oceanicity of climate everywhere than in any part of the European continental land mass except in the immediate coastal zones. This is reflected in the abundance of *Ulex europaeus* L., and *Primula vulgaris* Huds., which are general in lowland Britain. The relative development of different types of substrates, and hence the extent of different types of soils, differs much in the two regions; there is for example a far greater development of geological formations giving rise to sandy base-poor soils in Southeast England than in North France.

The history of land use also has been somewhat different in the two areas.

In spite of the potential interest of a comparative phytogeographical study of the two regions, remarkably little work has been attempted on the subject in the past. Good (1928) produced a short comparison of the floras of Kent and the Pas-de-Calais. Chevalier (1923), it is true, compared the flora of Normandy and Brittany with that of Britain, but this study concerned areas much further west than those we are now considering. Rose and Géhu (1960), (1964), and Rose (1965) have however attempted comparative studies of the two regions, and this present paper is an attempt to develop certain aspects of our work and to draw certain further conclusions from it.

The principal points that I wish to consider are:

(1) How closely do the vegetation and flora correspond between Southeast England and North France?
(2) What factors account for the differences that exist?
(3) To what extent is there evidence that the presence of a sea barrier since about 5500 B.C. has affected the colonization of Britain by plants from the continental land mass?
(4) Has the line of the Straits of Dover been a major point of entry for species into Britain?

GEOLOGICAL STRUCTURE OF THE TWO REGIONS AND THE ENVIRONMENTAL BACKGROUND

Geologically, Southeast England and the coastal region of the Pas-de-Calais form a unit, breached as we have seen on the latest occasion only some 7000–8000 years ago by an arm of the sea. This region is dominated by the Wealden anticlinal system, in which successively older layers of rocks have become exposed by erosion towards the centre of the anticline. The whole system is ringed by chalk with an escarpment that faces inwards around the whole system; while beneath the chalk the older rocks form a series of escarpments, more or less continuous, facing inwards. Those on the south side of the ellipse-shaped Wealden system are in general mirror-images of those to the north, until the oldest rocks form now the central ridge of the dome.

The long axis of this system runs W.N.W.–E.S.E. from near Petersfield in Hampshire to some 15 miles (24 km) east of Boulogne in the Pas-de-Calais, France. The British part of the system, ringed by its horse-shoe shaped chalk escarpment, is called the Weald; the French part, the Boulonnais.

The Boulonnais shows the same sequence of Cretaceous rocks as are represented in the English Weald, but the Lower Greensand and Wealden formations are there much thinner and cover less extensive areas; the greater part of the exposed floor of the Boulonnais consists of Jurassic strata, mostly Oxford and Kimmeridge Clays, but including also some limestones and sandstones, while in the north of the Boulonnais, Carboniferous, and even Devonian, limestones are exposed at the surface over limited areas. The Wealden and Lower Greensand times in the Boulonnais were characterized by much feebler deposition of the appropriate sedimentary rocks than in the deeper basin of the English Weald, with the result that erosion since the Oligo-Miocene uplift phase that formed the anticline has been able in the Boulonnais to expose over a wide area there the underlying synclinal system of Jurassic (and even Palaeozoic) rocks; while in the English Weald the Cretaceous rocks dominate the scene entirely at the present surface, except for some tiny, insignificant inliers of Jurassic Purbeck Beds exposed in the central Weald by faulting. These geological and geomorphological differences account for the very different extent of certain vegetational formations in the two areas of sandy acid forest and heath country on Lower Greensand and Hastings Sands.

Further afield, much of Northern France south and east of the Boulonnais is composed of a great chalk plateau, that of Picardy and Artois. This is in some ways the analogue of the Wessex chalk plateau, but unlike it is largely covered with extensive deposits of fertile loess, and hence has been far more intensively exploited and settled by man in the recent past than has the relatively thinly populated Wessex chalk plateau, though the reverse was probably true in Neolithic times. To the North of the Boulonnais lie the plains of Flanders with extensive exposures of clays and loams; one has to go much further afield to the Belgian Campine to find anything comparable ecologically with the Bagshot and other Eocene sand formations of the Thames and South Hampshire Basins.

There are profound differences in the coastline environments of the two regions. There are chalk cliffs of limited extent at Cap Blanc Nez west of Calais, (and far more extensive, to the south-west of the estuary of the Somme); but these chalk cliffs face north, providing a relatively cool environment for plants, and are subject to continual marine erosion. The chalk cliffs of Southeast England are not only much more extensive than those of France north of the Somme, but the majority of them are of southerly aspect, providing a much more sheltered and warmer microclimate, and in the case of those east of Folkestone, are protected from rapid erosion to some extent by a shingle bar developed in the lee of the Lower Greensand headland there.

Much of the low coastline of northern France, however, from the Belgian border to west of Calais and especially from south of Boulogne to the mouth of the Somme, is fringed with huge dune systems, magnificently developed both in height and in breadth, with several parallel dune ridges and extensive freshwater slacks in between them. This phenomenon is related to the fact that the coastline south of Cap Gris Nez faces down channel directly into the prevailing wind direction and the direction of greatest fetch of the waves and currents, up Channel from the Atlantic. These circumstances produce ideal conditions for large-scale dune development, similar to those on the west coast of Wales and at Braunton Burrows in Devon. Pure shingle beaches are very limited and are only developed to any extent to the south of the Somme estuary.

Along the southeast coast of Britain, however, the development of dunes is very limited; they lie on a coastline parallel to the prevailing winds and up channel currents, and so receive relatively little sand for their growth. Shingle beaches, however, are very well developed in contrast, above all at Dungeness, the largest feature of its kind in West Europe.

All these topographical and geological features have a considerable significance in relation to the development of the vegetation, and hence of the flora, in Southeast England and in North France respectively.

Many parts of the Wealden Sand, Lower Greensand, and Bagshot Sand areas of Southeast England were selected by early man for extensive clearance presumably because of their light readily cultivated soils, and later degraded into heathlands of an extent unparalleled in North France east of the Cotentin peninsula. Valley bogs also developed on these heaths, and as a result the species of acid heath and bog vegetation are far better developed than in North France, where areas of heathland are few and very limited in area owing to the lack of suitable substrates.

Many oceanic species, such as *Ulex minor* Roth., *Erica cinerea* L., *Viola lactea* Sm., and *Narthecium ossifragum* (L.)Huds. either extend much further east in Southeast England than in France, or if they occur in North France (such as *E. cinerea*) are far rarer than in Southeast England.

The sandstone area of the Central Weald with its still extensive forests on acid sandy loams, its deep ravines, and its sandrock cliffs, has no parallel in the adjacent continent nearer than the Ardennes to the east or the Cotentin to the west. Only in the small area of the Forêt de Desvres in the Boulonnais can any approach to this type of vegetation be seen, and there only to a limited extent.

Probably as a result of this, such oceanic species as *Hymenophyllum tunbridgense* (L.)Sm., *Dryopteris aemula* (Ait.) O.Kuntze (Fig. 1), *Wahlenbergia hederacea* (L.)Rchb., *Sibthorpia europaea* L., *Cicendia filiformis* (L.) Delarbre and a large number of bryophytes (Rose, 1964) occur in the Kent and Sussex Weald but are absent from North France east of the Cotentin.

The flora of the dunes of North France, on the other hand, contains some oceanic species lacking in Southeast England, such as *Viola kitaibeliana* Schult., and also other species of wider Northwest European distribution which do not appear to find suitable habitats in Southeast England, such as *Viola curtisii* E.Forst., *Centaurium littorale* (D. Turner) Gilmour, and *Corynephorus canescens* (L.)Beauv.; while the development of subclimax scrub of *Ligustrum* and *Hippophaë* reaches levels unparalleled on British coasts.

The extensive dune slacks on the North French coast support permanent fen communities and even permanent lakes, which contain rich floras including species absent now in Southeast England, such as *Teucrium scordium* L. and *Liparis loeselii* (L.)Rich. in quantity, and one species now unknown in the British Isles, if it did in fact ever occur—*Carex trinervis* Desf. On the other hand *Juncus acutus* L., a species of Atlantic-Mediterranean distribution in Europe as a whole, occurs in dune slacks at Sandwich in Kent, while it is unknown in France east of the Cotentin. The flora of the fixed dunes in Southeast England also contains oceanic elements unknown in North France such as *Trifolium glomeratum* L. and *T. suffocatum* L., and *Medicago hispida* Gaertn.

The greater development of shingle beaches in Southeast England is probably responsible for the far greater abundance there of *Glaucium flavum* Crantz and *Crambe maritima* L. than on the French coastline; while *Lathyrus japonicus* Willd. (Fig. 4), now extinct in its only French locality

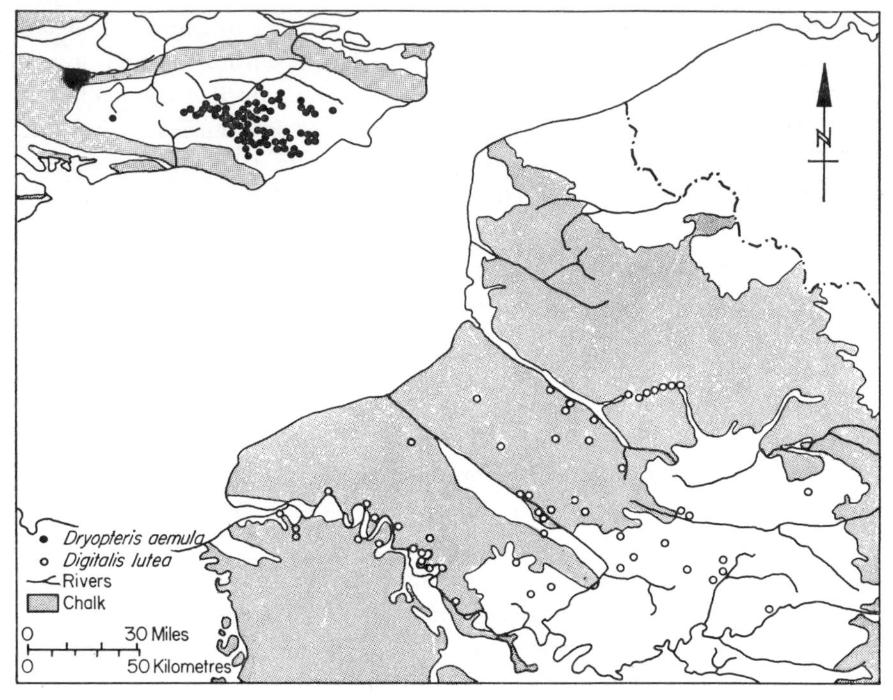

FIG. 1. (a) *Dryopteris aemula*. An "oceanic" calcifuge species that occurs in abundance as far east as the Weald on the English side of the Channel, but is absent on the French side east of the Cherbourg peninsula; this is a distribution apparently controlled by edaphic factors and microclimatic factors related to topographic relief.
(b) *Digitalis lutea*. A "continental" calcicole species that is frequent as near to Britain as the Somme Valley but does not reach the Channel coast nor does it occur in Britain.

on shingle west of the mouth of the Somme, is locally abundant and even appears to be spreading on the southeast English shingle beaches. In contrast, however, another northern-atlantic species of muddy shingle, *Halimione pedunculata* (L.)Aell., persists in several places on the estuaries of the North French coast but has become extinct in Britain.

The greater development of muddy or sandy estuaries in Southeast England accounts probably for the better development of the salt-marsh flora; *Frankenia laevis* L. and *Inula crithmoides* L. are examples of species of the drier southeast English salt marshes which are unknown in France east of the Cotentin.

The flora of the chalk cliffs in Southeast England is far richer than that of the opposite coast habitats, probably for the reasons mentioned above; well developed scrub communities occur with species of Southwest

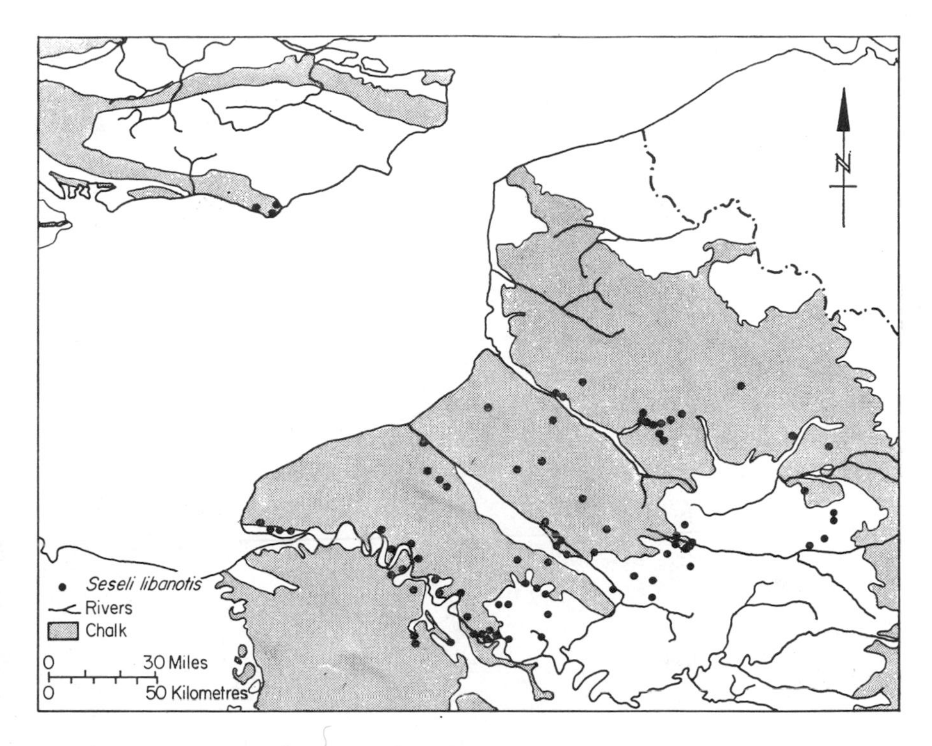

FIG. 2. *Seseli libanotis*. A "continental" calcicole species which reaches the Somme Valley in North France in local abundance but in South England is only found near the Sussex coast very locally on chalk.

European distribution such as *Rubia peregrina* L. and *Iris foetidissima* L., absent in North France east of Normandy.

The chalk grassland vegetation of the two areas has much in common in its basic components, but there are important differences in species. *Bromus erectus* Huds., which is an important dominant of chalk grasslands in Southeast England is very rare in North France, though *Brachypodium pinnatum* (L.)Beauv. is equally common on chalk within 20–30 miles of the coast in both areas. *Gentianella amarella* L., so characteristic of English chalk grasslands, is replaced entirely in its North French counterparts by *G. germanica* Willd. The chalk grasslands of East Kent and East Sussex have rich orchid floras that are largely lacking in the Pas-de-Calais; *Ophrys sphegodes* Mill., for example, so abundant locally on the English side near the coast, is only known in one site in North France north of the Somme Valley.

Other chalk grassland species which are fairly frequent or even common

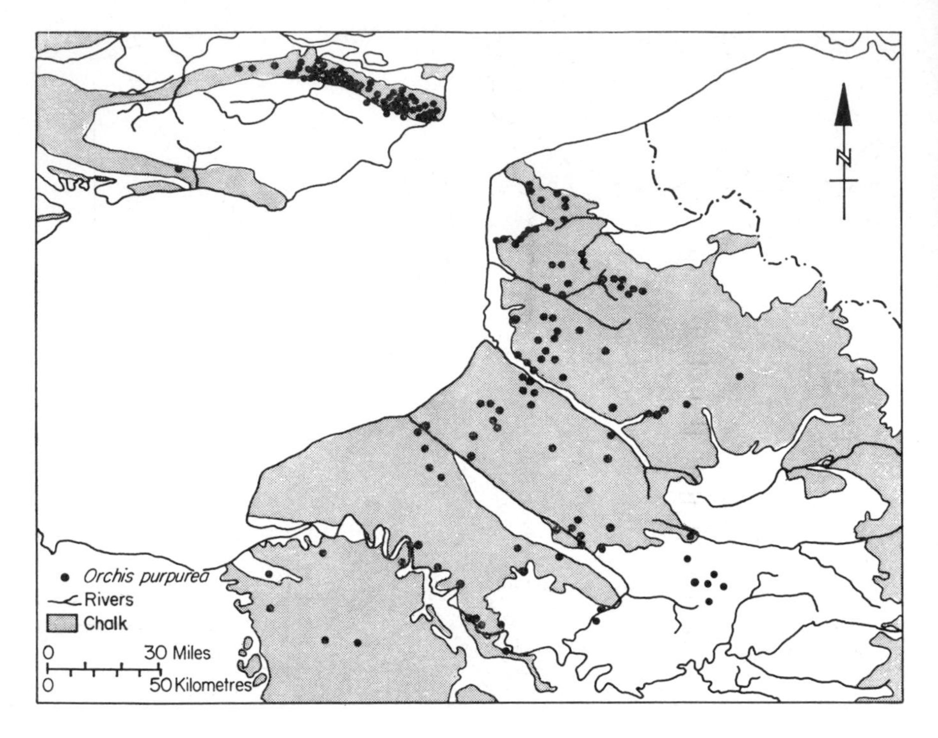

FIG. 3. *Orchis purpurea*. A "southern continental" calcicole species whose area of abundance just reaches across from North France into Kent but no further into Britain.

in parts of Southeast England, but very scarce or absent in North France, include:

Campanula glomerata L.; *Filipendula vulgaris* Moench.; *Anacamptis pyramidalis* (L.)Rich. (absent north of Somme Valley); *Phyteuma tenerum* R.Schulz (absent north of Seine Valley); *Senecio integrifolius* (L.) Clairv. (absent in all France); *Aceras anthropophorum* (L.)Ait.f.

The chalk grasslands of the Pas-de-Calais have a cooler, wetter climate [with over 40″ (100 cm) of rain per annum], than those of Southeast England, apparently because of the position of this region which in effect blocks off the end of the "funnel" of the English Channel.

In woodland and scrub on chalk (and also on sandy soils) in Southeast England *Taxus baccata* L. and *Sorbus aria* (L.) Crantz are very common; the former is absent in North France north of the Seine and west of the Meuse; the latter occurs only in a very limited area on chalk in the Boulonnais, within North France.

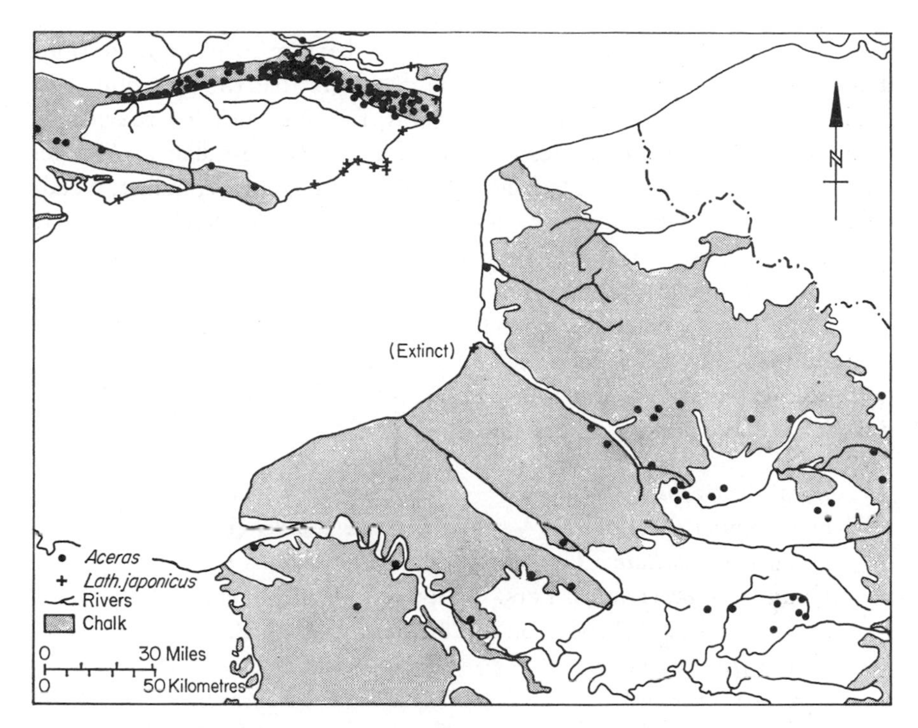

FIG. 4. (a) *Aceras anthropophorum*. A "southern continental" calcicole, abundant on the North Downs in Southeast England, but extremely rare in France, north of the Seine and Somme Valleys.
(b) *Lathyrus japonicus*. A "Northern Atlantic" species of coastal shingle, frequent as far south as the Southeast coast of England, but now extinct in its only French locality.

A NUMERICAL COMPARISON OF THE FLORAS

What information can we obtain from numerical comparisons of the floras of Southeast England and the north of France? Good (1928) attemped this for Kent and the Pas-de-Calais. Rose and Géhu (1960) attempted a comparison of the flora of Kent and Sussex with that of the Pas-de-Calais.

We found that some 114 species are recorded for Kent and Sussex that are unknown in the Pas-de-Calais, and 129 species are recorded for the Pas-de-Calais that are unknown in Kent and Sussex; 50 of the latter are not recorded as native in Britain. At first sight, this suggests a considerable difference in the flora of the two areas, though even on these figures it suggests that many species have been able to reach Southeast England other than via the Pas-de-Calais; and it suggests further that the

Channel *as a whole* has not proved a serious barrier to plant migration. However, if one analyses the data, one discovers that the situation is in reality rather different from what it seems, for the following reasons.

At least thirty of the species recorded for the Pas-de-Calais by Masclef (1886) and since, which are lacking in Southeast England are, in fact, weeds of cultivation or naturalized or even casual aliens, which have owed their presence there to human activity and to methods of farming practised in the region. Many of these are now in any case dying out or extinct, and should perhaps be excluded from the comparison, as should also some of our southeast English species of doubtful status.

In this category are the following Pas-de-Calais species recorded by Masclef (1886), and by others since.

Adonis aestivalis L.	*Adonis flammea* Jacq.
Braya supina (L.)Koch.	*Alyssum alyssoides* (L.)L.
Holosteum umbellatum L.	*Gypsophila muralis* L.
Cerastium brachypetalum Pers.	*Spergula morisonii* Boreau
Spergula segetalis (L.)G.Don fil.	*Herniaria hirsuta* L.
**Suaeda fruticosa* Forsk.	*Malva alcea* L.
Geranium phaeum L.	*Impatiens nolitangere* L.
Trifolium patens Schreb.	*Seseli montanum* L. (probably an error)
Orlaya grandiflora (L.)Hoffm.	*Polygonum bellardii* All.
Verbascum thapsiforme Schrader	*Verbascum phlomoides* (L.)Schrader
Veronica acinifolia L.	*Linaria supina* Desf.
Stachys annua L.	*Plantago indica* L.
Gagea arvensis (Pers.)Dumort.	*Muscari comosum* (L.)Miller
Briza minor L.	*Festuca heterophylla* Lam.
Bromus tectorum L.	*Bromus inermis* Leysser

A considerable number (at least 50) of the other species, in the Pas-de-Calais list are recorded only from the far interior of the department about Arras, some 50–60 miles (80–96 km) from the sea, and can hardly be regarded as species which have reached the Channel coast and been unable to cross over. Many of these, mostly more Continental species in a phytogeographical sense, occur elsewhere in Britain (e.g. *Pulsatilla vulgaris* Miller, *Veronica verna* L.) and may well be limited by the much more oceanic climate of the regions close to the Channel from occurring

* Now known to have been introduced in Pas-de-Calais.

in either the maritime part of the Pas-de-Calais or in Kent or Sussex. If they contribute to the argument at all, they merely support the hypothesis that present-day climate, and the distribution of suitable habitats, in the two regions, is more important than the effect of the barrier of the Channel in the past in controlling the present flora.

A more realistic approach is to compare the flora of the more natural habitats of Kent, Surrey and Sussex, or indeed that of the whole of lowland Britain, with that of the more maritime parts of the Pas-de-Calais (and indeed those of the adjacent Somme and Nord departments as well). Are there many species in the coastal zone of North France near the Channel coast which are, so to speak, "waiting on the brink of the sea" and unable to cross?

This approach leads to the production of a much reduced list of species.

There are only 12 species of the coastal zone of North France, up to 20 miles (32 km) inland between the mouth of the Somme and the Belgian border, which are unknown as natives in Great Britain. These are:

**Cirsium oleraceum* (L.)Scop	*Euphorbia dulcis* L.
**Veronica teucrium* L.	*Euphorbia palustris* L.
**Senecio spathulaefolius* DC.	**Myosotis arenaria* Schrad.
**Orchis palustris* Jacq.	*Epipactis muelleri* Godfrey
Cornus mas L.	**Carex trinervis* Desf.
Potentilla splendens Ram.	*Lactuca perennis* L.

Those marked with an asterisk reach the actual coastline. Of all these species, *Cirsium oleraceum* is the sole case of a species which is *common* in North France and unknown in Great Britain.

If we then list those species in the north French coastal zone which are unknown in Southeast England in the Wealden region, but occur, or have occurred, elsewhere in lowland Britain, we can add the 22 species of Table I.

There are thus only some 34 species which *might* be considered as having been unable to enter Southeast England via north France across the Straits of Dover while they were dry land.

In a total flora for Southeast England (Kent, Sussex and Surrey) of some 1090 accepted native macrospecies, this is not a large number, and one is led towards the conclusion that the Channel has not proved an important barrier to plant migration in the post-glacial period. In addition, some of the above *may* well have formerly existed in Southeast England, and have become extinct, before systematic recording began.

TABLE I.

Species	Nearest locality to coast
Thalictrum minus L. ssp. *arenarium* (Butcher) Clapham	coastal dunes
Actaea spicata L.	below Abbeville
Viola kitaibeliana Schult.	dunes
Alchemilla xanthochlora Rothm.	Boulonnais
Parnassia palustris L.	Boulonnais
Bunium bulbocastanum L.	Boulonnais
Selinum carvifolia (L.)L.	Somme Valley
Primula elatior (L.)Hill	general in Pas-de-Calais
Centaurium littorale (D. Turner) Gilmour	Ambleteuse dunes
Gentianella germanica (Willd.) E. F. Warburg	general on chalk
Melampyrum cristatum L.	near Hesdin
Utricularia intermedia Hayne	coastal fens
Ornithogalum pyrenaicum L.	Boulonnais
Polygonatum odoratum (Mill.)Druce	dunes south of Boulogne
Maianthemum bifolium (L.)Schmidt	west of Abbeville
Spiranthes aestivalis (Poir.)Rich.	near Le Touquet
Stratiotes aloides L.	east of Calais
Schoenus nigricans L.	dune slacks—just reaches N.W. Surrey in S.E. England
Eriophorum gracile Roth	fen near Le Touquet—reaches N.W. Surrey like the last
Carex limosa L.	fen near Le Touquet
Carex lasiocarpa Ehrh.	fens near Le Touquet
Corynephorus canescens (L.)Beauv.	dunes

This conclusion is heightened by the evidence of the important roles that topography, soil types, and land-use history appear to have played in the development in the flora of Southeast England, particularly with regard to the rich development there, as compared with North France, of (1) oceanic and oceanic-southern elements, in the floras of the Weald and the coastal zones of Kent and Sussex, and (2) the flora of the chalk.

What species are there that occur in Kent (or in Kent and East Sussex) and North France that appear to have been unable (on the trans-Straits of Dover migration theory) to have spread further into Britain? There are very few of them. They seem to comprise only the following five:

Orchis purpurea Huds. (Fig. 3)	common on N. French chalk; almost restricted to Kent in Britain.
Ophrys fuciflora (Crantz) Moench	rare (perhaps now extinct) in N. France north of Somme valley; restricted in Britain to E. Kent
Polygala austriaca Crantz.	not known in France north of the Somme Valley; restricted in Britain to Kent (and formerly E. Surrey)
Orobanche caryophyllacea Sm.	frequent on N. French dunes, restricted in Britain at least until recently to E. Kent coastal dunes and chalk
Phyteuma spicatum L.	rare in N. France, restricted in Britain to the E. Sussex Weald

It is noteworthy that orchids and *Orobanche* have very light seeds, probably capable of transport by wind; some of the species above may have arrived in Britain by wind dispersal relatively recently since the Channel was formed.

There is one last, very important point to consider. As there are not large numbers of European species occurring in the maritime region of North France which are now unable to cross into Britain because of the sea barrier, how near to Britain do the more Continental elements come that form so large a part of the flora of Central Europe (and of lowland, non-Mediterranean France), and what has been the real factor preventing their spread? To understand this we must look a little further south and east than the region of France we have been considering, to the Somme Valley, above, and to some extent just below, Amiens. Here the Somme has carved out a series of steep river-cliff escarpments in the chalk during and after the Pleistocene. Some of these escarpments, which are too steep for cultivation today, and perhaps even for the development of closed forest in prehistoric times, bear now scrub and grassland communities on skeletal calcareous soils which carry a very rich flora, particularly on south or southwest slopes, including many species of mainly Central European distribution. All those in the following list (Table II) are unknown in the more maritime parts of France north of the Somme Valley.

Delvosalle (1964) presents a full discussion of the distribution in North France and Belgium of these, and other, apparently "thermophilous" (summer warmth-demanding) calcicole species of open habitats.

TABLE II. Species which occur on calcareous escarpments of the Somme valley near Amiens, but which are absent from the more maritime regions to the north of the valley

*_Acinos arvensis_ (Lam.)Dandy
*_Ajuga chamaepitys_ (L.)Schreb.
*_Campanula glomerata_ L.
*_Hypericum montanum_ L.
*_Iberis amara_ L.
*_Lonicera xylosteum_ L. (an open scrub species)
*_Ophrys sphegodes_ Mill. (one locality a little north of Somme Valley) (B)
*_Polygala austriaca_ Crantz (C)
*_Seseli libanotis_ (L.)Koch (C) (Fig. 2)
*_Teucrium botrys_ L.
*_Teucrium chamaedrys_ L. (C)
* _Verbascum lychnitis_ L.
†_Epipactis atrorubens_ (Hoffm.)Schultes (C)
(extends slightly north of Somme Valley)
†_Orchis militaris_ L. (C) (French records for Pas-de-Calais dubious)
†_Phleum phleoides_ (L.)Karst. (C) (extends slightly north of Somme Valley)
†_Potentilla tabernaemontani_ Aschers. (C) (extends slightly north of Somme Valley)
†_Pulsatilla vulgaris_ Mill. (C)
†_Sesleria albicans_ Kit. (C)
(open habitat species, but on _north_-facing slopes in Somme Valley)
†_Stachys germanica_ L. (one old Kent record, perhaps adventive there)
†_Thymus serpyllum_ L. (C)
‡_Ajuga genevensis_ L. (only as an introduced species in Britain)
‡_Anthericum ramosum_ L. (C)
‡_Bupleurum falcatum_ L. (C) (probably only introduced in Britain)
‡_Digitalis lutea_ L. (C) (Fig. 1)
‡_Genista sagittalis_ L.
‡_Gentiana ciliata_ L.
‡_Globularia vulgaris_ L. (C)
‡_Kentrophyllum lanatum_ L.
‡_Linum tenuifolium_ L.
‡_Polygala comosa_ Schkuhr (C)
‡_Prunus mahaleb_ L. (C)
‡_Seseli montanum_ L. (C)
‡_Teucrium montanum_ L. (C)
‡ _Vincetoxicum officinale_ Moench (C)

* Found in warmer habitats in Southeast England on the chalk.
† On calcareous soils in mid or east England, but unknown as natives in Southeast England.
‡ Unknown in Great Britain.
(B) Discovered recently however (1972) in chalk turf E. of Colembert in the Boulonnais.
(C) Locally abundant in the Somme Valley above Amiens; those not so marked are rarer.

The Somme Valley appears to be the north-western limit of penetration, towards the Channel coast and Britain, of many of the species of this ecological type. Further south, in the Valley of the Seine, and further east, in the Laonnois and in Champagne, similar but even richer, floras occur, until a more typical, full "continental" type flora is found in eastern France and in the lower parts of the Jura.

The Somme valley about Amiens appears to be however of particular phytogeographical significance. Not only does it lie far enough inland to have a significantly more continental climate (with warmer summers) than the regions to its north and north-west; the steep, largely south-facing slopes accentuate this effect markedly in microclimatic terms.

An equally important point historically, however, is that the steep slopes along this part of the Somme Valley are likely to have remained open during much of post-glacial time, while the loess-covered plateau to the north and the south of the Somme Valley would have been far less suitable for many open-habitat calcicole species to grow, even before the development of forest on it in Boreal times.

Hence the Somme Valley, pointing like a dagger towards Southeast Britain from Continental Europe, may well have acted as a major channel of migration for open-habitat calcicole species north-westward from the interior of France, even *before* the present bed of the English Channel was inundated some time in the late Boreal period. The discussions in Pigott and Walters (1954), in Rose (1957), and in Stott (1971) are very relevant to this hypothesis. I would suggest that the *positive* role of the Somme Valley route has been more important, in leading Continental species towards and into what is now Britain up to the close of the Boreal period, than the *negative* role of the Channel as a barrier to migration in later post-glacial times. Present distributions of species within Britain are not likely in all cases to indicate migration routes, as so many changes in climate, and above all in human land use, have occurred in later post-glacial times, with consequent mass destruction of areas of former distributions of species—a process we can see at work today.

To conclude, I would like to try to answer the four questions I raised near the beginning of my paper.

(1) The vast bulk of the flora (and of the natural vegetation) in South-east England and in North France shows very great correspondence; the same associations in general can be recognized in what is in many ways one region climatically, geologically and even geomorphologically.

(2) The factors that account for the differences in flora and vegetation are primarily those of the differing proportions of the various habitats and

soil types within the two regions, and secondarily those due to different human land use patterns; for example the absence of *Taxus baccata* on chalk in North France north of the Seine and west of the Meuse is almost certainly due to anthropogenic factors.

(3) There is little evidence that the formation of the Channel since about 5500 B.C. has affected the colonization of Britain by plants from the Continent; this colonization was probably largely completed before that date.

(4) There is little evidence visible *today* that the Straits of Dover route has been a major point of entry into Britain, though human changes *may* have obliterated such evidence. Rather it appears more likely that the Somme valley may have been a major route of approach to Britain for more thermophilous calcicole species in the late stages before the Channel was inundated, perhaps even into Boreal times. The major route of entry of more oceanic species is more likely to have been further to the west, from the more suitable terrain of the Armorican Shield in Brittany and the Cotentin, into Devon and the E. Dorset–S. Hampshire basin, linking up with the more extreme Lusitanian elements in a continuum of progressive dispersal patterns.

ACKNOWLEDGEMENTS

I wish to thank Professor J. M. Géhu for his constant help in what has been very much a joint effort of research; Mr. P. A. Stott for many useful discussions, both in the laboratory and the field and for his valuable help in the compilation of the maps; and the Central Research Fund of London University for financial assistance with travel costs.

REFERENCES

Chevalier, A. (1923). *Bull. Soc. bot. Fr.* **70**, 598–623.

Delvosalle, L. (1964). *Bull. Soc. bot. Fr.* **111**, 90e session extraordinaire, 83–114.

Good, R. d'O. (1928). *J. Bot., Lond.* **66**, 253–264.

Masclef, A. (1886). "Catalogue raisonné des plantes vasculaires du département du Pas-de-Calais". Paris.

Pennington, W. (1969). "The History of British Vegetation". London.

Pigott, C. D. and Walters, S. M. (1954). *J. Ecol.* **42**, 95–116.

Pinchemelle, P. (1954). "Les Plaines de Craies du Nord-Ouest du Bassin Parisien et du Sud-est du Bassin de Londres et leur bordure". Paris.

ROSE, F. (1957). *In* "Progress in the Study of the British Flora", (J. E. Lousley, ed.), pp. 61–78, Arbroath.

ROSE, F. (1964). *Bull. Soc. bot. Fr.* **111**, 90e session extraordinaire, 209–238.

ROSE, F. (1965). *New Scient.* (July 15), 158–161.

ROSE, F. and GÉHU, J. M. (1960). *Bull. Soc. bot. N.Fr.* **XIII**, no. 4, 125–139.

ROSE, F. and GÉHU, J. M. (1964). *Bull. Soc. bot. Fr.* **111**, 90e session extraordinaire, 38–70.

STOTT, P. A. (1971). *Vegetatio* **23**, 61–70.

24 | Computational Methods in the Study of Plant Distributions

N. JARDINE

King's College Research Centre, Cambridge, England

Applications of computers to the analysis of the geographical distribution of plants and animals have rarely been described. Exceptions are the papers of Hagmeier and Stults (1964), Sneath (1967), Proctor (1967), Fisher (1968), Holloway and Jardine (1968), and Kikkawa and Pearse (1969). This is at first sight surprising. The usefulness of automatic data-processing methods in the preparation of distribution maps is generally recognized (see Soper and Perring, 1967; Cadbury *et al.*, 1971). And the use of automatic classification and ordination methods to investigate the factors which determine the *local* distributions of plants and animals is standard practice amongst ecologists (see, e.g. Grieg-Smith, 1964; Lambert and Dale, 1964; van der Maarel, 1969; Moore *et al.*, 1970). Perhaps one reason why computers have so rarely been used by phytogeographers is uncertainty about the nature of the numerical methods which could be profitably used. I shall therefore discuss in fairly general terms some of the methods for analysis of distribution maps which may be useful in the study of the ecological, geographical and historical factors which determine plant distributions. For mathematical and computational details of the various methods the reader is referred to Sokal and Sneath (1963) and Jardine and Sibson (1971). Technical details are given in footnotes when they are of practical importance to the non-mathematically inclined investigator.

I shall discuss first some techniques for analysis of affinities between distributions of taxa, and then some techniques for analysis of the floristic affinities of geographical areas.

AFFINITIES BETWEEN DISTRIBUTIONS OF TAXA

1. *Floristic Elements*

When one inspects the dot maps in Hultén's "Atlas över Kärlväxterna i Norden" or in the "Atlas of the British Flora" it becomes clear that certain patterns of distribution tend to recur. This recurrence is the intuitive basis for recognition of floristic elements or types. Many authors have speculated on the reasons for recurrence of distribution patterns. For example, Hultén related the elements of the arctic and boreal flora to the action of a limited number of areas of refuge from the glaciations as centres of dispersal. Such striking floral elements as the Lusitanian element of the European flora have been variously explained in terms of per-glacial refuges, past land-bridges, or the limited climatic tolerances of certain species (see Heslop-Harrison, 1953). Matthews (1937) in his classic paper on the elements of the British flora emphasized the dangers of seeking a single kind of explanation, whether historical or ecological (cf. also Walters, 1956, discussion of the origin of the British mountain flora).

There are certain pitfalls in grouping species distributions into floristic elements by eye. First there is the tendency of the human eye to discover groupings even when none is present. Secondly there is the tendency to select particular distributions as "types" and then to cluster the remaining distributions around them. And finally there is the tendency unconsciously to bias the classification because of preconceptions about the factors which determine the distributions of plants. The value of methods of automatic classification lies in their to some extent avoiding these pitfalls.

Automatic classification may be applied as follows in the study of floristic elements. First an overall area is selected and partitioned into unit areas. The unit areas may be grid-squares or such arbitrary areas as the British Watsonian vice-counties. Next a measure of association between each pair of species distributions is calculated. One appropriate measure is: number of squares in common/number of squares in which either or both occur.* Then a cluster method is applied to the matrix of

* This measure can be converted to a dissimilarity measure by subtraction from 1. Note that number of squares in common, and proportion of squares in common, are unsatisfactory measures because they exaggerate the similarities of widespread species. Note also that measures which are functions of χ^2 may be inappropriate because measuring similarity is unrelated to testing for independence. Numerous alternative measures of similarity are reviewed by Goodman and Kruskal (1954), Dagnélie (1960), and Cormack (1971). One

similarity values. The usual hierarchic cluster methods produce what is known as a dendrogram—a hierarchy in which the groups have associated numerical levels*. The final stage is the selection, by the investigator this time, of clusters which may be recognized as floristic elements. The investigator should beware of too eagerly recognizing as floristic elements clusters obtained by automatic classification. Cluster methods may sometimes find clusters even when the data show no significant cluster structure. For example, clusters may be generated when a cluster method is applied to the distances between pairs of points in a random 2-dimensional array (Sneath, 1966). It is therefore prudent to calculate a measure of the accuracy with which the system of clusters obtained represents the similarity matrix (see Hartigan, 1967; Jardine and Sibson, 1971). If the system of clusters turns out to represent the similarity matrix very inaccurately, a statistical test for cluster structure should be applied (see Calinski and Harabasz, 1972; Jardine, 1971).

Proctor (1967) pioneered this use of automatic classification in a study of distribution patterns amongst British liverworts. Figure 1 shows the distributions of six floristic elements of *Hieracium* s.s. in Britain. These floristic elements were obtained by the procedure described above.

It is important to consider carefully the limitations of this procedure. Like all the other methods described in this paper it presupposes an

* Choice of an appropriate cluster method is a controversial matter: see Jardine and Sibson (1971). The single-link method is the only known hierarchic method which is stable in the sense that small changes in similarity values cannot produce large changes in the system of clusters. However, it may fail to find clusters when intermediates between otherwise well-marked groups occur. This defect is avoided both by the unstable average-link and centroid methods and by certain stable non-hierarchic methods which allow limited overlap between clusters. These non-hierarchic cluster methods are, in principle, the most appropriate for study of floristic elements, but unfortunately they are at present computationally laborious. Using single-link up to *ca* 2000 species distributions can be clustered, whereas the non-hierarchic methods can handle only *ca* 100.

---

feature of all such measures of similarity is crucial for the subsequent selection of numerical methods. The absolute numerical values, and ratios of values of similarity measures are rarely significant in isolation because there is rarely an empirical reason for using them as they stand rather than, for example, their logarithms or square roots. If methods of automatic clustering or scaling are to be applied to them it is appropriate to use techniques such as single-link cluster analysis and non-metric multidimensional scaling which take account only of the ranking of similarity values. Such techniques as average-link cluster analysis and principal component analysis which take account of absolute values or ratios of values may be inappropriate.

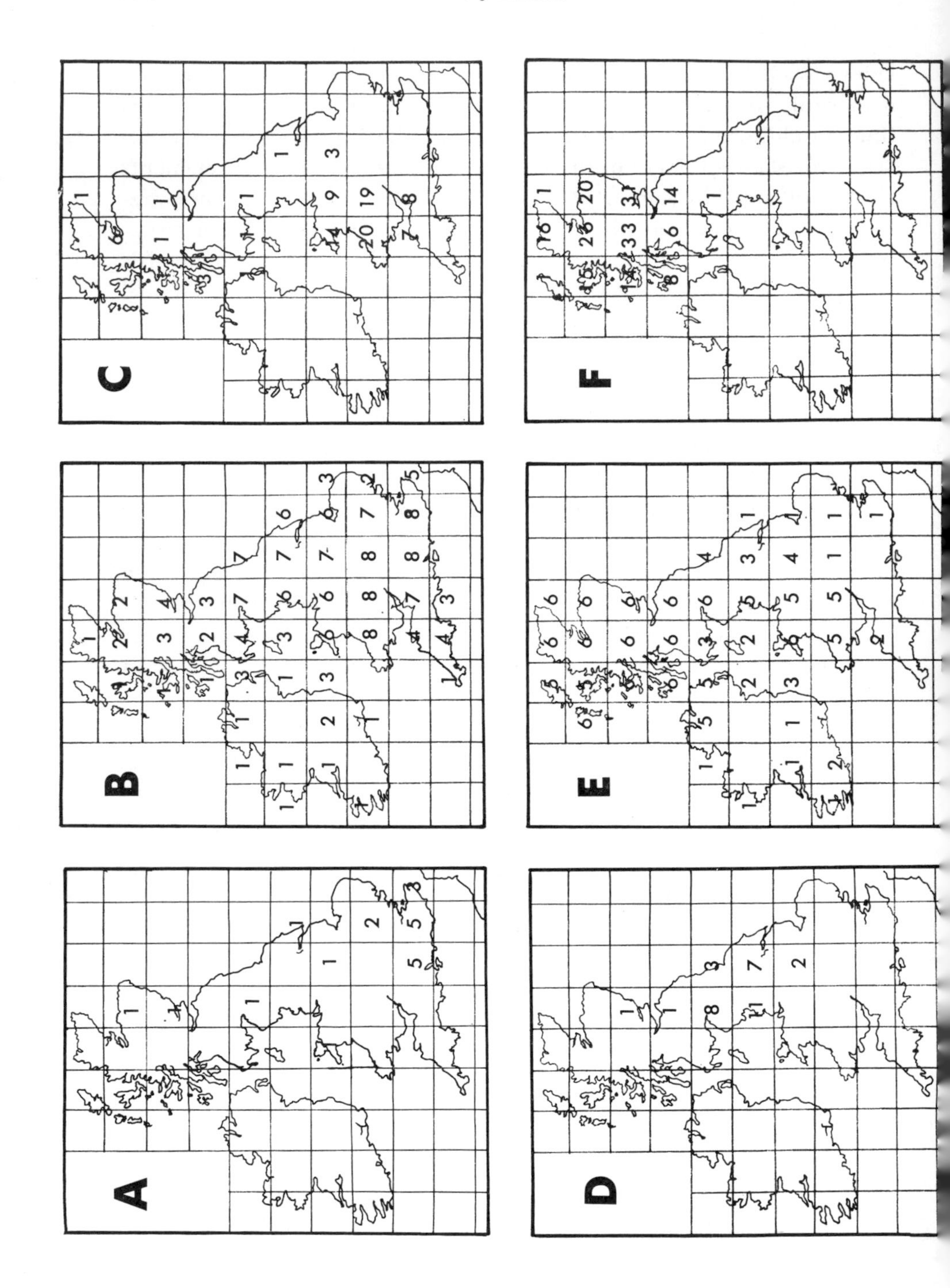

FIG. 1. Six floristic elements obtained by automatic classification of the *Hieracium* microspecies distributions mapped in the "Critical Supplement to the Atlas of the British Flora" (Perring and Sell, eds, 1968). Elements **C** and **D** contain many calcicoles. The sequence of elements **A—B—(C, D)—E—F** shows progressive displacement along a S.S.E.–N.N.W. axis. There are in addition many locally endemic elements. It is of interest to note that the distributions of *H. subplanifolium* and *H. angustisquamum* each form a single element cluster at the highest level in the system of clusters obtained by automatic classification. P. D. Sell (personal communication) has suggested that the former may have originated polytopically from *H. dicella* and the latter polytopically from *H. caledonicum*. This would explain their anomalous distributions. The microspecies in each floristic element are as follows:

A *H. virgultorum*
H. cantianum
H. rigens
H. surrejanum
H. cambricogothicum
H. acamptum

B *H. perpropinquum*
H. umbellatum subsp. *umbellatum*
H. vagum
H. strumosum
H. diaphanum
H. eboracense
H. calcaricola
H. trichocaulon

C *H. stenolepiforme*
H. britanniciforme
H. vagense
H. pseudoleyi
H. carneddorum
H. cambricum
H. holophyllum
H. snowdoniense
H. lasiophyllum
H. asteridiophyllum
H. pachyphylloides
H. radyrense
H. triviale
H. dissimile
H. solum
H. riddelsdellii
H. discophyllum
H. neocoracinum

C *H. rectulum*
H. subminutidens
H. linguans
H. stenstroemii
H. cacuminum
H. angustatiforme
H. cuneifrons
H. pulchrius
H. eustomon
H. subbritannicum
H. cyathis
H. leyanum
H. subamplifolium
H. nidense
H. repandulare
H. cinderella
H. substrigosum
H. scabrisetum
H. submutabile

D *H. pycnotrichum*
H. ornatilorum
H. mirandum
H. rhomboides
H. maculosum
H. angustatum
H. auratiflorum
H. crebridentiforme
H. cymbifolium
H. decolor
H. pseudostenstroemii

E *H. argenteum*
H. vulgatum

FIG. 1. Continued.

E *H. caledonicum*
H. rubiginosum
H. subcrocatum
H. strictiforme

F *H. subumbellatiforme*
H. petrocharis
H. eximium
H. caesiomurorum
H. subhirtum
H. pictorum
H. chloranthum
H. dasythrix
H. centripetale
H. senescens
H. subtenue
H. lingulatum
H. globosiflorum
H. hanburyi

F *H. alpinum*
H. piligerum
H. hyparcticoides
H. jovimontis
H. dipteroides
H. breadalbanense
H. isabellae
H. tenuifrons
H. aggregatum
H. pseudoanglicoides
H. nigrisquamum
H. cremnanthes
H. atraticeps
H. pseudocurvatum
H. langwellense
H. clovense
H. marshallii
H. grovesii

adequate taxonomy and its reliability is directly related to the completeness of the distribution records and to the reliability with which records of introductions have been eliminated (cf. Sneath, 1967). The initial choice of unit areas raises problems. If too coarse a grid is used valid floristic elements may be amalgamated. For example, use of 100km grid squares for the British *Hieracium* microspecies distributions failed to detect the several distinct elements within the "Scottish" element which would be obtained using 20km squares or by selecting as unit areas mountain blocks. Conversely, use of too fine a grid may fail to reveal major floristic elements and may generate many spurious small elements when recording has been uneven. There is also a difficulty of scale. There is no guarantee that the results obtained when a relatively small overall area is considered will remain valid when a larger area is considered. Thus, Watson's (1832) classification into floristic elements within Britain does not correspond well with Matthews' (1937) allocation of British species to floristic elements on the basis of their entire European distributions. *Hieracium* microspecies were chosen as a small-scale example because a high proportion are endemic to Britain. Ideally studies of floral elements should be on a world-wide scale.

The greater objectivity of floristic elements derived by automatic classification does not make their interpretation any the less problematical.

The explanations of arctic and boreal floristic elements by Hultén (1937) in terms of dispersal from per-glacial refuges are probably over-simplified. However, Hultén's demonstration that many widespread "Linnaean" species show segregation into several taxa of lower rank each of which has a distribution centred on one of the postulated refuges is highly suggestive. Similarly, Holloway showed that certain butterfly species are differentiated into subspecies or varieties whose ranges correspond to the ranges of the faunal elements derived by automatic classification of butterfly generic distributions by Holloway and Jardine (1968). This likewise suggests that centres of species concentration of particular faunal elements can be related to past areas of isolation in which differentiation occurred and which acted as centres of dispersal.

Explanations of floristic elements in terms of hypothetical past land-bridges are definitely unwise. The theory of plate tectonics is at present enabling geophysicists to reconstruct and date precisely past continental positions and may eventually enable them to reconstruct past climates. This will render the palaeogeographical speculations of biologists redundant. It is arguable that overmuch speculation about historical factors is unwise until these reconstructions are available and until much more is known about the present-day climatic and edaphic tolerances of plant species and about the dispersal capacities of plants. Sneath (1967) emphasized the importance of considering plant dispersal capacities when attempting to relate their distributions to past continental arrangements. In this connection Mollison's work (forthcoming paper and personal communication) on the relations between probability of long-distance dispersal of objects, their terminal velocities, and meteorological conditions is of interest. He has shown that for objects with terminal velocities below a few centimeters per second, probability of long-distance dispersal is nearly independent of terminal velocity: with greater terminal velocities the probability falls rapidly. By extrapolation from information about the fall-out of Saharan dust in Barbados he has shown that whilst long-distance dispersal of small flowering plant seeds by the trade-winds may occur too rarely to be likely to be observed, it may nevertheless be appreciable over long periods of time. Wind is one of many dispersal mechanisms, but it is of particular interest because it could account for major floristic links between areas separated by large water gaps. This possibility should be remembered in all attempts to relate plant distributions to past continental arrangements. The fact that the Hawaiian Islands have been isolated by large water gaps since their volcanic origin serves as a warning against too facile an appeal to past land connections to explain flowering plant distributions. Theoretical studies of the kind carried out by Mollison on the dispersive power of winds and currents may eventually lead to an important application of computers in phytogeography. By computer simulation of the dispersal of seeds it may be possible to obtain quantitative estimates of the number of seeds transported from one place to another over a given period of time. A simulation of a similar kind was carried out by Levison *et al.* (1969). Using information about Pacific winds and currents they estimated probabilities of accidental dispersal of men on primitive rafts and canoes between Pacific islands.

2. *Centres of Dispersal and Speciation of Higher Taxa*

Rothmaler (1955) described ways of mapping the geographical distribution of relative morphological diversity and of relative abundance of species within genera and families. Such methods may provide clues to past centres of speciation and dispersal. Holloway (1969) developed Rothmaler's ideas by using automatic classification to investigate the extent to which Indian butterfly genera can be assigned to groups with similar patterns of species distribution. He found that the hypothetical centres of dispersal of the Indian butterfly fauna mainly lie outside India itself and he related this to possible depletion of the Indian flora and fauna during its relatively rapid northward drift from Antarctica in the late Mesozoic and Tertiary. Similar methods could profitably be applied to floristic data. Such methods however depend on the assumption that the genera and families studied are monophyletic, and even where the taxonomy is quite adequate this assumption may be dubious.

3. *Trends in Species Distributions*

A tendency for species distributions to fall into clusters may provide clues to past areas of isolation and subsequent dispersal. But this is only one of the kinds of structure which may be indicated by a matrix of similarities between species distributions. Suppose that a group of species has dispersed from some area of isolation, or has immigrated into a region from some source. Under these circumstances one might hope to find a trend to present distributions of species showing progressively wider areas of dispersal, or distributions progressively dispersed in some direction. Hultén's theory of "equiformal progressive areas" postulates trends of the first kind. Often such trends may be revealed by drawing on a map the boundaries of the distributions of particular species as in Hultén (1937). The technique of non-metric multidimensional scaling provides a more objective way of seeking such trends.

Suppose that we have a matrix of similarities between pairs of objects (species distributions in this case). The non-metric multidimensional scaling algorithm of Kruskal (1964a, b) seeks a disposition of points representing the objects in two (or more) euclidean dimensions, in which the interpoint distances represent the original similarities as accurately as possible.* A trend amongst species distributions is shown up by non-metric multidimensional scaling as an array of points showing a linear tendency. Figure 2 shows the disposition of points representing the distributions of 67 of the more widespread British *Hieracium* micro-species.

The disposition of points shows a slight linear tendency, which represents progressive displacement of species distributions along a S.S.E.–N.N.W. axis in the British Isles. This trend is also evident in the floristic elements shown in Fig. 1. It may be related, speculatively, to invasion from Europe following the creation of open habitats in the wake of the retreating glaciation, or more prosaically to the fact that Britain is elongated with more open rocky habitats in the north and north-west.

FLORISTIC AFFINITIES OF GEOGRAPHICAL AREAS

1. Floristic Regions

Just as cluster analysis can be applied to measures of similarity between species distributions to generate floristic elements, so it can be applied to measures of similarity in species composition to generate floristic regions. Hagmeier and Stults (1964) used automatic classification of unit areas in a study of mammalian faunal regions in N. America; and similar applications of automatic classification have been described by Proctor (1967), Fisher (1968) and Holloway and Jardine (1968).

The main purpose of classifications into floristic regions and sub-regions is, I presume, to provide a convenient framework for teaching purposes (cf. Good, 1964), and for this purpose it does not matter if the classification is somewhat arbitrary. For purposes of description in plant geography an indication (however approximate) of the composition of each area in terms of floristic elements is preferable to a rigid classification

* The measure of goodness-of-fit which is optimized is a monotone regression measure.

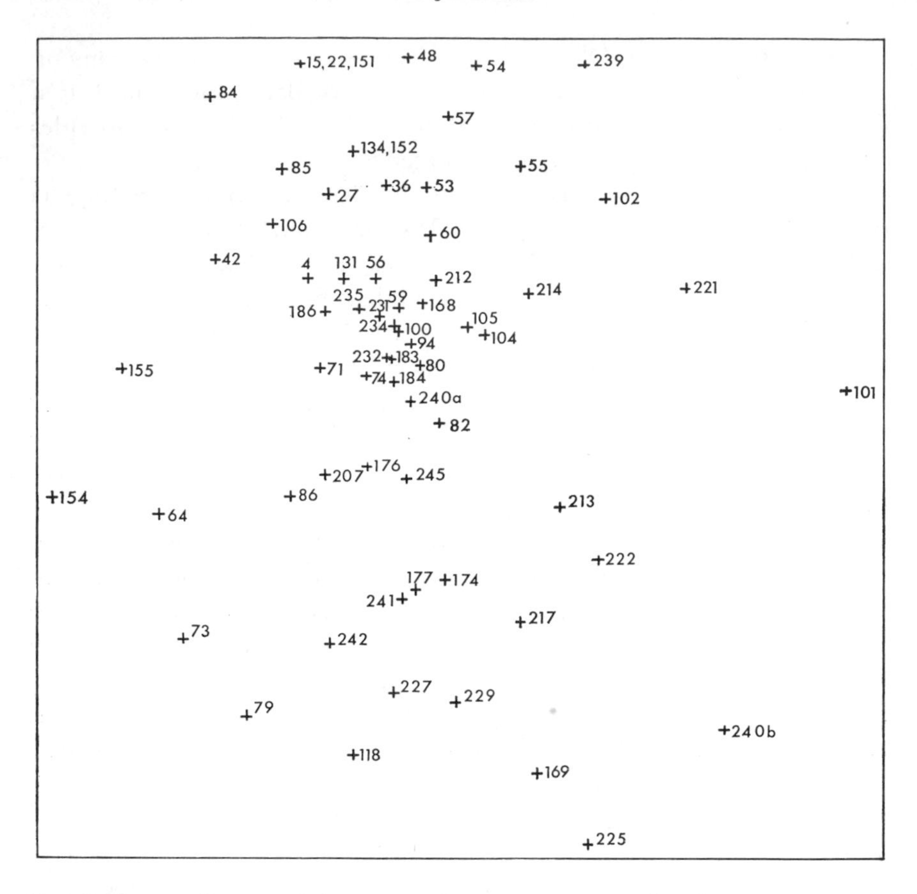

FIG. 2. The two-dimensional disposition of points representing distributions of the more widespread British *Hieracium* microspecies obtained by non-metric multidimensional scaling. The linear tendency indicates progressive displacement of distributions along a N.N.W.–S.S.E. axis in the British Isles. The numbers are those used in the "Critical Supplement to the Atlas of the British Flora."

of areas into floristic regions. However, the boundaries between floristic regions may in some cases be related to past or present barriers to dispersal. Of considerable interest are the relations between discontinuities in floristic composition of areas, their geographical history, and the taxonomy of particular groups of plants. For example, Strid (1970) gave a fascinating account of the relations between differentiation in the *Nigella arvensis* complex, the geological history of the Aegean, and the phytogeographical classification of the Aegean by Rechinger.

2. *Floristic Affinities and Geographical Disposition*

It is obvious that there is usually a partial dependence of similarity in the floristic composition of areas on their geographical separations. In particular, suppose that the greater the geographical separation of unit areas the less their similarity in species composition. Under this condition non-metric multidimensional scaling of the matrix of similarities of unit areas in species composition yields a two-dimensional disposition of points very similar to the geographical disposition of unit areas (provided, of course, that only a relatively small part of the earth's surface is considered). The differences between the two dispositions may be related to past and present barriers to dispersal. For example, Sneath (1967) applied a method related to multidimensional scaling to measures of similarity in conifer composition of unit areas of the entire world, using data published by Florin (1963). The disposition of points on the surface of a sphere which he obtained showed a marked aggregation of unit areas in the southern continents, which he related tentatively to continental drift. Holloway and Jardine (1968) applied multidimensional scaling to measures of faunal dissimilarity between land-masses in the Indo-Australian area. The 2-dimensional disposition showed marked aggregation of the land-masses which were united during periods of low sea-level in the Pleistocene and greatly exaggerated the isolation of Australia, which could be related to its northward drift during the Tertiary. It is reasonable to hope that such techniques could with profit be applied to floristic data.

CONCLUSIONS

It is arguable that progress in understanding the factors which determine plant distributions is more likely to come from detailed taxonomic and phytogeographical investigations of particular small groups of closely related taxa than from sledgehammer methods which analyse the distributions of large numbers of more distantly related taxa. I agree with van Steenis' suggestion at this conference that a genus such as *Nothofagus* with a striking geographical distribution, a good fossil record, and limited powers of dispersal, is more likely to provide clues to the relations between dispersal of plants and past position of land-masses, than are genera which lack these desirable features. It is unfortunate that the methods which I have described would all attach the same weight to the distribution of *Nothofagus* as to the distributions of other genera. Nevertheless, I believe that the two approaches are complementary. One example concerns the role of continental drift in the evolution and dispersal of the flowering plants. Both the Restionaceae and the Proteaceae show "Gondwana" distributions. At this conference Cutler suggested that the S. American and Australian Restionaceae are more closely related phylogenetically than is either group to the African representatives. The same may be true in the Proteaceae (see Johnson and Briggs, 1968). It is plausible to relate this to the fact that Africa broke away from Gondwanaland at least 50 million years before Australia broke away (Jardine and McKenzie, 1972). It would be an interesting complementary study to apply computer methods to investigate the extent to which the temporal order of break-up of Gondwanaland is reflected in the present-day floristic compositions of India and the southern continents.

ACKNOWLEDGEMENTS

I thank Mr P. D. Sell and Dr S. M. Walters for useful discussions and Dr R. Sibson for help with computing. I thank also all those who participated in the informal discussion at the conference on applications of computers in phytogeography, especially Mr D. L. Wigston whose more detailed application of computer methods to British *Hieracium* microspecies distributions will be described in a forthcoming paper.

REFERENCES

Cadbury, D. A., Hawkes, J. G. and Readett, R. C. (1971). "A Computer-Mapped Flora", A Study of the County of Warwickshire. Academic Press, London and New York.

Calinski, T. and Harabasz, J. (1972). *Biometrics* (In press).

Cormack, R. M. (1971). *J. R. statist. Soc. B.* **134**, 321–367.

Dagnélie, P. (1960). *Bull. Serv. Carte phytogéogr.* (*Sér. B*) **5**, 7–71; 93–195.

Fisher, D. R. (1968). *Syst. Zool.* **17**, 48–63.

Florin, R. (1963). *Acta Horti Bergiani.* **20**, 122–312.

Goodman, L. A. and Kruskal, W. H. (1954). *J. Am. statist. Ass.* **49**, 732–764.

Greig-Smith, P. (1964). "Quantitative Plant Ecology", 2nd Edition, Butterworth, London.

Hagmeier, E. M. and Stults, C. D. (1964). *Syst. Zool.* **13**, 125–155.

Hartigan, J. A. (1967). *J. Am. statist. Assoc.* **62**, 1140–1158.

Heslop-Harrison, J. (1953). *In* "The Changing Flora of Britain" (J. E. Lousley, ed.), B.S.B.I., London.

Holloway, J. D. (1969). *Biol. J. Linn. Soc. Lond.* **1**, 373–385.

Holloway, J. D. and Jardine, N. (1968). *Proc. Linn. Soc. Lond.* **179**, 153–188.

HULTÉN, E. (1937). "Outline of the History of Arctic and Boreal Biota during the Quaternary Period." Bökforlags Aktiebolaget Thule, Stockholm.

HULTÉN, E. (1950). "Atlas över Kärlväxterna i Norden", Stockholm.

JARDINE, N. (1971). *Phil. Trans. R. Soc. B.* **263**, 1–33.

JARDINE, N. and MCKENZIE, D. P. Continental drift and the dispersal and evolution of organisms. (In preparation).

JARDINE, N. and SIBSON, R. (1971). "Mathematical Taxonomy". Wiley, London and New York.

JARDINE, N. and SIBSON, R. (1971). *Comput. J.* (In press).

JOHNSON, L. A. S. and BRIGGS, B. G. (1963). *Aust. J. Bot.* **11**, 21–61.

KIKKAWA, J. and PEARSE, K. (1969). *Aust. J. Zool.* **17**, 821–840.

KRUSKAL, J. B. (1964a). *Psychometrika* **129**, 1–27.

KRUSKAL, J. B. (1964b). *Psychometrika* **129**, 115–129.

LAMBERT, J. M. and DALE, M. B. (1964). *Adv. ecol. Res.* **2**, 59–99.

LEVISON, M., FENNER, T. I., SENTANCE, W. A., WARD, R. G. and WEBB, J. H. (1969). *In* "Information Processing 68" (A. J. H. Morrell, ed.), North Holland Press, Amsterdam.

MATTHEWS, J. R. (1937). *J. Ecol.* **25**, 1–90.

MOORE, J. J., FITZSIMMONS, P., LAMBE, E. and WHITE, J. (1970). *Vegetatio* **20**, 1–20.

PERRING, F. H. and WALTERS, S. M. (eds) (1963). "Atlas of the British Flora". Nelson, London.

PROCTOR, M. C. F. (1967). *J. Ecol.* **55**, 119–135.

ROTHMALER, W. (1955). "Allegemeine Taxonomie und Chorologie der Pflanzen", 2nd Edition, Jena.

SNEATH, P. H. A. (1966). *Class. Soc. Bull.* **1**, 2–18.

SNEATH, P. H. A. (1967). *Nature, Lond.* **215**, 467–470.

SOKAL, R. R. and SNEATH, P. H. A. (1963). "Principles of Numerical Taxonomy". W. H. Freeman, London and San Francisco.

SOPER, J. H. and PERRING, F. H. (1967). *Taxon* **15**, 1–21.

STRID, A. (1970). *Op. bot. Soc. bot. Lund.* **28**, 1–169.

VAN DER MAAREL, E. (1969). *Vegetatio* **19**, 21–46.

WALTERS, S. M. (1956). *In* "Mountain Flowers". Collins, London.

WATSON, H. C. (1832). "Outlines of the Geographical Distribution of British Plants". Edinburgh.

Section VI

DISCUSSION

25 | Questions Answered and Unanswered

A. R. CLAPHAM, F.R.S.

The Parrock, Arkholme, Carnforth, Lancashire, England

Professor A. R. Clapham, in opening the final discussion, said he had found the Conference extremely interesting and wished to congratulate Professor Valentine and those associated with him for bringing together so distinguished a group of contributors and also to thank the contributors for a most stimulating series of papers.

The subject under discussion has been the Taxonomy and Phytogeography of Higher Plants in Relation to Evolution. It was clear that no progress could be made until we were in possession of the relevant facts of the situation as far as they could be learnt without the expenditure of excessive time and labour. And we needed both taxonomic and phytogeographical facts: what taxa, specified as precisely as was reasonably possible, were distributed in what way, the distribution also being specified with precision. Evolutionary relationships could only be inferences from the taxonomic and phytogeographical facts with the help of certain additional information and hypotheses.

What had emerged from the Conference was the impressive degree of refinement that had been achieved in the specification of taxa whose distribution and evolutionary history were being studied. Attention was being directed increasingly to small-scale morphological and cytological variability, to anatomical, biochemical and genetic features, to pollination mechanisms and breeding systems and to evidence of hybridity. There were many examples in the recent literature of this increasingly refined specification, and amongst papers contributed to the Conference those by Professors Favarger, Greuter, Hara and Harlan Lewis and others all reported investigations in which cytological information had been both abundant and illuminating. Drs Cutler and Stace had shown the value of anatomical data, Dr Morley had drawn attention to the importance of the sizes of breeding populations and of the territories of bird pollinators, and Professor Böcher had emphasized the value of growing representatives of different populations in controlled environments so as to reveal relative richness in biotypes as a significant additional item in specification. The question that now arose was how far the collection of all these types of data must be regarded as an essential routine in any biosystematic investigation. Was it the view of those attending the Conference that the present degree of precision in taxonomic specification was more or less adequate? There was clearly no end to what might be thought potentially relevant, but there must be some selection of items most likely to be of real value.

The precise specification of phytogeographical facts was perhaps largely a matter of hard work in the field by those few competent to collect distributional data, but here again there was scope for recording information of a more detailed kind than heretofore.

In a general discussion it was suggested that habitat data should always be recorded and should be as full as possible, with reference not only to precise location but also to altitude, aspect, substratum, etc., and that phenological information should be added where possible. It was also agreed that some phytosociological information would always be valuable. The need for more experimental data on growth under controlled conditions was emphasized in relation to climatically determined distribution-limits, though there was some hesitation about approving the concept of "potential distributions". An appeal was made for the collection of seeds and other propagules in a condition enabling plants to be raised for cytological investigation. A special point was made of the rapidly increasing urgency of recording the distribution of the many plant species in danger of elimination from some or all of their present range. There were already schemes for recording rare plants in Europe and elsewhere but much more needed to be done. Herbaria were vital for assembling data and making maps and apart from making additions to herbaria there was a great need for better information about the contents of our existing collections in herbaria all over the world.

Professor Clapham pointed out that a corollary of all this was that it became even more important than in the past to devote sufficient time and thought to the selection of areas in which biosystematic team-investigations were to be carried out. He went on to refer to the obvious value of detailed knowledge of the geological history, with firm dates, of areas selected for biosystematic study. It was pointed out in discussion that relevant information was often available but was usually not readily accessible to biologists unless with the assistance of geological specialists. Where adequate geological information was lacking it might be best to defer a full-scale biosystematic study until it was forthcoming or to include one or more appropriately qualified geologists in the investigating team. There were possibilities of considerable developments in quaternary studies as a result of recent progress in micromorphology, in particular through the use of the scanning electron microscope. Other dating techniques were also becoming available and could be of assistance to biosystematists.

Another interesting point of the Conference was the indication that further criteria might emerge for the determination of the direction of evolutionary changes, at least within fairly narrow circles of affinity. Use had long been made of the presumption that polyploid forms were derivative from diploids in inferring migration-routes, and evidences of hybridity could be similarly informative. Professor Stebbins, in a stimulating contribution, had given examples of sympatric species believed to have come together after divergent adaptative evolution from a common ancestral type. If we could learn more about the modes of control of developmental patterns we might more safely infer the directions of progressive adaptational change and thence the chronological history of current distributional patterns.

There was a final discussion of the evolutionary significance of apomixis, during the course of which it was emphasized that apomixis should not be regarded as an evolutionary "dead-end": there was scope, through various cytological mechanisms, for appreciable rates of evolutionary change.

Author Index

Numbers in italic indicate the page on which the reference is listed.

O

P

Q

R

S

Subject Index

C

H

Q

R